"Energy production has had a major impact on landscape transformation everywhere on the globe. This important collection of articles draws our attention to the environmental damage done, but also to the impact of energy production on work and leisure patterns. The case studies presented here are brimming with insights into changes to socio-economic structures, migration patterns and tourism as well as heritage initiatives. The book is a must-read for anyone interested in the history and transformation of energy landscapes."

Stefan Berger, *Professor of Social History, Ruhr University Bochum, Germany*

"This carefully edited volume provides a rich and engaging account of lives entangled with energy infrastructures, in numerous European regions across the former East-West divide. The book convincingly shows how hydropower plants and dams, coal mine fields and oil rigs produced wider landscapes of hope and home, loss and resignation, and, in their afterlives, of ambivalent heritage and unsettled tourist attractions."

Anna Storm, *Professor of Technology and Social Change, Linköping University, Sweden*

Cultural Landscapes of Energy

This volume explores the contested heritage of landscapes impacted by energy production. It offers a comparative perspective across Europe on different energy resources and reveals the hidden histories behind current efforts to revalorise the industrial heritage of energy production.

Including case studies from across the European continent, this volume adds a crucial historical perspective to current debates on energy transition and the future of Europe's landscapes, which have been deeply impacted by energy production. Coal mining, oil drilling, peat extraction, and the construction of large-scale infrastructure, such as dams, have shaped 'cultural landscapes of energy' in present-day Europe. The exploitation of natural resources served economic development and established new industrial work cultures, but it also destroyed settlements through excavation and flooding. This volume brings together conflicting histories around work, habitation, and leisure in contemporary landscapes across Europe. Drawing on archival records, interviews, and fieldwork, the chapters in this volume combine perspectives on the productive and destructive sides of energy. They address the tensions emerging from heritage-making processes, which focus on the end of energy production despite ongoing and future commissioning projects.

This volume contributes new insights to the fields of energy and environmental history, heritage studies, memory studies, landscape architecture, and sustainability science. It provides rich materials on energy landscapes across Europe for researchers as well as policymakers and practitioners interested in energy transition, (post-)industrial heritage, and cultural tourism.

Corinne Geering holds a tenure-track position at the Department of Economic, Social, and Environmental History and the Linz Institute for Transformative Change (LIFT_C) at Johannes Kepler University Linz. Specialising in nineteenth- and twentieth-century history from a transnational and global perspective, her wider research interests include the use of the past in regional and urban development.

Torsten Meyer is Senior Scientist at the Deutsches Bergbau-Museum Bochum/Leibniz-Research Museum for Geo-resources. His main research interests include environmental history, the history of technology (eighteenth to twentieth century), and the use of industrial heritage in regional planning.

Critical Heritages of Europe
Series Editors: Christopher Whitehead and Susannah Eckersley, both at the University of Newcastle, UK

The *Critical Heritages of Europe* series seeks to explore the cultural and social politics of the European past in the present. Bridging theoretical and empirical research, the series accommodates broad understandings of Europe – a shifting and historically mutable entity, made both of internal tensions and exogenous encounters, re-imaginings and influences. 'Heritage' too is taken as an expansive paradigm, made in myriad practices where the past is valorised for the present, from folk traditions to museums and memorials, the management of historic sites and traditions, and everyday matters such as education, political discourse, home life, food consumption and people's relations with place. Consequently, the series spans a broad array of foci, disciplinary approaches and ways of investigating and questioning the diverse meanings of European heritages today.

The following list includes only the most-recent titles to publish within the series. A list of the full catalogue of titles is available at: https://www.routledge.com/Critical-Heritages-of-Europe/book-series/COHERE

Diversity of Belonging in Europe
Public Spaces, Contested Places, Cultural Encounters
Edited by Susannah Eckersley, Claske Vos

Viking Heritage and History in Europe
Practices and Re-creations
Edited by Sara Ellis Nilsson and Stefan Nyzell

Revoicing Intangible Cultural Heritage
Perspectives from the Margins of Europe
Edited by Laura Hodsdon, Valts Ernštreits, Kadri Koreinik and Sjoerd-Jeroen Moenandar

Cultural Landscapes of Energy
Constructing Histories of Power, Prosperity, and Decline in Europe
Edited by Corinne Geering and Torsten Meyer

Cultural Landscapes of Energy

Constructing Histories of Power, Prosperity, and Decline in Europe

Edited by Corinne Geering and Torsten Meyer

Routledge
Taylor & Francis Group
LONDON AND NEW YORK

First published 2026
by Routledge
4 Park Square, Milton Park, Abingdon, Oxon OX14 4RN

and by Routledge
605 Third Avenue, New York, NY 10158

Routledge is an imprint of the Taylor & Francis Group, an informa business

British Library Cataloguing-in-Publication Data
A catalogue record for this book is available from the British Library

ISBN: 978-1-032-75336-2 (hbk)
ISBN: 978-1-032-75344-7 (pbk)
ISBN: 978-1-003-47353-4 (ebk)

DOI: 10.4324/9781003473534

Typeset in Times New Roman
by Newgen Publishing UK

Work on this book was supported by the Leibniz Research Alliance "Value of the Past". The department for Regional Development and Minority Protection is financially supported by the Foundation for the Sorbian People, which receives grants from the Federal Ministry of the Interior based on the approved budgets of the German Bundestag.

Contents

Figures

Table

Contributors

Andrew Demshuk is Professor of History at American University. His five books and multiple articles explore German expellee memory politics, East Germany's black-market economy, transnational architectural politics after Nazism, and the politics of ecology in East Germany's industrial South before and after 1989. His current project compares the devastation of villages, recultivation practices, environmental movements, and memory politics in lignite pit mining regions in East and West Germany before and after 1989.

Petra Dolata is Associate Professor at the University of Calgary. A former Canada Research Chair in the History of Energy, who held previous academic positions at the Freie Universität Berlin in Germany and at King's College London in the UK, her research examines the 1970s energy crises, transatlantic energy relations, and the historical connections between deindustrialisation and energy transitions.

Rune Frandsen graduated in architecture from the École Polytechnique Fédérale de Lausanne (EPFL) in 2015 and earned his doctorate from the Swiss Federal Institute of Technology (ETH) Zurich in 2023. His research focuses on integrating the histories of territory, construction, and labour and exploring their intersections within architectural contexts.

Corinne Geering holds a tenure-track position at the Department of Economic, Social, and Environmental History and the Linz Institute for Transformative Change (LIFT_C) at Johannes Kepler University Linz. She is the author of *Building a Common Past: World Heritage in Russia under Transformation, 1965–2000* and co-editor of *Heritage under Socialism: Preservation in Eastern and Central Europe, 1945–1991*. Her wider research interests include the use of the past in regional and urban development.

Ewan Gibbs is Senior Lecturer in Economic and Social History at the University of Glasgow. He is a historian of energy, industry work, and protest in the United Kingdom. His research focuses on how the evolution of energy economies mould people and place through shaping and rearranging politics and power.

Jenny Hagemann is a cultural anthropologist and works in the Department 'Regional Development and Minority Protection' at the Sorbian Institute in Bautzen/Cottbus, Germany. Her research focuses on minority rights in industrial and regional planning, minorisation processes and collective memories, regional heritage, and popular cultures.

Fabian Jacobs is an ethnologist and works in the 'Cultural Studies Department' at the Sorbian Institute in Bautzen/Cottbus, Germany. From the perspective of comparative minority studies, he is primarily concerned with questions of cultural diversity and sustainable development in regions of Lusatia affected by lignite mining.

Marin Kuijt is a PhD candidate at the University of Amsterdam. He studies the power of fossil fuel corporations in the past and present. His dissertation focuses on the colonial and postcolonial history of oil in the Netherlands and its empire. Besides this, he is interested in studying how fossil fuel conglomerates shape our understanding of sustainability and the future of energy in museums.

Lutz Laschewski is a rural sociologist and economist and has been dealing with structural change and transformation processes, especially in rural areas in eastern Germany. Since 2022, he is the Head of the Department 'Regional Development and Minority Protection' at the Sorbian Institute in Bautzen/Cottbus, Germany.

Hanna Lempinen is a senior scientist at Natural Resources Institute Finland (Luke). Her research has focused on social and cultural sustainability of extractive industries and justice considerations associated with the green energy transition. She has worked with peatlands, Finnish peatland politics, and the nations' peat workers for the last 15 years.

Saara Mildeberg is a junior research fellow at the Centre for Landscape and Culture at Tallinn University, Estonia. She holds a bachelor's degree in photography and a master's degree in ethnology. Her current research focuses on people and landscapes, difficult heritage, and cultural tourism.

Merve Neziroğlu holds a BA in History and Cultures of the Middle East/Islamic Studies and an MA in European Studies. She was a research associate in the project 'Contested Waterway. Governance and Ecology on the Lower Danube, 1800–2018' funded by the Leibniz Association. Her research interests include Southeast European history, cultural and environmental history, and memory studies.

Gertjan Plets is Associate Professor in Heritage Studies at Utrecht University. He studies the musealisation and heritagisation of fossil fuels through exhibitions and industrial heritage. He is particularly interested in how corporate actors mobilise heritage as a vehicle for promoting their understandings and discourses about energy transitions and sustainability.

Leyla Sayfutdinova is Senior Lecturer in Critical Area Studies at the University of Glasgow. Her research has focused on the outcomes of the post-Soviet industrial restructuring in Azerbaijan. Her latest project, 'Turning Oil into Stone' examines the co-production of oil industry and urban space in Baku.

Sarem Sunderland is a landscape architectural researcher based in Zurich and Munich. He is writing a doctoral thesis on the project of hydroelectric landscapes in Switzerland at the Swiss Federal Institute of Technology (ETH), Zurich. He is also a practicing landscape architect and co-founding member of the architecture collective la-clique.

Acknowledgements

The story of this book goes back to 2021, when we first began to have a conversation about landscapes impacted by energy production. We jointly coordinated a research lab on the commodification and valorisation of the past as part of the Leibniz Research Alliance 'Value of the Past.' We are immensely grateful to the Leibniz Research Alliance for granting us financial support to prepare, edit, and publish this book. Working on this book in Leipzig and Bochum, post-mining landscapes were right on our doorsteps. We share an interest in combining the histories of mining districts and moving beyond national borders. As our professional and private lives led us further afield to other landscapes, such as hydropower reservoirs in the Alps or former oilfields in contemporary Poland and Ukraine, we sought to move beyond a focus on just one energy resource. What can we learn from shifting our focus to the people living in these landscapes, affected by coal mining, oil drilling, peat extraction, and hydropower?

In September 2022, we convened with invited colleagues in the Mining and Technology Park in the *Neuseenland* (New Lake District), a recultivated post-mining landscape south of Leipzig, to discuss and explore this question. We would like to thank Thorsten Hinz, Martin Baumert, Manuela Gründel, and the entire team at the Mining and Technology Park for hosting us in this post-mining landscape amidst excavators and technological remnants, a venue that was so inspiring for our discussions. While most of the contributions in this volume are based on the papers that were presented at this gathering, we also invited some more scholars to contribute to this publication in order to reflect the European dimension of this topic more fully. We are indebted to all of the authors for bringing together spaces across the continent and for their continuous hard work on their chapters. It has been such a pleasure working with them.

We would like to thank Susannah Eckersley and Christopher Whitehead for including this volume in their series, 'Critical Heritages of Europe' and Heeranshi Sharma from Routledge for her thoughtful editorial assistance during the publishing process. We would also like to thank Steffen Butzkus and Lena Zirkel for their support in preparing this volume. In a very special way, we would like to express our gratitude to Radhika Johari, whose careful and precise copy editing gave the texts the 'finishing touches.'

We hope that this book will inspire further conversations, and we look forward to having them.

Corinne Geering and Torsten Meyer,
Leipzig/Bochum in autumn 2024

Introduction

Cultural Landscapes of Energy and the Conflicting Temporalities of Energy Transition

Corinne Geering

In January 2023, more than 10,000 people came to Lützerath in the state of North Rhine-Westphalia in Germany to protest against the hamlet's imminent destruction. Like other recent protests against extractivism and large-scale energy infrastructure, the demonstrations in Lützerath were covered in the media across Europe, transforming the small settlement into a site of transnational activism. Almost 20 years earlier, the energy company RWE had initiated resettlement, court-ordered expropriation, and compensation of approximately 100 villagers to make way for the expansion of an open-cast lignite mine. Some of the residents had tried to avert the destruction of their homes, but Lützerath's fate had been sealed by a German Federal Constitutional Court ruling in 2013.[1] After several days of stand-off, the police cleared the protesters, and soon after, like hundreds of other localities in Germany during previous decades,[2] the hamlet was swallowed up by the open-cast mine to make way for energy production.

The timing of the destruction challenges the temporalities associated with the heritagisation of mining in the Ruhr district and the German government's coal phase-out plans. The last active colliery located around 70 km from Lützerath had closed in 2018, marking the end of a decades-long deindustrialisation process in the Ruhr district. Over the past few decades, this region has emerged internationally as a leading example of industrial heritage-making, with prominent images of redeveloped industrial remnants looking quite different from landscapes still undergoing extraction.[3] Moreover, just one month before the protests, the German federal government had adopted a law preponing the planned ending of lignite mining in North Rhine-Westphalia by eight years to 2030 and had committed to doubling the generation of renewable energy from wind power during this time.[4] Nonetheless, alongside the demolition of houses, the local church, and the clearance of trees in Lützerath, the energy company began to disassemble a wind farm situated adjacent to the hamlet to make way for lignite mining.[5] Thus, in what is now an open-cast mine—with future plans to transform it into a lake once coal extraction ends—the multiple layers of the history of this energy landscape were rendered visible for a brief moment (Figure 0.1).

This landscape is just one of many sites across Europe that have been impacted profoundly by the production of energy. The extraction of natural resources

DOI: 10.4324/9781003473534-1

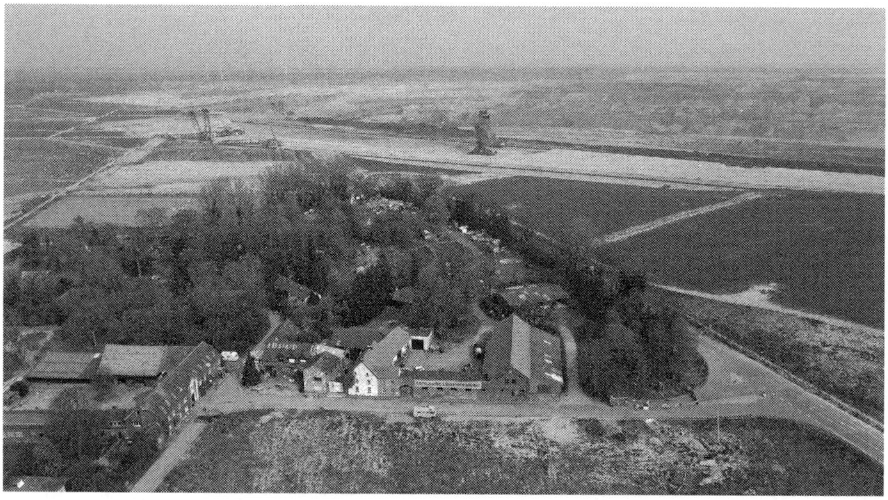

Figure 0.1 View of Lützerath where protesters had moved to a farm, holding out against the advancing excavator. Both the village and the wind park in the background were destroyed by the adjacent open-cast lignite mine soon after this picture was taken. Photograph by Dirk Jansen, 2022.

through coal mining, oil drilling, and peat production and the construction of large-scale infrastructure for electricity generation, such as dams, power plants, and wind farms, have shaped what we call 'cultural landscapes of energy' in this volume. Different histories ranging from an economic boom and promises of a prosperous future to sociocultural dislocation, loss, and ecological decline shape the understanding of these landscapes. Today these stories are largely told in the wake of the closure of mines and the dismantling of infrastructure, thus positioning them firmly in the past. However, as the example of Lützerath vividly illustrates, the past, present, and future of energy production is not a linear process. It generally involves different types of energy resources, both fossil fuels and renewables, simultaneously. Consequently, the process of energy transition is characterised not only by shutdowns, but also by deferral and new commissioned projects. Moreover, for the people interacting with these landscapes—workers, planners, residents, or visitors—the built environment of energy production is still an important part of their present-day experiences. In many places, the monostructural orientation towards energy production has had an enduring impact on the local society, reflected in outmigration and unemployment after these industries are shut down. Regional development initiatives have sought to counteract these tendencies by promoting other economic sectors, in particular tourism and leisure activities, which also actively draw on the past through heritagisation. Here service provision, souvenir design, museum exhibitions, and itineraries are all engaged in the construction of histories for the present and future of the affected landscapes.

In these places undergoing change, heritage is not just about history; it is also a future-making practice.[6] It evokes the past in order to serve present-day concerns, such as viable economic development, and to craft the desired future of sustainability and the achievement of net zero emissions. Region branding through heritage initiatives impacts processes of identity formation, be they centred on class, ethnicity, or gender, as industrial work cultures end with pit closures and new opportunities emerge in what has been termed the 'post-industrial society.'[7] Some regions are viewed as particularly successful examples in using their industrial heritage to project a future following deindustrialisation, such as the aforementioned Ruhr district in Western Germany.[8] 'Born on coal,'[9] as a popular Ruhr slogan states, regional heritage politics became a shared venture across the political spectrum through cooperation with energy companies for transforming the closed-down industrial infrastructure into new business, science, leisure, and industrial ventures in the 1990s.[10] Many more examples can be found across Europe where the cessation of coal and other fossil fuels extraction in the late twentieth century became an important part of the history written by these initiatives. Energy has thus been positioned in the past, seemingly detached from ongoing struggles over national energy security and global climate targets. The contributions in this volume clarify that heritage-making does not start after the industry has ended; rather, it is initiated along the way or at times even before the extraction of energy resources or the construction of energy infrastructure begins. This is particularly relevant for current debates on the energy transition from fossil fuels to renewables, achieving net zero emissions and mitigating the impacts of climate change.

This volume sheds light on tensions over the past in contemporary Europe, where energy production is again at the forefront of political debates in view of climate change, armed conflicts, and a renewed focus on environmental and social justice. Tying into research on 'the politics of heritage,'[11] it also provides new perspectives on how the past is mobilised in the present by different stakeholders as energy production is rearranged, extended, or even revived in response to shortages, addressing pressing economic, social, and environmental challenges. While most publications on energy, both in the past and present, deal with one particular energy resource, the focus on 'cultural landscapes of energy' in this volume enables us to bring together conflicting histories surrounding work, habitation, and leisure that are shared by these landscapes across different types of energy. Written by a multidisciplinary group of authors, the chapters position the stories of coal, oil, hydropower, and peat in the respective landscapes within a wider context of energy politics. They critically assess the past—and thereby the future—constructed for the regions in Europe located between the North Sea, the Alps, the Carpathian Mountains, and the Caspian Sea. Heritage-making can thus be analysed as an ongoing process involving competing narratives of gain and loss centring on the coexistence of various energy resources and the transition from one to another. In this setting, histories are constructed not only to make sense of lived experiences, deal with difficult pasts, and find solutions for environmental remediation but also to support future energy production policies. The energy sector is—and will continue to be—a major contributor to economic development and employment in regions across Europe.

This introductory chapter outlines the volume's comparative framework across states and energy resources. By providing an overview of the processes shaping the 'cultural landscapes of energy' discussed in this volume, it situates them within current debates on industrial heritage, environmental and energy history, and energy transition. The first section of this chapter foregrounds the tensions between productive and destructive elements in the history of energy landscapes in Europe, tracing the emergence of a new 'landscape of power' in the twentieth century. The second section then examines practices of heritage-making in landscapes. While this body of scholarship is extensive, the discussion here focuses on (post-) industrial landscapes to highlight the distinctive traits of the 'cultural landscapes of energy.' Finally, the third section explains the structure of this volume, which is organised around the themes of creating, living with, and sharing landscapes.

How Energy Shapes Landscapes: Productive and Destructive Histories

Several researchers have noted that there was no energy *transition*, conceived in a narrow sense, but rather energy *additions*.[12] The use of older resources always continued, with additions of new sources of primary energy, thus leading to an overall increase in energy production. Therefore, what occurred was merely a shift in the proportions of different resources in the energy mix, with the overall required energy supply continuing to expand.[13] From the early twentieth century, with the advancing industrialisation of production and electrification of homes and workplaces, energy demands rose drastically. While energy production has a long history, particularly through the burning of biomass like wood or the use of hydropower, populations now began increasingly turning to fossil fuels. In ever-increasing quantities, coal fuelled railway and maritime transportation, and it fired new power plants, which generated electricity for both workplaces and private homes. Between the 1940s and 1970s, several technologies replaced coal with oil and gas that also ran the engines of vehicles in an age of increasing mobility by car and aeroplane.[14] The demand for fossil fuels extended far beyond energy production, thus boosting coal and oil extraction to ever greater degrees. For example, coal is used in steelmaking, and petrochemicals derived from oil are used to manufacture thousands of products that are part of our everyday lives.[15] The consumption of coal continued to increase even after its replacement by new technologies. Against this background, what may be considered the past in energy production continues in the present and extends into the future.

The production of energy reshaped societies, economies, and political governance and had a transformative effect on landscapes worldwide.[16] While these processes date back centuries, evidenced, for instance, in land clearing and water management, what emerged in the twentieth century was a new 'landscape of power' whose modernist infrastructure continues to shape popular images of energy heritage today.[17] This landscape comprised 'immense constructions of the oil refineries, the nuclear reactors and power stations, and, radiating from them, the great latticed towers and overhead lines of the electricity grid, forming a network which covers the whole country.'[18] Notwithstanding general trends, the impact of energy production on a specific landscape largely depended on local geographical conditions and the prevalent energy resource, thereby revealing longer historical trajectories. This can be illustrated in the European regions explored in the chapters of this volume (Figure 0.2), which are representative of many more landscapes across the continent. The peatlands concentrated in the Ostrobothnian regions in Finland continue in the present to play a central role in the country's economy and energy mix. Coal and lignite deposits have been mined to produce energy in both western and eastern Germany. Oil shale mining in Ida-Virumaa and oil drilling on the Abşeron Peninsula began during a time when both regions were part of the Russian Empire and continue to be dominant industries in contemporary

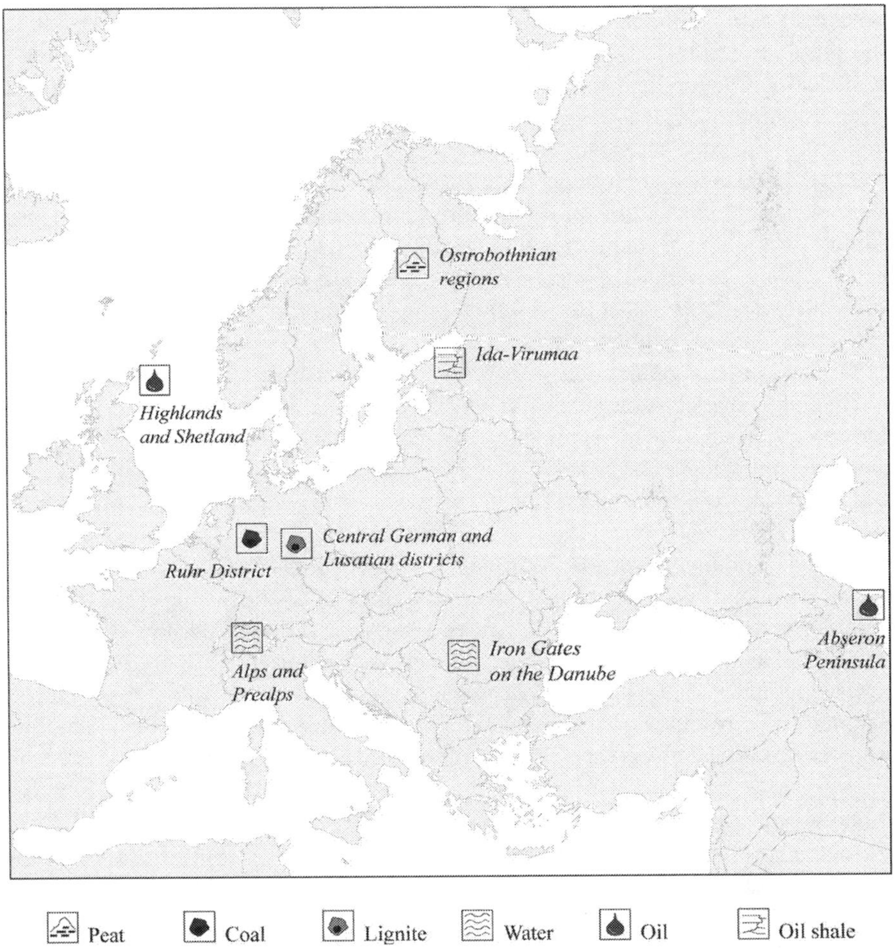

Figure 0.2 The 'cultural landscapes of energy' explored in this volume. Map by Silke Haps.

Estonia and Azerbaijan, respectively; and the petroleum reservoirs in the North Sea were discovered several decades later. Finally, the alpine mountainous environment and the Danube River provided the groundwork for the erection of dams with adjacent water reservoirs and the flooding of the valleys and their settlements to produce hydropower. However, while wandering through these and other ubiquitous 'energy landscapes' across Europe, it might not always be readily discernible to the visitor that the hills or lakes they see are the outcomes of energy production and the mitigation of its impact.[19]

The use of these landscapes has been subject to contestation and political struggles over the last century. The territoriality of energy production was crucial to this contestation: it combined the horizontal and vertical dimensions of landscapes by adding, through the extraction of fossil fuels, a subterranean energy regime to the already existing surface energy regime.[20] The transition to renewable energy resources will not ease these tensions; instead, as Matthew T. Huber and James McCarthy anticipate, it is likely to further amplify earlier political debates. The production of wind and solar power is more land-intensive compared with fossil fuel extraction, and it 'will entail the *restructuring* of land-based struggles in the constitution of the energy–society relations.'[21] For instance, recent protests by Sámi indigenous activists in Norway against the construction of a wind farm, which will impede their reindeer herding, seem to confirm this restructuring.[22] Moreover, notwithstanding protests against fossil fuel extractivism, which are gaining pace globally, industries that depend on it, such as petrochemical manufacturing, have experienced sustained growth and are projected to make up for oil companies' declining fossil fuel production in the coming decades.[23] As a result, a new territoriality of extraction emerges from the diverse sites involved in the production and processing of energy resources and downstream industries.

Thus, the 'planetary mine' is not spatially limited to a specific site; rather, it can be conceived as a transnational network of infrastructures and technologies integrated by a logistical system, which spans the entire globe.[24] Using a comparative framework focused on Europe allows for the connection of the planetary consequences of energy production and the transregional networks to national frameworks and finally specific locations. This is particularly important, given that energy production was connected to programmes of social engineering, nation-building, and geopolitical interests, all of which reflect back on how the past is constructed today.[25] In some instances, energy even provided a foundation for imperial region-making, in which particular landscapes were used for the economic exploitation of resources. The Donbas (a portmanteau for Donets Basin), a region in eastern Ukraine, which was named with reference to the large underground coal deposits, first fuelled the industries of the Russian Empire and subsequently those of the Soviet Union.[26] Moreover, with the emergence of the Soviet Union as one of the main exporting states of oil and gas, the pipeline systems built from the fields in the Idel-Ural and Western Siberia leading to Europe became an important means of consolidating the socialist bloc during the Cold War.[27] Russia's ongoing war against Ukraine clearly demonstrates how this history reverberates in the present and will continue to impact the future.

Against the backdrop of political and economic interests, local perspectives on cultural landscapes of energy across Europe have tended to fade into the background of these overarching narratives. In recent years, critical heritage studies have aimed to 'rethink heritage in precarious times,' addressing current crises such as the climate emergency and growing inequality in an effort to craft more inclusive futures.[28] In this context, landscapes have been a focus of research on the environmental and social impacts of extractivism. When heritage-making is sustained under these transformative processes, it becomes less about the conservation of stable entities and more about 'maintaining continuity with the past' in an unstable future.[29] The topic of energy speaks directly to these questions, which have been explored mostly in Anglophone regions, including post-settler and postcolonial contexts.[30] The chapters in this volume expand on this research by zooming in on marginal spaces in Europe—from the offshore oil fields to open-pit mines, mountains, and peatlands—and revisiting prominent cases of industrial heritage such as the Ruhr district. They challenge established temporalities that position heritage-making as succeeding energy production.

Energy has transformed and continues to transform the lives of those residing and working in landscapes affected by large-scale infrastructure and extraction, forcing them to readjust to their new environments and adapt their daily practices.[31] New workers' settlements have had a lasting impact on places even decades later when the respective worker communities no longer exist.[32] The social and cultural dimensions of large structural transformations, such as pit closures, reveal unresolved challenges and, at times, conflicting responses on the part of those facing unemployment and an uncertain future.[33] Even in countries like Switzerland, where public votes were held on the construction of energy infrastructure, which would cause the destruction of villages through flooding, poverty and economic pressures compelled the voters to concede to offers of compensation by the concerned energy company and accept the dam project.[34] Examples such as this one highlight the ambivalent ways in which processes on the ground unfolded. Particularly in places where the majority of the population worked in the energy sector, individual biographies integrated both the productive and the destructive sides of energy production. The account of a young lignite miner in an East German documentary from 1983 portrays this ambivalence very well. The film traces the destruction of the young miner's village to make way for an open-cast lignite mine. The filmmaker first attends the miner's wedding-eve party in the village, and two years later, he accompanies him on an excavator moving across the new open-cast mine. The young miner is asked how he feels about driving by the location where his home used to be and whether he ever thinks of it. He gives a shrugging response: 'Oh, thinking. Shall we say, what is gone is gone.'[35] He continues to operate the excavator while the devastated landscape, which can be seen in the background, provides him with employment and serves the national objectives of a planned economy.

The drastic interventions in the environment relating to energy production gave rise to the idea, in the twentieth century, that these landscapes needed active planning. Essentially, 'a new landscape' was to be created in place of the former one that the industry was destroying, an idea that reverberates in practices of conservation and preservation.[36] Architects, designers, and planners attempted to integrate the new energy infrastructure seamlessly into the environment. For instance, already at the turn of the twentieth century, when substations had to be installed for the transmission and distribution of electricity, they were designed to fade into the background of the historical surroundings, for instance, as timber framing architecture in rural settings.[37] In regional planning, these approaches have continued into the present to minimise adverse 'scenic impacts' from renewable energy infrastructure.[38] Furthermore, when energy extraction and production end, some landscapes have been recultivated in ways that hardly anything reminds current visitors who come to these places for recreation today of the earlier presence of heavy industries. This is particularly true for open-cast mining and other forms of resource extraction that leave behind large pits.[39] After the excavators had left the former site of the destroyed village featured in the East German documentary mentioned above, the pit was flooded as part of recultivation measures. The former open-cast mine is today covered by one of the post-mining lakes in what is now the *Neuseenland* lake district south of Leipzig.

Landscape transformation does not, however, imply that references to the past are absent—quite to the contrary. Some recultivation measures even explicitly incorporate the past through practices of heritage-making. The destroyed church tower from the young miner's devastated village in the aforementioned East German documentary was recreated as a floating platform for hosting events (Figure 0.3). Named 'VINETA' after a mythical place swallowed up by the sea, the location actively combines fact and fiction from medieval legends and contemporary history. Similar to an iceberg, the floating platform hides a large construction

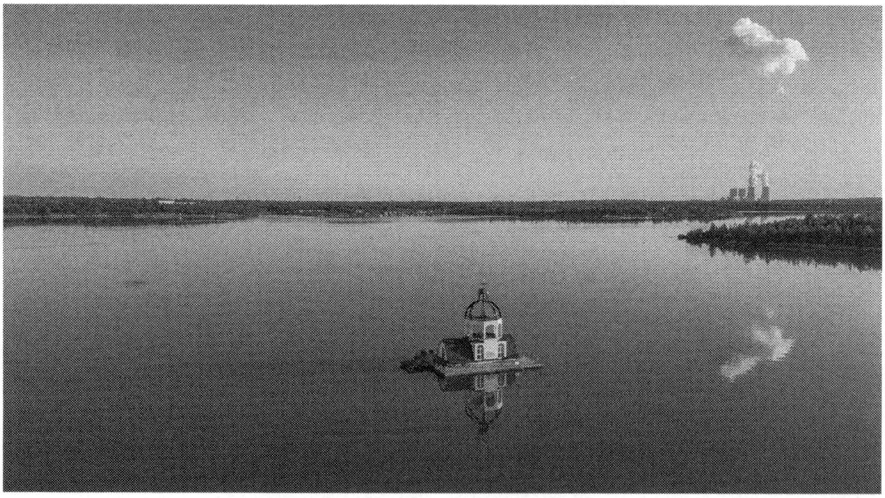

Figure 0.3 The floating 'VINETA' platform located on Lake Störmthal in the Leipzig *Neuseenland* post-mining landscape. Photograph by Carsten Müller, May 2022.

below the water surface in the seemingly pristine appearance of the landscape. It is attached by 86-metre-long iron chains to four containers filled with concrete that had been installed at the bottom of the open-cast lignite mine at the church's former location before it was flooded by the lake.[40] What is gone, as the young miner said in the documentary, has returned as heritage.

Heritage-Making: Conflicting Pasts, Presents, and Futures of Energy

Many of the landscapes shaped through energy production in the twentieth century no longer look as if they are pointing the way to the future. Power plants have been decommissioned, abandoned derricks are rusting, cracks are showing in the concrete of hydropower dams, and open-cast coal mines have been recultivated as lakes. What were once viewed as 'new lives' and 'new landscapes' now oftentimes represent a history of environmental degradation and pollution as well as declining industry and unemployment.[41] Studies on deindustrialisation have focused in particular on Western Europe and North America, whose former industrial heartlands experienced major structural transformations in the 1970s and 1980s. The transformation of the Steel Belt into the Rust Belt in the United States can be seen as emblematic of this transition, leading to a multitude of industrial ruins and millions of displaced workers. While a more general global trend of a transition to a 'post-industrial society' can be observed, reflecting a shift from the production of goods to the provision of services, deindustrialisation processes are characterised by 'spatial and temporal unevenness' across the globe and by the varying responses of people to the ruination of their lived environments.[42] In the context of Europe, which is the focus of this volume, the experience of decline and degradation across the continent has been fragmented. The end of state socialism and of planned economies in 1989–1991 led to the dismantling of large state-owned industries in Central and Eastern Europe, thus further intensifying the process of deindustrialisation already underway. Consequently, industrial ruins today may be found in various parts of Europe, but the reasons for their appearance and their interpretations may differ radically.

The declining regions with their industrial ruins were to be transformed across the continent into 'blooming landscapes' to live and work in.[43] The decommissioned industrial built environment was to be integrated into redevelopment initiatives, facilitated by the recent trend of official recognition of industrial heritage by national and international bodies. While the preservation of technical artefacts can be traced back to the turn of the twentieth century, it was not until the 1960s–1970s that the field of industrial archaeology emerged and the first international congresses on the conservation of industrial monuments were held in different European cities. Historical scholarship reveals that these efforts spanned both the Western and Eastern blocs during the Cold War, thus providing a relevant international framework for virtually all of the cases explored in the chapters of this volume.[44] The official recognition of energy infrastructure also gained ground during this period, which is why in this volume energy infrastructure is discussed in the overarching context of industrial heritage rather than as a distinct phenomenon, such as mining heritage or 'nuclear cultural heritage.'[45] For example, the Lake Sihl reservoir in Switzerland explored by Sarem Sunderland in this volume was given protected status in 1962, and the first colliery in the Ruhr district was recognised as industrial heritage in 1969. The reservoir has continued to operate, whereas the colliery has been decommissioned. Therefore, it is evident from these two sites that there is no clear correlation between deindustrialisation and heritagisation, and one is not a prerequisite for the other.

Notwithstanding continuing industrial operations, and even their expansion, most of the literature on heritage-making processes in the context of energy production has focused on what comes after the industries end. They generally frame specific buildings, sites, and landscapes as well as local histories as 'post-industrial,' thus focusing on ruins rather than ongoing operations. A particular focus has been on processes of memorialising the industrial past as an anchor of identity in deindustrialised regions, given the repurposing of former factories and production sites and the reshaping of forms of belonging.[46] When it comes specifically to landscapes, affective and embodied encounters, especially by means of autoethnography, have played a central role as ways of approaching the 'post-industrial landscape.'[47] In environmentally degraded landscapes, 'toxic heritage,' for instance from nuclear waste, raises the question of how unwanted pasts can be managed and thus contained.[48] Landscapes have been described as wastelands that are 'reclaimed, restored, and redeveloped'[49] and as 'places of latent potential,'[50] offering a canvas for creative practice to make new sense of the memories associated with abandoned industrial sites.[51] In many places, the post-industrial heritage has been operationalised for tourism to generate new employment and an income flow from outside the site, though generally below the level provided by the previous industry.[52]

By building on traditions of landscape architecture and planning, or alternatively, by analysing recent activities in these fields, scholars have sought to direct attention to what Anna Storm described in *Post-Industrial Landscape Scars* as 'ambiguous, painful, and damaging pasts, as well as to possible healing processes.'[53] In another article written with Krister Olsson, Storm has explored the local meanings and narratives centring on a continuously growing pit in the former centre of Malmberget, a northern Swedish mining town, showing how the local population have connected their values to the pit's physical changes over time.[54] These authors argue that the 'landscape scar' can be deployed as a cultural tool: to the local population, the pit represents the past mining boom as well as experiences of forced relocation, decline, and loss of employment. It is also a special landmark regularly shown to visitors, distinguishing the town from other places, thus enabling its emergence as a heritage site in its own right. Storm and Olsson are two of many researchers who have sought to expand the concept of heritage, emphasising the perspectives of local stakeholders on the historic remains of an industry and its impact on landscapes. Alice Mah, a sociologist who expressed criticism of celebratory accounts that ignore the lives of the inhabitants of a site, has focused on the lived process of industrial ruination and decline to understand how local communities relate to the physical legacy of deindustrialisation and the changes that it reflects.[55] More recently, as a researcher interested in post-industrial ruins, Mah has also pointed to the coexistence of decline and booming industry as an unexpected aspect of her earlier research.[56] Examples like these highlight the many processes of heritage-making associated with local industrial pasts that are taking place outside of the space of official recognition.

Our understanding of 'cultural landscapes of energy' builds on this research and challenges linear temporalities of deindustrialisation. Instead, we seek to consider the coexistence of phase-out and ongoing energy production, expanding the analysis of heritage-making, which previous scholarship has typically confined to the period following the end of energy production. The concept of a 'cultural landscape' is well suited for pursuing this approach, as it has been used for enquiries into the intimate relationship between cultural and physical environments.[57] The concept first emerged in the late nineteenth century to refer to the formative influence of human activity on natural landscapes, later inspiring methodologies of 'reading' landscapes.[58] From the 1970s onwards, preservationists and heritage site managers adopted the cultural landscape concept, who were interested in developing new approaches focusing on holistic entities conceived as a series of layers linking past, present, and future rather than as individual monuments viewed in isolation. This approach encompassed not only the natural environment and the countryside but also conservation of the urban landscape,[59] analyses of which have often sidetracked the sensory and environmental dimensions of place.[60] Against this backdrop, the cultural landscape concept has been used to critically assess and overcome the traditional split between nature protection and monument preservation.[61] New inter- and transdisciplinary approaches were being developed in the recent past, bringing together conservation with anthropology, archaeology, geography, ecology, architecture, and several other disciplines.[62] These approaches enable the analysis of the cultural landscape as a process, encompassing complex social-ecological systems, a place inventoried in a heritage conversation register, or a practice carried out by state authorities and international bodies.[63] Against the backdrop of energy transition policies, recent discussions on transformative conservation have shifted the emphasis on the entanglement of nature-based and people-centred approaches. Ongoing initiatives in many landscapes shaped by energy aim to restructure systems, promote the transition to net-zero emission economies, and secure sustainable use of natural resources.[64]

The valorisation of landscapes shaped by energy production is a relatively recent practice, as industrial activity long remained outside the focus of the 'Authorized Heritage Discourse' for a long time.[65] For example, when the concept of a cultural landscape gained wider acceptance internationally among heritage professionals through its inclusion in the Operational Guidelines of the United Nations Educational, Scientific and Cultural Organization (UNESCO) World Heritage Convention in 1992, its official definition initially excluded industrial landscapes. Emerging from competing visions based on different understandings of landscapes rooted in specific geographical contexts,[66] this definition instead encompassed a specific set of human-made landscapes whose protection was seen to support sustainable land use and the maintenance of biological diversity.[67] Arguably, compared to agriculture and crafts, industrial landscapes were deemed less aligned with these objectives. Moreover, they were perceived to lack the traditionally conceived aesthetic qualities projected through the genre of landscape painting. However, advocates for the recognition of industrial landscapes have argued that factories, transport routes, waste disposal areas, workers' housing, and other infrastructure equally represent stages of interaction in the evolution of a historic landscape.[68] The European Landscape Convention of the Council of Europe, adopted in 2000, reflects this more holistic approach to landscapes. Its conception of landscape includes both places of beautiful scenery and those devastated by resource extraction, such as open-cast lignite mines or the growing pit in Malmberget mentioned earlier.[69] Consequently, in 2011, the urban mining landscape of Carbonia, a city in Sardinia, received the Landscape Award for its exemplary approach to sustainable development. UNESCO has also drawn inspiration from these developments, and several sites currently recognised as cultural landscapes have been shaped by extractive industries, including energy resources, such as coal in northern France and South Wales and uranium in the German–Czech border region.

However, what all of these landscapes have in common is the end of mining and the abandonment of their industrial infrastructure, often accompanied by economic decline, which preceded their redevelopment and international recognition. Mining activities, for instance the planned lithium mining in the German Ore Mountains, are unfolding adjacent to the protected territory and are monitored by international bodies to mitigate potential impacts on heritage management.[70] In case of extractive activities or the construction of large-scale infrastructure in the protected landscapes, heritage management becomes unfeasible, and this can lead to the removal of the heritage status of these landscapes. Given that extractive industries and the energy sector are still expanding, the way the local impact of industrialisation and energy production is valorised entails at times a tension between past and present. For example, the Nord-Pas de Calais Mining Basin in northern France is recognised not only for the technical installations and factories associated with coal extraction but also for the 'remarkable example of worker solidarity' in the landscape shaped by the 'continuous mono-industrial occupation of coal mining over almost three centuries.'[71] While this mono-industry is lauded for preserving an authentic and integral mining landscape devoid of other industries, it has also resulted in enduring social challenges in the region. On the one hand, the decommissioning of mines led to high unemployment rates in the basin, which were still at 20–30% at the time of the UNESCO World Heritage inscription.[72] On the other hand, the mining heritage is being used to revitalise the region as a tourist destination. Tourists can hike up the slag heaps, glide down an artificial ski slope, and drink 'Charbonnay' wine (a wordplay on *charbon*, the French word for coal) from the new vineyards.[73] These practices of heritage-making firmly situate energy production in the past, much like the *Neuseenland* lake district south of Leipzig discussed earlier.

The above examples underline the need for more integrated perspectives, which take account of both official heritage-making processes and those unfolding outside protected territories. This approach renders visible the conflicting pasts, presents, and futures of cultural landscapes of energy in which some industries may have ended while others have continued or recently commenced. Depending on the perspectives of various stakeholders, radically different readings of the histories of the landscapes emerge. A focus on the end of energy production paints only a partial image of the past and how people relate to it in the present, thus projecting desired futures. This selective view also contributes to obscuring how histories are being constructed to support and mobilise specific energy policies. It is crucial to consider here that energy production is not only ongoing but also that it is on the rise. As highlighted by the cases explored in this volume, energy production remains a present and future concern, especially when talking about the past.

Creating, Living with, and Sharing Landscapes

The past of cultural landscapes of energy is integral to the ways in which people interact with their environments in the present, how they make sense of it, and how they articulate desired futures. By exploring the meanings and uses of landscapes, the contributions in this volume, authored by scholars from history, heritage studies, architecture, sociology, anthropology, and sustainability science, reveal the conflicting temporalities of energy transition. Coming from different disciplines and academic traditions, and drawing on a variety of methods, the authors shed light on the 'socio-political complexities that enmesh heritage.'[74] Fieldwork and oral history interviews conducted in the Scottish Highlands and Finnish North reveal the lasting impact of historical energy regimes, which local communities uphold even as states seek to decommission production. The analysis of museums and media in the Swiss Prealps and at the Romanian-Serbian border highlights how heritage-making, in fact, preceded energy production and even played an integral part in the construction of the local energy landscape. Memories in post-mining landscapes are conflicted, and a careful uncovering of forgotten, overwritten, and suppressed voices of resettlers and minorities contributes to broadening the processes of heritage-making. And finally, the analysis of energy infrastructure, including the traces of temporary workers' settlements next to high-rise dams, raises new questions for conservation practices.

The chapters in this volume shed light on the diverse uses of landscapes that have been undergoing changes. This includes the domain of work that may experience a transition from the industrial to the service sector, such as tourism. They also enquire into the landscapes as places where people reside, possibly because they moved there for new work opportunities, which they later lost, or because their families had already lived in the region and experienced the changes over a longer period of time. Many of these landscapes are also now places of leisure for people living elsewhere who come into the environment shaped by energy production possibly without being aware of this. They bathe in post-mining lakes or hike on hills recultivated from waste heaps. In other places, the energy infrastructure is a touristic attraction in its own right with platforms erected for visitors to observe the scenery comprising power plants or open-cast mines. Visitors can also register for one of the many activities offered by local entrepreneurs, such as bungee jumping off high-rise dams. Thus, the ways in which the cultural landscapes of energy are currently used are manifold, and they can be seen as places where many different perspectives intersect, often involving conflicting aspirations.

The case studies explored in this volume bring together the perspectives of diverse stakeholders, including inhabitants, workers, excursionists, government officials, museum staff, and energy companies, thus revealing the multifaceted histories of prosperity and decline associated with the landscapes under study. They illuminate new visions for landscapes, at times even preceding the industry by decades, and the new industrial work cultures that were shaped by the exploitation of energy resources. While the production of energy was introduced to advance economic development, it not only changed the landscape but also destroyed villages and small towns in the vicinity as a result of excavation and flooding. Against this backdrop, current perceptions of the material remains of energy industries and the new landscapes they created are highly ambivalent. They range from appreciation of technological innovation, workers' pride, and artistic repurposing of industrial infrastructure to critiques of enduring negative ecological impacts and the contested legacies of displacement and hazardous working conditions. And finally, the story of energy production leading to a prosperous future features personal histories of loss or gain, and sometimes both.

This volume is structured into three parts that trace the processes shaping the cultural landscapes of energy, while paying close attention to the changes for people and the environment and the conflicting interpretations of the past, present, and future. The three parts do not follow a traditional chronology, thus reflecting the fact that the histories of power, prosperity, and decline associated with the landscapes are not linear. Instead, as the cases reveal, the temporalities of creation, interaction, and decommissioning can overlap and impact each other, sometimes repeatedly. The first part titled 'Creating Energy Landscapes' carves out the making of energy landscapes in the twentieth and twenty-first centuries. The second part, 'Living with Energy Landscapes,' then zooms in to explore how changes in the landscape have affected the environments and the people who live in them. The third and final part, 'Sharing Energy Landscapes,' looks at how current heritagisation projects approach and disseminate the histories of energy landscapes. These three parts reveal that the landscapes discussed in this volume were made and continue to be remade by energy. The ten chapters of the volume are followed by a concluding commentary by the renowned energy historian, Petra Dolata, who reflects on common themes explored across regions and energy sources and points to new questions and directions for future research.

In the first part of this volume, three chapters trace the emergence of cultural landscapes of energy, revealing how they were shaped by aesthetic traditions, the interests of political economy, and processes of nation-building. They highlight commonalities across energy sources by looking at hydropower, peat, and oil. The chapter by Sarem Sunderland titled 'The Cultural Construction of a Hydroelectric Landscape: The Project of Lake Sihl in Switzerland, 1897–1937' zooms in on Lake Sihl, a reservoir in the Swiss Prealps fuelling a dam power plant, which began operations in 1937. Approved through a public vote, the new lake flooded the valley, including peatlands, and destroyed the homes of more than 1,000 inhabitants. Sunderland's account reveals that by drawing on traditions of landscape painting and alpine visual culture, especially in tourism, various stakeholders started imagining the energy landscape long before it came into existence. While the human-made lake reflected the Swiss national image of pristine nature, the chapter by Leyla Sayfutdinova titled 'Land of Fire, Temples of Extraction: Azerbaijan's Geo-Architectural Assemblage of Oil and Nation' looks at the politics surrounding oil in Azerbaijan. By focusing not on the periphery but on the capital city of Baku, and by exploring the urban landscape, Sayfutdinova highlights the relevance of vertical territory. She shows how, over the course of more than a century, energy imaginaries were attached to Baku's built environment and extended to cover the entire territory of Azerbaijan. The chapter by Hanna Lempinen titled 'Energy Transition as Cultural Trauma: The Making and Unmaking of the Finnish Peat Industry' also examines the role of energy in forging a national cultural identity through an exploration of the continuing role of peatlands in Finland's energy mix. Based on fieldwork and interviews with peat workers, Lempinen reveals not only the communities' economic dependence on peat but also their intimate relations to the resource and landscapes shaped by its extraction. The ongoing energy transition in Finland, which seeks to move away from fossil fuels, including peat, is experienced as a cultural trauma by communities living near peatlands, which has been used within political campaigns to influence decisions on energy policies.

The second part of this book focuses on living with cultural landscapes of energy as workers and residents—both temporary and more permanent. Enquiring into ephemeral infrastructure, work cultures, and displacement, the three chapters in this part reveal how big visions of energy production intersected with personal histories. The chapter by Ewan Gibbs titled 'Northern Scotland's Late Oil-Fuelled Industrialisation: Labour Mobility and Community Transformation since the 1970s' traces the dramatic changes in the Scottish Highlands and Shetland following the discovery of oil in the North Sea. Drawing on a large number of interviews, this chapter tells the stories of workers migrating to northern Scotland as well as those of erstwhile inhabitants, showing how they experienced changes from the initial boom to the eventual decline. The account connects offshore extraction with the onshore oil terminal, thus revealing an extensive cultural landscape of energy. The topic of labour is also explored in the following chapter by Rune Frandsen titled 'The Cultural Landscapes of Dam Building in Switzerland: Secondary Infrastructure and its Territorial Archive at the Grande Dixence Dam, 1950–1965.' His analysis, focusing on temporary workers' settlements established for the construction of energy infrastructure, addresses a crucial challenge in the heritagisation of energy production, namely, the focus on individual technical monuments—in this case, the 285-metre-high dam—while ignoring the 'secondary infrastructure' and the logistics required for its construction. Frandsen follows the traces left behind by temporary workers, who were largely seasonal 'guest workers' from Italy, in this high-altitude site. Traces of earlier settlements are also the focus of the final chapter in the second part by Andrew Demshuk titled 'Erasure as Heritage: Two Villages between Restoration and Destruction on an East German Lignite Moonscape.'

Focusing on the Central German Mining District south of Leipzig, this chapter traces two memorial archetypes that emerged following the destruction of settlements to make way for open-cast lignite mining. One village, which was submerged under water in what is now the *Neuseenland* lake district, has been commemorated as an idealised past. The other village experienced extreme pollution but ultimately survived, and its ecological remediation was embedded in the preservation of the historical built environment following German reunification.

The third and final part of this volume titled 'Sharing Energy Landscapes' examines how the past is disseminated through contemporary heritagisation initiatives to foster cultural tourism, support corporate and political objectives, or recentre marginalised perspectives. The four chapters in this part also point to blind spots and desiderata in present-day heritage practices when it comes to places and people whose histories have been closely intertwined with energy production. The chapter by Saara Mildeberg titled 'A Post-Industrial "Adventure Land?" Challenges for Cultural Tourism Development in the Estonian Oil Shale Region' focuses on ongoing efforts to support the economic diversification of northeastern Estonia, whose economy has been shaped by carbon-intensive oil shale extraction. The region of Ida-Virumaa, formerly an industrial hub during the Soviet period, has been experiencing economic decline and unemployment in the recent past. Drawing on ethnographic fieldwork and interviews with policymakers and stakeholders in tourism, Mildeberg explores how these actors are attempting to reinvent the region as an 'adventure land.' The chapter by Marin Kuijt and Gertjan Plets titled 'Curating a Future for Coal and Petrochemicals: Ruhrkohle AG's Corporate Influence on the Zeche Zollverein Heritage Site' also looks at efforts to promote an industrial built environment as a cultural tourism site. This chapter revises earlier histories of the 'success story' of Zeche Zollverein in Germany's Ruhr District, shedding light on the involvement of corporations and the combination of mining and downstream processing of coal. By drawing on archival company records, the authors show how the narrative of energy transition was co-written by the energy company, thereby positioning coal in the past and obscuring their ongoing business activity. Along similar lines, highly selective accounts presented in contemporary museums and heritage sites are also the subject of the chapter by Merve Neziroğlu titled 'Representing What Has Been Destroyed: The Sunken Island of Ada Kaleh in Museums of the Iron Gates Region in Romania.' Visiting the region on the fiftieth anniversary of the dam's construction, Neziroğlu focuses on the island of Ada Kaleh, a former Ottoman exclave inhabited by a Turkish minority, which was submerged underwater following the construction of the largest dam on the Danube. It has resurfaced in the present in several museums in the region, including the technological museum of the dam, and depictions of life on the island continue to sustain an orientalist tradition that emerged more than a century ago with the onset of tourism in the region. The final chapter of this volume reflects the shared perspective of three scholars who were involved in a recent regional heritagisation project. Written by Jenny Hagemann, Fabian Jacobs, and Lutz Laschewski, the chapter titled 'Integrating Minority Perspectives into the Heritagisation of Post-Mining Landscapes in Lusatia, Germany' addresses a fundamental problem, namely, the separation of the industrial history and the history of Lusatia as a region inhabited by Sorbians, one of Germany's recognised minorities. Their insight into a failed nomination for UNESCO World Heritage listing draws attention to competing needs and learning outcomes during the participative process at the local level.

Taken together, the chapters in this volume outline the conflicting histories surrounding work, habitation, and leisure in landscapes across Europe, which were affected by coal mining, oil drilling, peat extraction, and the construction of hydroelectric power reservoirs. Moreover, they highlight the ambivalent attitudes towards 'cultural landscapes of energy' where changes brought about by energy production are lauded by some as markers of progress and bemoaned by others for the destruction of traditional livelihoods. Drawing on archival records, interviews, and fieldworks, many of the accounts in these chapters combine both perspectives on the productive and destructive sides of energy, addressing the tensions emerging from heritage-making narratives that focus on the end of energy production despite ongoing and future commissioning projects. Thus, the authors provide novel insights not only into the multifaceted and complex ways in which the landscapes they describe were shaped by energy in the past but also into how they continue—and will continue—to be shaped by it.

Notes

1 "Power Trumps Property."
2 See Berkner, *Bergbau und Umsiedlungen*.
3 See Berger, Wicke, and Golombek, "Burdens of Eternity?"
4 Deutscher Bundestag, "Braunkohleausstieg im Rheinischen Revier"; Landesregierung Nordrhein-Westfalen, "Wir machen Tempo."
5 Müllender, "RWE reißt Windräder ab."
6 Harrison, "Heritage as Future-Making Practices."
7 Bell, *The Coming of Post-Industrial Society*.
8 Wicke et al., *Industrial Heritage*.
9 This is a translation of 'auf Kohle geboren.' All translations are by the author.
10 Berger, "Representing the Industrial Age," 23.
11 Harrison, *Understanding the Politics of Heritage*; Bozoğlu et al., *The Routledge International Handbook of Heritage and Politics*.
12 York and Bell, "Energy Transitions or Additions?"; Bonneuil and Fressoz, *The Shock of the Anthropocene*.
13 For data on the current global energy mix, see the Energy Institute, "Statistical Review."
14 See Melsted and Pallua, "The Historical Transition."
15 See Chandler, *Shaping the Industrial Century*.
16 Gross and Needham, *New Energies*.
17 Ross et al., *New Lives*.
18 Crowe, *The Landscape of Power*, 9–10.
19 Pasqualetti, "Reading the Story of Energy," 12.
20 Huber and McCarthy, "Beyond the Subterranean Energy Regime?," 655.
21 Huber and McCarthy, "Beyond the Subterranean Energy Regime?," 666.
22 "Sámi Rights Activists."

23 Mah, *Petrochemical Planet*, 73.
24 Arboleda, *Planetary Mine*, 4–5.
25 See Mitchell, *Landscape and Power*.
26 Zabirko, "The Donbas." For an exploration of the ways in which this industrial emphasis continues in selective present-day representations of the region's past, see Sklokina and Kulikov, "Industrial Heritage."
27 Flade, "Creating a Common Energy Space."
28 Shepherd, *Rethinking Heritage*.
29 DeSilvey, "Ruderal Heritage."
30 See Zhu, "Putting the 'critical' in heritage studies."
31 Parr, *Sensing Changes*, xiii.
32 See, for example, the "Workers' Villages" project on the reimagining of a former peat worker settlement as part of Creative Ireland's Climate Action Fund at https://reimagineplace.ie/fieldwork/workers-villages/.
33 For an account that illuminates individual responses to large structural transformation, see Gibbs, *Coal Country*.
34 Mark, *Ein Bergdorf geht unter*, 29.
35 'Ach, denken. Sagen wir mal, was weg ist, ist weg.' This quote appears in Tetzlaff's documentary, *Erinnerung an eine Landschaft*, 1:10:03.
36 Fairbrother, *New Lives*, 382.
37 An example of a substation in timber framing from 1912 is preserved in the Erzgebirge Open Air Museum in Seiffen, Germany.
38 See Apostol et al., *The Renewable Energy Landscape*.
39 Meyer, "'Biofacts.'"
40 Lausitzer und Mitteldeutsche Bergbau-Verwaltungsgesellschaft mbH, "Espenhain," 28.
41 Anderson and Kelly, "What Happens," 258.
42 High et al., "Introduction," 4. See also MacKinnon and High, "Deindustrialization." For a broader perspective on deindustrialisation history in the United States, see Cowie and Heathcott, *Beyond the Ruins*.
43 Translated from 'blühende Landschaften.' For an early reference, see Kohl, "Der entscheidende Schritt," a speech delivered in 1990 by German chancellor Helmut Kohl on the monetary union of West and East Germany. The metaphor was later taken up and used in election posters of his party, the Christian Democratic Union of Germany (CDU) in the 1990s.
44 See Lehmann, "Socialism and the Rise of Industrial Heritage."
45 See Ross, "Nuclear Cultural Heritage."
46 Berger, *Constructing Industrial Pasts*; Orange, *Reanimating Industrial Spaces*; Storm, *Hope and Rust*.
47 Jaramillo and Tomann, *Transcending the Nostalgic*.
48 Wollentz et al., "Toxic Heritage," 295.
49 Kirkwood, *Manufactured Sites*, xiv.
50 McLean, *Transformative Ground*, 1.
51 Heeney, *The Post-Industrial Landscape*.
52 Rhodes et al., *Geographies of Post-Industrial Place*.
53 Storm, *Post-Industrial Landscape Scars*, 3.
54 Storm and Olsson, "The Pit."
55 Mah, *Industrial Ruination*, 10–12.
56 Mah, *Petrochemical Planet*, x.
57 See Jones, "The Concept of Cultural Landscape."

58 Sauer, *The Morphology*; Lewis, "Axioms.".
59 Taylor, "Connecting Concepts."
60 Hayden, *The Power of Place*, 18.
61 Taylor and Lennon, "Cultural Landscapes."
62 Jacques, "The Rise of Cultural Landscapes."
63 Brown and Goetcheus, "Introduction," 2–3.
64 Fougères et al., "Transformative Conservation."
65 Smith et al., "Introduction: Class Still Matters," 2; Kisiel, "Unwanted inheritance?"
66 Gfeller, "Negotiating the Meaning," 495.
67 This comprised landscapes like gardens, which were intentionally created, as well as those that evolved from social, economic, administrative, and/or religious imperatives, or those with religious, artistic, or cultural associations, mostly lacking material evidence. United Nations Educational, Scientific and Cultural Organization (UNESCO), "Operational Guidelines," 7–8.
68 Fowler, *World Heritage Cultural Landscapes*, 57.
69 Ermischer, "Walking the Landscape," 246.
70 UNESCO, "Current Development Projects."
71 International Council on Monuments and Sites (ICOMOS), "Nord-Pas de Calais Mining Basin," 214, 216.
72 Mission Bassin Minier, *Portrait socio-économique*, 22.
73 Winston Nicklin, "This Former Coal Mining Region."
74 Winter, "Clarifying the Critical," 533.

Bibliography

Anderson, Ben, and Matthey Kelly. "What Happens When Rural Modernity Ceases to be Modern?" In *New Lives, New Landscapes Revisited: Rural Modernity in Britain*, edited by Linda M. Ross, Katrina Navickas, Matthew Kelly, and Ben Anderson, 258–280. London: The British Academy, 2023.

Apostol, Dean, James Palmer, Martin Pasqualetti, Richard Smardon, and Robert Sullivan, eds. *The Renewable Energy Landscape: Preserving Scenic Values in our Sustainable Future*. London/New York: Routledge, 2017.

Arboleda, Martín. *Planetary Mine: Territories of Extraction under Late Capitalism*. London: Verso, 2020.

Bell, Daniel. *The Coming of Post-Industrial Society: A Venture in Social Forecasting*. London: Heimann, 1974.

Berger, Stefan. "Representing the Industrial Age: Heritage and Identity in the Ruhr and South Wales." In *The Invention of Industrial Pasts: Heritage, Political Culture and Economic Debates in Great Britain and Germany, 1850–2010*, edited by Peter Itzen and Christian Müller, 14–35. Augsburg: Wißner, 2013.

Berger, Stefan, ed. *Constructing Industrial Pasts. Heritage, Historical Culture and Identity in Regions Undergoing Structural Economic Transformation*. New York/Oxford: Berghahn, 2020.

Berger, Stefan, Christian Wicke, and Jana Golombek. "Burdens of Eternity? Heritage, Identity, and the 'Great Transition' in the Ruhr." *The Public Historian* 4, no. 39 (2017): 21–43.

Berkner, Andreas. *Bergbau und Umsiedlungen im Mitteldeutschen Braunkohlerevier.* [*Mining and Resettlement in the Central German Lignite Mining District*]. Beucha/Markkleeberg: Sax-Verlag, 2022.

Bonneuil, Christophe, and Jean-Baptiste Fressoz. *The Shock of the Anthropocene: The Earth, History, and Us*. London/New York: Verso, 2016.

Bozoğlu, Gönül, Gary Campbell, Laurajane Smith, and Christopher Whitehead, eds. *The Routledge International Handbook of Heritage and Politics*. Abingdon/New York: Routledge, 2024.

Brown, Steve, and Cari Goetcheus. "Introduction." In *Routledge Handbook of Cultural Landscape Practice*, edited by Steve Brown and Cari Goetcheus, 1–13. Abingdon/New York: Routledge, 2023.

Chandler, Alfred D. *Shaping the Industrial Century: The Remarkable Story of the Evolution of the Modern Chemical and Pharmaceutical Industries*. Cambridge, MA/London: Harvard University Press, 2005.

Cowie, Jefferson, and Joseph Heathcott, eds. *Beyond the Ruins: The Meanings of Deindustrialization*. Ithaca: Cornell University Press, 2003.

Crowe, Sylvia. *The Landscape of Power*. London: The Architectural Press, 1958.

DeSilvey, Caitlin. "Ruderal Heritage." In *Deterritorializing the Future. Heritage in, of and after the Anthropocene*, edited by Rodney Harrison and Colin Sterlin, 289–310. London: Open Humanities Press, 2020.

Deutscher Bundestag. "Braunkohleausstieg im Rheinischen Revier wird auf 2030 vorgezogen." ["The Lignite Phase-out in the Rhenish Mining District Will Be Pushed to 2030"]. Last modified December 2, 2022. www.bundestag.de/dokumente/textarchiv/2022/kw48-de-braunkohleausstieg-923096.

Energy Institute. "Statistical Review of World Energy 2024." www.energyinst.org/__data/assets/pdf_file/0006/1542714/EI_Stats_Review_2024.pdf.

Ermischer, Gerhard. "Walking the Landscape." In *Landscape Mosaics. Thoughts and Proposals for the Implementation of the Council of Europe Landscape Convention*, edited by Council of Europe, 236–255. Strasbourg: Council of European Publishing, 2022.

Fairbrother, Nan. *New Lives, New Landscapes*. Victoria: Penguin Books, 1970.

Flade, Falk. "Creating a Common Energy Space: The Building of the Druzhba Oil Pipeline." In *Cold War Energy. A Transnational History of Soviet Oil and Gas*, edited by Jeronim Perović, 321–344. Cham: Palgrave Macmillan, 2016.

Fougères, Dorian, Mike Jones, Pamela D. McElwee, Angela Andrade, and Stephen R. Edwards. "Transformative Conservation of Ecosystems." *Global Sustainability* 5 (2022): 1–14. https://doi.org/10.1017/sus.2022.4.

Fowler, Peter J. *World Heritage Cultural Landscapes 1992–2002*. Paris: UNESCO World Heritage Centre, 2003.

Gfeller, Aurélie Elisa. "Negotiating the Meaning of Global Heritage: 'Cultural Landscapes.' in the UNESCO World Heritage Convention, 1972–92." *Journal of Global History* 8 (2013): 483–503.

Gibbs, Ewan. *Coal Country: The Meaning and Memory of Deindustrialization in Postwar Scotland*. London: University of London Press, 2021.

Gross, Stephen G., and Andrew Needham, eds. *New Energies: A History of Energy Transitions in Europe and North America*. Pittsburgh: University of Pittsburgh Press, 2023.

Harrison, Rodney. "Heritage as Future-Making Practices." In *Heritage Futures: Comparative Approaches to Natural and Cultural Heritage Practices*, edited by Rodney Harrison, Caitlin DeSilvey, Cornelius Holtorf, et al., 20–50. London: UCL Press, 2020.

Harrison, Rodney, ed. *Understanding the Politics of Heritage*. Manchester: Manchester University Press, 2010.

Hayden, Dolores. *The Power of Place. Urban Landscapes as Public History*. Cambridge, MA: MIT Press, 1997.

Heeney, Gwen. *The Post-Industrial Landscape as Site for Creative Practice: Material Memory*. Newcastle upon Tyne: Cambridge Scholars Publishing, 2017.

High, Steven, Lachlan MacKinnon, and Andrew Perchard. "Introduction." In *The Deindustrialized World: Confronting Ruination in Postindustrial Places*, edited by Steven High, Lachlan MacKinnon, and Andrew Perchard, 3–22. Vancouver: UBC Press, 2017.

Huber, Matthew T., and James McCarthy. "Beyond the Subterranean Energy Regime? Fuel, Land Use and the Production of Space." *Transactions of the Institute of British Geographers* 42, no. 4 (2017): 655–668. https://doi.org/10.1111/tran.12182.

International Council on Monuments and Sites (ICOMOS). "Nord-Pas de Calais Mining Basin (France) No 1360." *UNESCO World Heritage Convention*, 2012. https://whc.unesco.org/document/152468.

Jacques, David. "The Rise of Cultural Landscapes." *International Journal of Heritage Studies* 1, no. 2 (1995): 91–101.

Jaramillo, George S., and Juliane Tomann, eds. *Transcending the Nostalgic. Landscapes of Postindustrial Europe beyond Representation*. New York/Oxford: Berghahn, 2022.

Jones, Michael. "The Concept of Cultural Landscape: Discourse and Narratives." In *Landscape Interfaces: Cultural Heritage in Changing Landscapes*, edited by Hannes Palang and Gary Fry, 21–51. Dordrecht: Springer, 2003.

Kirkwood, Niall, ed. *Manufactured Sites: Rethinkinig the Post-Industrial Landscape*. London/New York: Spon Press, 2001.

Kisiel, Piotr. "Unwanted inheritance? Industrial past as the EU heritage." *International Journal of Heritage Studies* 26, no. 7 (2020): 652–666.

Kohl, Helmut. "Der entscheidende Schritt auf dem Weg in die gemeinsame Zukunft der Deutschen. Fernsehansprache des Bundeskanzlers zum Inkrafttreten der Währungsunion am 1. Juli 1990." ["The Decisive Step on the Way to the Common Future of Germans. Televised Speech by the Chancellor on the Coming into Force of the Monetary Union on July 1, 1990"]. *Bulletin des Presse- und Informationsamtes der Bundesregierung* 86 (1990): 741–742. www.chronik-der-mauer.de/system/files/dokument_pdf/58697_cdm-900701-bulletin-kohl.pdf.

Landesregierung Nordrhein-Westfalen. "Wir machen Tempo beim Windenergieausbau!" ["We Are Accelerating the Expansion of Wind Energy"]. Accessed September 4, 2024. www.windenergieausbau.nrw.de/.

Lausitzer und Mitteldeutsche Bergbau-Verwaltungsgesellschaft mbH. *Espenhain*. Senftenberg: n.p., 2018. www.lmbv.de/wp-content/uploads/2021/04/doku-02_Espenhain.pdf.

Lehmann, Nele Hendrijke. "Socialism and the Rise of Industrial Heritage: The Preservation of Industrial Monuments in the German Democratic Republic." In *Heritage under Socialism: Preservation in Eastern and Central Europe, 1945–1991*, edited by Eszter Gantner, Corinne Geering, and Paul Vickers, 195–216. New York/Oxford: Berghahn, 2021.

Lewis, Peirce. "Axioms for Reading the Landscape." In *The Interpretation of Ordinary Landscapes*, edited by D. W. Meinig, 11–32. Oxford: Oxford University Press, 1979.

MacKinnon, Lachlan, and Steven High. "Deindustrialization." In *The Routledge Handbook to the Political Economy and Governance of the Americas*, edited by Olaf Kaltmeier, Anne Tittor, Daniel Hawkins, and Eleonora Rohland, 57–67. Abingdon/New York: Routledge, 2020.

Mah, Alice. *Industrial Ruination, Community, and Place: Landscapes and Legacies of Urban Decline*. Toronto: University of Toronto Press, 2012.
Mah, Alice. *Petrochemical Planet: Multiscalar Battles of Industrial Transformation*. Durham: Duke University Press, 2023.
Mark, Paul J. *Ein Bergdorf geht unter. Das Schicksal von Marmorera*. [*A Mountain Village Is Drowning. The Fate of Marmorera*]. Chur: Terra Grischuna, 2005.
McLean, Ross. *Transformative Ground: A Field Guide to the Post-Industrial Landscape*. Abingdon: Routledge, 2020.
Melsted, Odinn, and Irene Pallua. "The Historical Transition from Coal to Hydrocarbons: Previous Explanations and the Need for an Integrative Perspective." *Canadian Journal of History* 53, no. 3 (2018): 395–422. https://doi.org/10.3138/cjh.ach.53.3.03.
Meyer, Torsten. " 'Biofacts' – Recultivating the Post-Mining Landscape in the Anthropocene." In *Boom – Crisis – Heritage: King Coal and the Energy Revolutions after 1945*, edited by Lars Bluma, Michael Farrenkopf, and Torsten Meyer, 267–282. Berlin/Boston: De Gruyter, 2022.
Mission Bassin Minier. *Portrait socio-économique du Bassin minier*. [*Socio-Economic Characteristics of the Mining Basin*]. Oignies: Mission Bassin Minier Nord – Pas de Calais, 2018. www.missionbassinminier.org/wp-content/uploads/Portrait-socio-economique-du-Bassin-minier-2018.pdf.
Mitchell, William John Thomas. *Landscape and Power*. Chicago: University of Chicago Press, 1994.
Müllender, Bernd. "RWE reißt Windräder ab." ["RWE Is Demolishing Wind Turbines"]. *Die Tageszeitung (taz)*, August 28. 2023. https://taz.de/Foerderung-von-Kohle/!5953026/.
Orange, Hilary, ed. *Reanimating Industrial Spaces. Conducting Memory Work in Post-Industrial Societies*. London/New York: Routledge, 2015.
Parr, Joy. *Sensing Changes. Technologies, Environments, and the Everyday, 1953–2003*. Vancouver: UBC Press, 2010.
Pasqualetti, Martin. "Reading the Story of Energy in the Landscape." In *Sustainable Energy Landscapes: Designing, Planning, and Development*, edited by Sven Stremke and Andy van den Dobbelsteen, 11–44. Boca Raton/London/New York: CRC Press, 2013.
"Power Trumps Property." *Deutsche Welle*, December 17, 2013. www.dw.com/en/constitutional-court-clears-residents-complaint-on-garzweiler-ii-coal-plant/a-17304292.
Rhodes, Mark Alan, William R. Price, and Amy Walker, eds. *Geographies of Post-Industrial Place, Memory, and Heritage*. London/New York: Routledge, 2021.
Ross, Linda M., Katrina Navickas, Matthew Kelly, and Ben Anderson, eds. *New Lives, New Landscapes Revisited: Rural Modernity in Britain*. London: The British Academy, 2023.
Ross, Linda M. "Nuclear Cultural Heritage: From Energy Past to Heritage Future." *Heritage & Society* 17, no. 2 (2024): 296–315.
"Sámi rights activists in Norway charged over protests against wind farm affecting reindeer herding." *Euronews*, January 19, 2024. www.euronews.com/2024/01/19/sami-rights-activists-in-norway-charged-over-protests-against-wind-farm-affecting-reindeer.
Sauer, Carl. *The Morphology of Landscape*. Berkeley, CA: University of California Press, 1925.
Shepherd, Nick, ed. *Rethinking Heritage in Precarious Times: Coloniality, Climate Change, and Covid-19*. Abingdon/New York: Routledge, 2023.
Sklokina, Iryna, and Volodymyr Kulikov. "Industrial Heritage and its Multiple Uses in Donbas, Ukraine." *Region: Regional Studies of Russia, Eastern Europe, and Central Asia* 10, no. 1 (2021): 33–60.

Smith, Laurajane, Paul A. Shackel, and Gary Campbell. "Introduction: Class Still Matters." In *Heritage, Labour and the Working Classes*, edited by Laurajane Smith, Paul A. Shackel, and Gary Campbell, 1–16. Abingdon/New York: Routledge, 2011.

Storm, Anna. *Hope and Rust: Reinterpreting the Industrial Place in the Late 20th Century*. Stockholm: Division of History of Science and Technology, Royal Institute of Technology, KTH, 2008.

Storm, Anna. *Post-Industrial Landscape Scars*. New York: Palgrave Macmillan, 2014.

Storm, Anna, and Krister Olsson. "The Pit: Landscape Scars as Potential Cultural Tools." *International Journal of Heritage Studies* 19, no. 7 (2013): 692–708.

Taylor, Ken. "Connecting Concepts of Cultural Landscape and Historic Urban Landscape: The Politics of Similarity." *Built Heritage* 2 (2018): 53–67.

Taylor, Ken, and Jane Lennon. "Cultural Landscapes: A Bridge between Culture and Nature?" *International Journal of Heritage Studies* 17, no. 6 (2011): 537–554.

Tetzlaff, Kurt. *Erinnerung an eine Landschaft – Für Manuela*. [*Memory of a Landscape – For Manuela*]. DEFA, 1983. 1 hr., 24 min. www.youtube.com/watch?v=bRy9BEDEnqQ.

United Nations Educational, Scientific and Cultural Organization (UNESCO). *Operational Guidelines for the Implementation of the World Heritage Convention*. Paris: World Heritage Centre, 1999. https://whc.unesco.org/archive/opguide99.pdf.

United Nations Educational, Scientific and Cultural Organization (UNESCO). "Current Development Projects within the Nominated Property and Buffer Zone in the Saxon and Czech Component Parts of the 'Erzgebirge/Krušnohoří Mining Region.'" *UNESCO World Heritage Convention*. Last modified 2019. https://whc.unesco.org/document/168556.

Wicke, Christian, Stefan Berger, and Jana Golombek, eds. *Industrial Heritage and Regional Identities*. London/New York: Routledge, 2018.

Winston Nicklin, Mary. "This Former Coal Mining Region Now Fuels Green Tourism." *National Geographic*, March 9, 2022. www.nationalgeographic.com/travel/article/this-former-coal-mining-region-now-fuels-green-tourism.

Winter, Tim. "Clarifying the Critical in Critical Heritage Studies." *International Journal of Heritage Studies* 19, no. 6 (2013): 532–545.

Wollentz, Gustav, Sarah May, Cornelius Holtorf, and Anders Högberg. "Toxic Heritage: Uncertain and Unsafe." In *Heritage Futures: Comparative Approaches to Natural and Cultural Heritage Practices*, edited by Rodney Harrison et al., 294–312. London: UCL Press, 2020.

York, Richard, and Shannon Elizabeth Bell. "Energy Transitions or Additions? Why a Transition from Fossil Fuels Requires More than the Growth of Renewable Energy." *Energy Research & Social Science* 51 (2019): 40–43.

Zabirko, Oleksandr. "The Donbas: A Region and a Myth." In *Ukraine's Many Faces. Land, People, and Culture Revisited*, edited by Olena Palko and Manuel Férez Gil, 331–344. Bielefeld: transcript, 2023.

Zhu, Yujie. "Putting the 'critical' in heritage studies from non-Anglophone regions: China and beyond." *International Journal of Heritage Studies* 30, no. 12 (2024): 1476–1486.

Part 1

Creating Energy Landscapes

1 The Cultural Construction of a Hydroelectric Landscape

The Project of Lake Sihl in Switzerland, 1897–1937

Sarem Sunderland

From a cultural perspective, hydroelectric infrastructures can be viewed as a complex and ambiguous mosaic. The most imposing elements of these infrastructures, namely dams and reservoirs, are sources of both enthusiasm—as objects of technological fascination, leisure, and economic activity—and concern, given their environmental and social impacts. At the same time, other less conspicuous elements that lie underground, upstream, downstream, and along electric lines largely go unnoticed. This mosaic conditions how landscapes are perceived, valued, and handled in the face of their transformation for energy production. Moreover, it is currently a major source of conflict in projects to establish new hydropower plants as well as renovate existing ones. In these projects, the promoters and authorities tend to consider landscapes in terms of their previous states and project impacts. They conduct impact evaluations to define the degree of gravity of the impacts of each intervention and the required measures in response (the usual triad is to avoid, reduce, and compensate).[1] Yet in doing so, they neglect the possibility of a more nuanced impact, which includes problematic as well as beneficial aspects. Hydropower plants are thus seen to be exclusively destructive of landscapes (which, like any material intervention, they necessarily are) and not as constructive. In other words, their impacts on landscapes are viewed solely as negative and never as positive. This position considerably limits the range of possibilities and paralyses debates on current projects. Landscapes, seen from this perspective, become a key social and environmental issue in hydropower projects. This chapter excavates alternative views that existed in the past, with the aim of broadening perspectives and contributing to ongoing and future debates around hydropower works. It does so by offering a more nuanced and in-depth account of landscape formation in the case of a hydroelectric reservoir from the Swiss Prealps.

From water wheels used during the pre-Christianity era to initial explorations of hydroelectricity 150 years ago and the more recent boom in hydroelectric infrastructure during the post-World War II period, the production of hydroelectricity exemplifies the growth of anthropogenically initiated geomorphic activity. With their imposing and easily identifiable appearance, dams have become the symbols of this production. It is therefore understandable that researchers examining the impacts of hydropower on landscapes have tended to focus on dams and other

DOI: 10.4324/9781003473534-3

conspicuous built elements of hydroelectric systems (e.g., power plants, power lines, and substations) along with the expanded scales of electric and hydrological networks.[2] Yet, among the components of hydroelectric infrastructure, reservoirs often have the most extensive spatial imprint. Despite this importance, few works have examined reservoirs in the depth as for dams. In most cases, reservoirs are referred to briefly, if at all. They are perceived as lake-like landscapes of minor importance in comparison to dams. The formation of these landscapes is either considered accidental, determined by purely functional parameters, or guaranteed to be beautiful and taken for granted.

Some rare works have paved the way for a more substantive understanding, such as the study by the cultural geographers Denis Cosgrove, Simon Rycroft, and Barbara Roscoe on British reservoirs.[3] Their analysis of the Ladybower Reservoir Project (completed in 1945) and the Rutland Water Project (completed in 1975) in the UK illuminates the concern and conflicts that were generated by the idea of these new water bodies. In response, planners worked on the layout of new roads, the siting of leisure facilities, the plantation of trees and shrubs along the shoreline, and the design of the dam, with the aim of integrating these reservoirs into the existing landscape and producing a picturesque setting. The resulting landscapes were then mobilised by project promoters to legitimise their infrastructural works. Endorsing this political perspective, geographers Viviana Ferrario and Benedetta Castiglioni explored the social implications of landscapes within hydroelectric reservoir projects in the Piave watershed in Italy.[4] They showed how discreet actions, such as referring to a reservoir as a lake and presenting it as an embellishment of the landscape, or hiding it in visualisations of a dam, served to emphasise the positive contributions of reservoirs while hiding their problematic implications. These discourses and depictions revealed some of the insidious 'politics of invisibility'[5] that can underlay the design and accommodation of reservoirs. The two examples, described above, along with other works,[6] have shown that far from being accidental or necessarily appealing outcomes, the landscapes of hydroelectric reservoirs may be actively produced by design and driven by political projects.

This chapter explores this intersection of hydropower, landscape, politics, and design. It presents a case study of the landscape of a hydroelectric reservoir, which was culturally produced decades before the valley was actually flooded. It further highlights the interplay of a hydroelectric project with landscape aesthetics and the work of visual artists in shaping the idea of the infrastructure and producing expectations of modernity and beauty. My aim is not to promote hydroelectric landscapes; rather, it is to trace and problematise the relationship between humans and their environment in the context of hydroelectric works. To do so, I examine the project of Lake Sihl, a reservoir in central Switzerland planned and built between 1897 and 1937. I begin by tracing the construction of an idea of landscape during the project phase, drawing on texts and images that promoted or criticised the future landscape. Next, I compare this early idea of landscape with present-day depictions of Lake Sihl. As this chapter shows, the landscape of Lake Sihl was actively projected and mobilised even before it had been built, with ideas of this future landscape presented enthusiastically and contested by a variety of actors (residents and outsiders, public authorities at different scales, experts, and artists). These positive images and their contestations continue to structure the ways in which the landscape is shaped and perceived today, resulting in ambiguous and sometimes contradictory ideas.

These outcomes shed light on the history of one of Switzerland's first large hydroelectric reservoirs and on early experiments in dealing with hydroelectric landscapes, which apply to many similar infrastructures. They show that the very idea of this landscape, even before the detailed planning of the infrastructure began, was not a given condition; rather, it was a cultural construction. The ways in which people imagined the future landscape of the reservoir were thus actively shaped, instrumentalised, and contested, and the effects of this construction are still at stake today, with people gradually beginning to overcome its limitations.

The Case of Lake Sihl

The case examined here focuses on Lake Sihl, the reservoir of the Etzelwerk, a hydroelectric plant located in the Prealps in the canton of Schwyz in north-eastern Switzerland. Initiated in 1897 and completed 40 years later in 1937, the project was part of a wider wave of hydropower projects in the early twentieth century, centred in (but not confined to) the Global North. This wave produced the first large dams and reservoirs and hydroelectricity, with the Etzelwerk as a paradigmatic Swiss example of this period. However, in the present, projects from the early twentieth century, such as the Etzelwerk, often remain in the shadows of the ones that followed. The post-war period brought another global wave of even larger hydropower plants in more countries, resulting in well-known examples, such as the works on the Tennessee and Columbia Rivers in the USA and on the Volga and the Dnipro Rivers in the USSR, the Aswan High Dam in Egypt, and alpine works, such as the Grande Dixence in Switzerland, discussed in Rune Frandsen's chapter in this volume. While the boom phase of new monumental dams has ended in areas of the Global North, such as in the Alps, the practice is flourishing in other regions of the world, with more recent and even more radical projects, such as the Three Gorges Dam in China. However, the technologies, economic structures, and cultural reception of these recent large-scale constructions were largely pre-established through earlier projects. Viewed in this light, projects such as the Etzelwerk are of interest as early laboratories for the confrontation of hydropower production with landscapes. They provide opportunities to observe the evolving dynamics shaping hydroelectric landscapes from their early days until the present.

The Etzelwerk targeted the water of the Sihl River, which starts in the canton of Schwyz and flows down to the city of Zurich, where it first joins the Limmat River, then the Aare River, and eventually the Rhine. The project proposed to impound the river in the Upper Sihl Valley, a wide prealpine valley located approximately 900 metres above mean sea level. Outside of the valley, this rural area was known primarily as a passage for pilgrims on their way to a large Benedictine abbey in the nearby town of Einsiedeln. The open and relatively flat valley floor was half covered with moorlands rich in peat. The project would bring the Sihl's water into a different watershed, 6 km further north and nearly 500 m lower than the valley, through galleries to pass through turbines and be released into Lake Zurich. Undoubtedly, the most ambitious aspect of the project was its reservoir. The proposed Lake Sihl would cover a surface area of over 1,000 hectares (10 km^2). Its unprecedented size and coverage would make it by far the largest potential reservoir in the country. The shifts that this project represented for hydropower practices at large were just as visible locally. With the proposal for its reservoir to replace an important peat extraction site, the Etzelwerk embodied an energy transition and modernisation of the region's ways of life.

The Maschinenfabrik Oerlikon (MFO), a large engineering company headquartered in Zurich, 30 km north-east from the Upper Sihl, stood behind this project. This company produced tools and machines, particularly for railways and hydropower plants, being one of the main proponents of the latter in the country. For the MFO, the Etzelwerk represented a source of energy for its operations, with additional energy to sell, as well as a strategic step toward the consolidation of the Swiss hydroelectricity industry.[7] The project was developed in close consultation with the canton and municipality of Zurich, which also required energy to supply their industries. Thus, the Etzelwerk was the work of the combined private and public interests of one canton outsourcing their energy supply to other cantons, as there were hardly any inland rivers left to impound—at least none that offered economic advantages.

On the other side of the negotiating table was a large panel of actors, as required by the Swiss political structure for decision making relating to such projects. Hydropower plants had to be approved by the cantonal authorities, who held water rights, and by the municipal or district authorities, who had jurisdiction over land use and its impacts on localities.[8] For municipalities and districts, such decisions were taken by holding open assemblies of male-only adult citizens who participated in a public vote. This meant that the project developers had to convince the cantonal authorities, the municipal or district authorities, and at least half of the local adult male population, as the population's decision could not simply be dismissed through expropriation—at least, not without a national-level intervention. Two districts were involved in the Etzelwerk project: Höfe, located by Lake Zurich and home to the future power plant, and Einsiedeln, located higher up in the Prealps and home to the future reservoir. For both districts, the infrastructure represented a leap into modernity and a source of income for the district authorities as a result of a number of provisions secured through negotiations. These included financial compensation, the supply of low-priced electricity, prioritisation of local companies and a local workforce for the works, construction of new roads, and channelling of rivers flowing into the reservoir, which was paid for by the hydropower operator. However, the material expenses differed in the two districts. In Höfe, the implementation of the power plant was an easily digestible idea, given its limited spatial imprint. In Einsiedeln, by contrast, the infrastructure's spatial implications were considerably more radical. The plan was to flood 10% of the district's surface area, completely submerging the homes and land of 274 residents and portions of the land and homes of a further 1,763 residents under the reservoir.

This affected approximately a quarter of the district's population in 1930.[9] These numbers broke another record in Swiss hydropower as the largest number of residents to have been displaced by a hydroelectric reservoir, a record that was far from being associated with heroic technical feats. These costs and benefits resulted in a variety of positions and perspectives within the district: proponents of the project were those who either gained direct or indirect benefits or held pro-modernist viewpoints, whereas its opponents were those who lost their land and homes or were in solidarity with them or held anti-modernist viewpoints. This diversity of opinions and the sheer number of actors involved in decision making made the Etzelwerk a complex and ambitious affair, with difficulties aggravated by the long-standing tension between the liberal and highly industrialised canton of Zurich and the more conservative and rural canton of Schwyz.[10]

Within this constellation, the first few years of negotiations proceeded smoothly. The population of Einsiedeln, which was most affected by the project, initially approved it by public vote in 1900, and negotiations with developers and the cantons made considerable progress. Yet after a decade, the MFO and the city and canton of Zurich found themselves unable to bring their proposal to fruition. The project required not only the approval of the canton and districts but also changes to a previous intercantonal agreement for the use of the Sihl's water. For the MFO and the Zurich authorities, the compensation demanded by Schwyz and its districts was considered unacceptable, making the Etzelwerk project financially unviable. After several attempts, the initiators eventually gave up, and in 1910, the recently nationalised Schweizerische Bundesbahnen (SBB), or Swiss Federal Railways, took over the project, given the organisation's quest to electrify its fast-growing railway network. This change brought about a major shift in the scale of interests behind the project, making it a matter of national interest. The SBB created synergies with another major actor, the Nordostschweizerische Kraftwerke (NOK), or North-Eastern Swiss Power Plants, a public sector energy supplier for the north-eastern part of the country. The project continued through a partnership of the two organisations over another two decades of negotiation, which was more successful this time. The population of Einsiedeln approved the project once again in 1926, construction started in 1932, and the hydroelectric complex was inaugurated in 1937. Some changes did occur after that, as the infrastructure was converted into a pump-storage hydropower system a few years later, and in 1987, the NOK withdrew, leaving the Etzelwerk project solely in the hands of the Federal Railways. At present, the complex produces 10% of the electricity that is annually consumed by the railways, and the reservoir remains the largest in the country according to its surface area, though its volume, fall height, and electricity production have long since been exceeded by those of other projects.

The Inception of a Landscape Idea

In the early years of the Etzelwerk project, its developers made no explicit mention of the infrastructure's landscape. On September 6, 1897, when the MFO submitted its first proposal for the Upper Sihl, it presented the project as a material and economic operation. The project would 'facilitate and rationalise the utilisation [*Nutzbarmachung*] of the Sihl,'[11] while also preventing floods by holding back storm water. It could therefore appeal to everyone: the district (Einsiedeln had no general electricity supply at the time), the industries of the surrounding cantons, and even the country at large, which would benefit from electricity, and the Lower Sihl Valley, which would be protected from floods. The project's spatial implications were explained in the same way as a technical plan based on quantifiable facts. The infrastructure was described piece by piece, with explanations provided on the location, size, construction, and capacity of each component.[12] One spatial point was emphasised, namely the vastness of the reservoir and the ambitious proportions of the infrastructure. In parallel with this emphasis on size was an emphasis on production. The indication of size made the potential production, which was otherwise abstract, tangible.

Apart from describing the size of the project and the individual constructions, the MFO made no mention of its concrete appearance in the Upper Sihl Valley and how it would impact residents' everyday lives. The developer hardly addressed the landscape, and this would remain the case in the MFO's subsequent communication. This way of presenting the infrastructure and its spatiality was not unusual. It was representative of the texts of many engineers at the time, including those of the project engineer, Louis Kürsteiner.[13] The emphasis lay on clearly identifiable built features (primarily dams and power plants and secondarily penstocks, power lines, substations, and bridges required for the project). There was no discussion of the larger landscape, including the reservoir, but other parties soon took up the matter.

The first textual thematisation of the infrastructure's landscape came from the local population. Residents were quick to question what the infrastructure would look like and how it would affect their everyday life. How these local communities related to the infrastructural project can be traced in the pages of the local newspaper, the *Einsiedler Anzeiger*. This newspaper delivered news while also providing an arena for debate, including the diverse and usually anonymous opinions of readers in addition to those of its editors and correspondents in its redaction. In a first brief mention of the project nine days after the MFO's submission of its project proposal to the authorities in 1897, the editor made an effort not only to describe the components of the infrastructure and their size, as the MFO did too, but also to envision the reservoir's presence in the valley. Adopting a positive yet caricatural tone, the editor described an 'alpine lake' thanks to which a 'series of guest houses would draw in foreigners enthusiastic of the view'[14] and imagined people fishing and bathing there. In the following weeks, articles on the Etzelwerk started to proliferate, some following the promotional logic of the MFO, while others expressed concern and mentioned resistance among the inhabitants of the Upper Sihl Valley. In terms of the landscape, the caricatural tone of the first depiction, which was possibly a reaction to the almost immeasurable size of the proposed reservoir, was dropped, but the positive tone remained, without any negative references. A short article described the public's positive reception of the project and promised 'a splendid image of a mountain landscape' and 'a proper image of a lake,' which would replace the 'boring rust-brown wetland' covering most of the Upper Sihl

Valley with the 'silver-sparkling mirror of the lake.'[15] Another article, published a month later, depicted a complete scene, imagining someone's account of the reservoir in the future:

> Some days ago, I stood on the slope of the Freiherrenberg. Further down lay Lake Sihl, its blue waves comfortably swirling against the shore in Böcklin-like light. Evening bells were ringing from the Wyler Zella [the Willerzell settlement in the Upper Sihl Valley]; some fishermen from Gross [another such settlement] cast their nets, and people in Gross were joyfully hanging their laundry by the shore. The bleeding sun was flashing over the valley and spread the magic of a unique, magnificent image of a high alpine valley.[16]

What is apparent from these initial accounts is that authorities or promoters did not openly tackle the issue of the Etzelwerk's landscape, whereas the district's inhabitants did. Local communities sought to engage with the idea of the infrastructure and to imagine what it would mean for them according to the plans, numbers, and somewhat technical descriptions of the promoters. In this context, local residents took up the landscape question, imagining and exchanging possible scenarios. Remarkably, they quickly reached a consensus. No one, in those early years, expressed concern publicly about the prospective landscape, and this consensus soon found support in the work of external experts assessing the project.

The Landscape of Experts

Among the first batch of experts who examined the Etzelwerk project proposal, several sanitary experts addressed the question of how this new landscape would be lived. During the initial years of the project, no fewer than four such reports were commissioned, driven by concern over possible deterioration of the conditions for life in the valley caused by the reservoir's water. This concern arose at a time when waterscapes, and especially moors and wetlands, were strongly considered insalubrious. Early in the process, the MFO commissioned Oskar Wyss, a Zurich-based doctor and professor to examine this issue. Wyss published two reports in 1898, with the second containing a remarkable passage. After the mandatory introductory paragraph, Wyss reviewed the current state of the valley, noting that its floor was a sparsely inhabited moor used for peat extraction. He painted a picture of a valley in a sad state, highlighting noxious smells attributed to variations in the humidity and a view that was 'not capable of uplifting the human spirit.'[17] He then went on to envision the new landscape:

> Let us ask ourselves how would the valley look if its floor were to be changed into a lake? The beautiful mountainsides, partly occupied by forest, partly by willows, houses, and huts, would attract more attention than [they do] now and would form a wonderful frame to the lake that would take up the entire valley. It is undoubted that no one would come to miss the currently desolate moor. The Upper Sihl Valley would turn into an alpine valley like other high-alpine valleys

> that are endowed with a lake and which visitors hold in high praise for their beauty. I am reminded of the lakes of the Berner Oberland; of those in Engadin. The beauty of the valley and therewith of the surroundings of Einsiedeln would significantly rise with such an artificial lake; there can be no doubt [of that].[18]

This view of the landscape of the Etzelwerk drew on the depictions portrayed by journalists and residents in the *Einsiedler Anzeiger* some months earlier. Wyss did indeed see Lake Sihl as an improvement, which could create a better albeit anthropogenic landscape. The reservoir would serve as an aesthetic element, which would raise the valley's scenic status and elevate it to the enviable category of a beautiful alpine valley endowed with a lake. However, this landscape was not just about a beautiful scene in the future; it was also about a less-than-ideal state in the present. Wyss turned this present state into a desolate one because of the moors that covered the valley floor. His appreciation was aesthetic as well as economic, considering the poverty in the valley and the altogether difficult living conditions. This negative depiction of the present landscape acted as a contrasting foil, juxtaposing ugliness with beauty, dark, brownish tones with lush greens and clear blues, and hardship with hope. These words had considerable resonance, cementing the early conception of the reservoir's landscape.

Wyss's statements were the first articulations of a scientific opinion on the Etzelwerk's landscape, offering a helpful base for others to build on, as the authors of later reports had access to Wyss's work.[19] Half a year later in December 1898, Schmid, the national health adviser in Bern, and Professor Roth in Zurich co-signed another report on the sanitary implications of the Etzelwerk. This time, there was no description of scenery resembling the one in Wyss's report; there was only a reference to Wyss's previous identification of this quality.[20] The text was written with noticeably more facts and references and less explicit value judgements than that in Wyss's report. Initially, the authors took a different position regarding the appearance of Lake Sihl. They were concerned about the variations in the reservoir's water levels and the resulting reservoir margins, which designate the portion of the lakebed that becomes a visible shoreline at drawdown (the period of low water level). This concern was rooted in the MFO's submission of 1897, which mentioned that the water would fluctuate and reach its lowest point in the winter months, subsequently filling up with snowmelt in the spring and remaining high until November.[21] Schmid and Roth examined the issue in greater depth for possible sanitary implications, such as the rotting of organic deposits in the early years and risks of malaria and other typical wetland illnesses. In this context, they feared that 'the sight of the half or almost entirely empty lakebed would substantially hamper the scenery of the valley.'[22] For the first time, doubt was openly expressed about the certainty of a beautiful landscape in times to come. Yet despite this concern, the authors still gave credence to Wyss's prognosis, concluding that landscape embellishment was a key benefit of the reservoir.[23] Although their report was more nuanced than the first report, their conclusion was the same, namely that the reservoir was an aesthetic improvement.

The idea of the reservoir as a beautifying intervention conveyed in the views of these authoritative experts eventually reached the district authorities and became part of their argumentation in support of the project. In advance of the public vote on the Etzelwerk project held on April 1, 1900, the district authorities in Einsiedeln distributed copies of the proposed contract and a 45-page explanatory booklet to all households. This booklet argued in favour of the project and quoted the prognosis of both reports regarding a beautiful landscape, while omitting the concern raised by Schmid and Roth regarding the low water level in that same landscape.[24] One out of nine key takes in the conclusion of this document stated that the lake would not have any negative climatic or sanitary implications and that 'on the contrary, the lake will have an improving effect on the climate and will raise the landscape beauty of the barren valley.'[25] The expert's imaginary of the reservoir as a beautiful scenic spot was thereby incorporated into the district's core logic and instrumentalised for its promotion. At that point, what had initially been a matter of discussion amongst Einsiedeln locals envisioning the infrastructure progressively culminated in an official political matter.

From Words to Images: Depictions of the Reservoir

In 1901, four years after the first submission of the Etzelwerk project proposal and one year after its approval in a public vote, the first images of the Etzelwerk taken on the ground and not in the plan appeared in an almanack, the *Einsiedler Kalender*. The almanac's 1901 edition featured an article about the Etzelwerk titled 'A new lake for Einsiedeln.' Five pages of text were devoted to praising the Etzelwerk and included some images: a plan of the lake, a sketch of Willerzell Village, and two views of the Upper Sihl Valley from a site named Roblosen, painted by a local artist from Einsiedeln named Wydler. With the inclusion of this visual material, the article differed from previous ones published in magazines and journals that had been limited to text, numbers, and technical plans. The two views from Roblosen were its main novelty. Both focused on the southern end of the valley, depicting the same agrarian scene in the foreground and the peaks of the Drüsberg mountain chain in the background. What differed in the images was the middle ground: the first image depicted the valley at the time with its pastures, peat fields, and the meandering Sihl River, and the second depicted the reservoir of the Etzelwerk covering all of this area (Figure 1.1). Despite its apparently idyllic setting, this last view illustrates a radical proposal. The Lake Sihl it depicted did not yet exist, and it would not do so for another 36 years. For the first time, however, the lake was represented not as an abstract, technical object but as something experienced through the senses. These images were the first published ones of a large reservoir in Switzerland at a time when no built examples existed. Lake Sihl thus became a laboratory for imagining such a landscape, and these images were among the first instances of this being done visually.

In line with previous textual descriptions of the Etzelwerk, the article depicted the valley's present state in a highly negative manner and presented the Etzelwerk as an improvement of the landscape.[26] Again, the depiction of a beautiful and

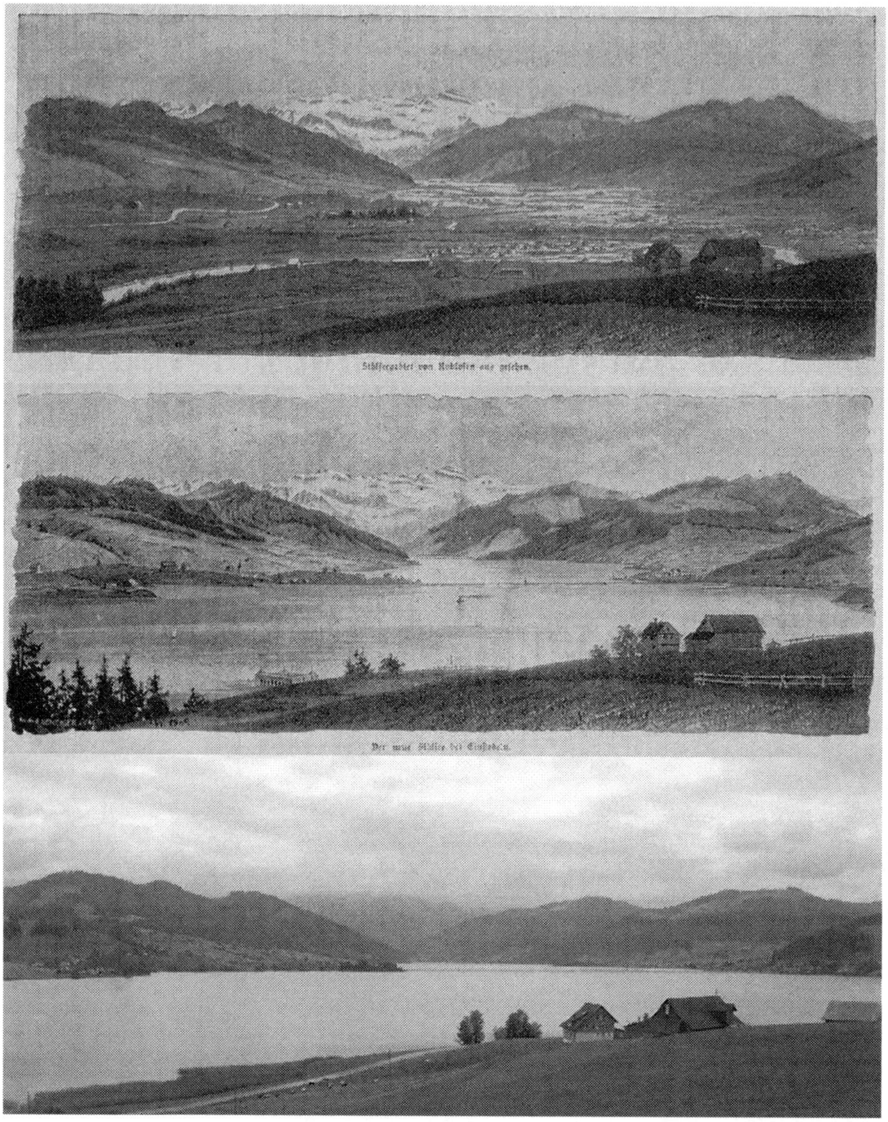

Figure 1.1 Drawings of the Upper Sihl Valley with and without Lake Sihl in the early days of the project, and a photograph taken by the author on July 27, 2022, from the same location.

Source: "Ein neuer See bei Einsiedeln," *Einsiedler Kalender*, 1901; Archiv Stiftung Kulturerbe Einsiedeln, Yb.34.

hopeful future went hand in hand with deprecation of the present. The image itself shows subtle signs of support for the depiction of the reservoir. The 'before' image is darker than the 'after' one, emphasising an undesirable present state. Conversely, the 'after' image includes features associated with leisure activities, such as a boat and bathing facility on the shore, denoting a certain bias in favour of the lake scenario. Less explicit, but equally powerful, is the selection of the viewing point. The view towards the south of the valley, subtly hiding the dam whilst framing the mountains of the Drüsberg chain, presented an appealing composition and a classic image of an alpine lake. Had the artist picked a view towards the north, there would have been no mountains, only a soft, slightly hilly relief lacking the same visual intensity. Thanks to these combined visual devices, a preference for the image without a reservoir was unlikely. Despite their apparently neutral appearances and accurate composition, the before-and-after depictions of the valley thus operated as a device for promoting the power plant.

An essential structuring logic for the images and texts considered here was the artistic tradition of alpine (natural) lakes. At the turn of the twentieth century, these lakes had become associated with a well-established landscape aesthetic. However, this had not always been the case, with perceptions during the Middle Ages being in stark contrast to these ones. Open waters, including lakes and rivers, were preferably located at a distance from settlements at the time and were despised and perceived as objects of disgust rather than of beauty or pleasure. People had functional interactions with these waters, which provided transport, water, and fish, but they were far from being celebrated in the ways that they currently are. From the late seventeenth century to the mid-nineteenth century, these older conceptions were reversed. Theological and scientific advances (mainly in geology and medicine) combined with the practice of travel and a literary genre celebrating picturesque voyages transformed mountains and water bodies into objects of fascination. The result was no less than a rediscovery of mountains and shorelines and, in terms of aesthetics, a reinvention of both.[27] This aesthetic reinvention of mountains and water bodies was directly reflected in lakes, which became objects of pleasure and beauty. The blend of lakes and mountains resulted in what the landscape theorist, Michael Jakob termed a 'double valorisation.'[28] The lake allowed for a sublime experience of the mountains, contrasting tall mountains with a wide-open surface in front and mirroring the scene, while the irregular and undulating shoreline offered a classically picturesque vista. These powerful effects offered a helpful reference to anyone seeking to imagine a large reservoir in the Swiss Prealps.

In the paintings and descriptions of the Lake Sihl landscape, these principles were transposed one-to-one: a water body in the centre of an appealing scene, framed by a picturesque shoreline in the foreground and dramatic mountains in the background. Those who sought to imagine Lake Sihl thus mobilised images of natural lakes and projected them onto the reservoir. In doing so, they turned the hypothetical future of a technical infrastructure into a scene that was both familiar and desirable. This borrowed aesthetic was highly effective: familiarity generated comfort and certainty by lending an aesthetic and appealing image to the infrastructure, erasing doubts and generating enthusiasm and even desire. For supporters of the Etzelwerk, the lake became a *tour de force*, simultaneously producing energy, beauty, economic stability, and flood security. The landscape was no longer just a matter of people wondering what the infrastructure would look like; it was instrumentalised and became a highly charged political issue.

The article in the almanack, illustrated with Wydler's paintings, was the first example of a wider artistic practice, which was soon to be followed by other works of mostly local painters. A large mural was commissioned by the gymnastic association of Einsiedeln as the background set of the main stage for a national gymnastics festival. The view of Lake Sihl made a strong impression on the audience, with the *Einsiedler Anzeiger* describing 'an enormous image of the future Lake Sihl and the green mountain pastures and the defiant mountains of the Sihl Valley.'[29] Another painting on the wall of a former local restaurant in Einsiedeln depicted the future Lake Sihl, combining the aesthetics of Wydler's drawing and the traditional format of an alpine panorama.[30] Furthermore, images adorned the pages of a touristic guide intended for visitors interested in the Etzelwerk, which was under construction at the time. Its pages were richly illustrated and included a before-and-after depiction almost identical to Wydler's composition, though the base was now a photograph and not a drawing (Figure 1.2). In parallel, Wydler's painting assumed a life of its own. In the 1920s and 1930s, it was featured in magazines for the general public within a specialised literature and in booklets produced for the inauguration of the infrastructure in 1937.[31] By this time, these futuristic images of Lake Sihl had been seen by a large public within and outside of Einsiedeln District before the reservoir had materialised. Both the visual and the textual logic of the three previous decades remained identical but were now widespread. The recurrence of this visual structure across such a wide variety of contexts shows that the image effectively embodied the thoughts of diverse authors and publishers, specialists, and laypeople alike.

Considered together, these artistic efforts reveal that the landscape idea of Lake Sihl became increasingly visual and established during the decades following the approval of the project in 1900. Engaging with this hypothetical landscape became a way not only to question the impact of the infrastructure but also to celebrate it as a positive impending change. Furthermore, the audience that received this idea had expanded from the readership of local newspapers and experts' reports to that of magazines and events with a national reach. These efforts also reveal how the landscape became an object of increasing interest for artists and 'consumers' of these works, extending from commissioners to a wider audience. Notably, for all these works, there were only positive expectations, with hardly any criticism expressed.

Contestations of a Landscape Idea

In the early days of the Etzelwerk project, voices hinting that the landscape of the reservoir might not be as ideal as promised were rarely heard. Schmid and Roth were the first to raise this issue, expressing concerns about the sight of the lake at low water levels in their report written in 1898. At first, their concern did not resonate widely. The reports and articles that followed picked up the reference to

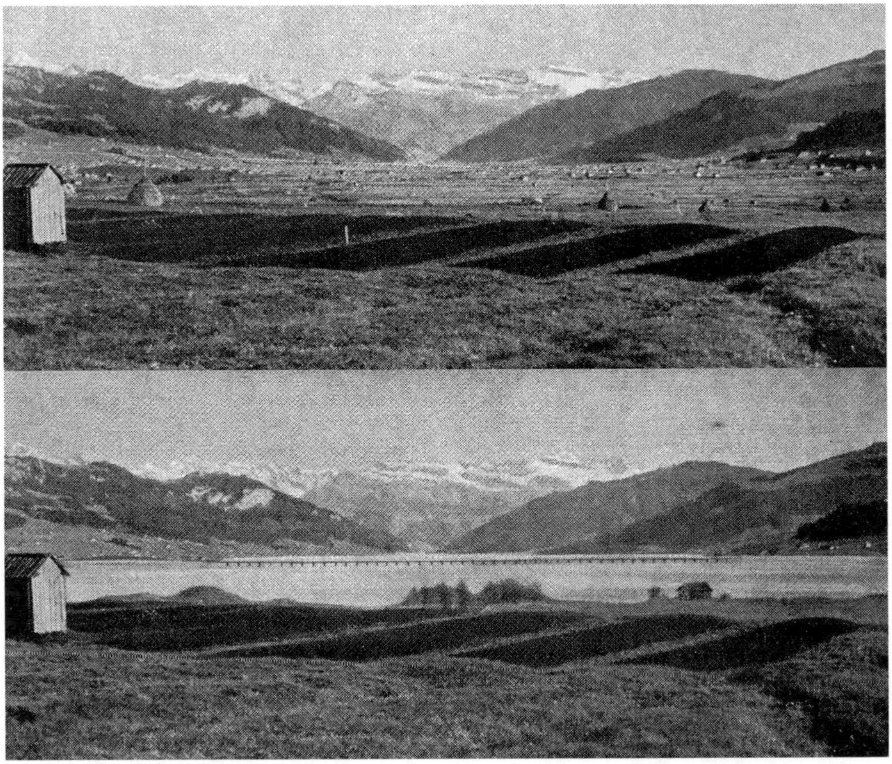

Figure 1.2 A photograph of the Upper Sihl Valley from a site close to Roblosen, and the same photograph, with Lake Sihl and the viaduct of Willerzell superimposed over the scene. The black surfaces in the foreground are planting beds, typically used to cultivate potatoes. In the centre are the peat fields, with scattered huts built by peat diggers and spikes used to stack peat extracts. In the second image, the most productive surface area remains, but the peat fields have been erased. The photographs were published in 1933.

Source: Salathé, *Wägital und Etzelwerk*, 7, 10.

the beautifying effect, but not the one about the poor display at drawdown. By the end of World War I, the Etzelwerk project had garnered national support stemming from the need for domestic energy production to counter reliance on coal imported from abroad. However, criticism of the project had also grown within Einsiedeln District. Contrasting with supporters at the national level, the local population's enthusiasm had waned, as the novelty of electricity had worn off and the project had not progressed since its approval two decades earlier. This climate of discontent and the threat of the construction of a much larger Lake Sihl than that originally proposed sowed new seeds of doubt, some of which concerned the promised landscape.

The second round of criticism of the proposed landscape came from Father Cölestin Muff, a Benedictine priest at the prominent Einsiedeln Abbey. In 1921, Muff alias Professor Frohmut wrote a booklet opposing the proposal of an 'extra-large Lake Sihl' and had it printed by one of Einsiedeln's publishers and sold for 60 centimes. The text was also published in the *Einsiedler Anzeiger* newspaper, thereby ensuring that it reached people throughout the district. In this work, Muff launched a frontal attack on the four key arguments of the supporters of the Etzelwerk: the promise of work generated by the construction, fair compensation for displacement, industrial development, and beautification of the Upper Sihl Valley. On this last point, Muff—or rather Frohmut—wrote the following:

> A reservoir with such flat and thus shallow shores, which remains almost or entirely empty for months, will never be capable of embellishing our valley, instead only appallingly deforming it. And we don't even want to speak of the bad vapours coming from the shores. A green and blue laughing lake, on which one may sail and swim with pleasure, is a false fantasy, a daydream, a Fata Morgana.[32]

The author thus made the same point, expressed more delicately in the report by Roth and Schmid two decades earlier, with greater vehemence and to a larger audience. There was also some exaggeration in his words, which suggested that the image of an alpine lake would never appear even at high water, though the hyperbole may have been needed to counter the earlier tsunami of positive descriptions.

This time, the critique must have resonated and reached the authorities because the official position of the Einsiedeln District authorities on the landscape evidenced a shift from the one that was articulated two decades earlier. The shift was evident in a booklet distributed by the district authorities in 1926 in preparation for a new general vote on the Etzelwerk project. As for the publication released prior to the vote held in 1900, this one aimed to present the issue to be voted on as well as the district's position on it. However, reacting to the climate of doubt that prevailed concerning the project at the time, the general tone of the text was far less celebratory. Gone were the references to a beautiful landscape to come, even if the conviction remained that the Etzelwerk offered a valuable opportunity for the district's development. Instead, the authors expressed their concern about the landscape, noting the 'unsightly' shallow shores at drawdown and explaining that this issue, along with possible sanitary risk, would lead to regulation of the water level, with a minimum water level set for the summer months.[33] Additionally, one article in the proposed contract requested the preservation of the landscape's image to the greatest extent possible.[34] The position on the landscape in the Etzelwerk project had thus begun to shift from being an unequivocally positive and promising one to growing concern and a view that beauty could not simply be expected but had to be actively secured.

From a contemporary standpoint, one would expect opposition to the project based on social and environmental concerns. Indeed, there were instances of such opposition stemming from outside of the valley at the annual assembly of the Swiss

Federation for Nature Protection (now Pro Natura) held in Zurich in 1927. Rudolph Gelpke, a civil engineer and national councillor, opposed the Etzelwerk, due to which 'the much-visited and praised pre-alpine valley, in which the wild-romantic pairs with charm, will be robbed of its living landmark.'[35] Gelpke did not reject hydropower in principle, as he supported the goal of 'a rational mobilisation of fortunate height differences for the needs of the domestic economy.'[36] However, he saw the Etzelwerk as illustrative of a tipping point, where unregulated hydroelectric development occurred at the cost of a '*disfiguration of our landscape's physiognomy.*'[37] However, Gelpke's position and concern were only shared by a minority of people in the early twentieth century. The Swiss Federation for Heritage Protection, one of the organisations that was most active in questioning the relevance of hydropower projects at the time, held the opposite view. In 1929, the topic of the Etzelwerk came up during the Federation's assembly in Einsiedeln and a clear consensus was reached:

> As it is known, heritage protection never acted against this project, since here a beautiful landscape will not be destroyed for mere speculation. … The valley floor by Willerzell, which will be put under water, is mostly a swampy surface which will someday become much better looking in the form of a lake. At the time of its lowest water levels, it is covered with snow. … The rural population, which loses parts or the entirety of its cultivated land, will be cared for in a downright exemplary way. A walk to the Vogelherd, which offers an open sight on the ground of the future reservoir, taught the heritage protecters that there is really no need to worry about the landscape of Lake Sihl.[38]

Remarkably, not even the issue of the low water level was of concern here. This position, shared in heritage protection circles, and Gelpke's isolated rejection of the Etzelwerk project reveal that concern emanating from outside the valley over heritage and nature was limited in the Etzelwerk area. Landscape positivism prevailed. Had the project been implemented a few decades later, things would have been different, with the first generalised wave of social and environmental opposition to new hydropower plants occurring during the late post-war period. The Etzelwerk project preceded this wave of opposition and possibly played a role in triggering it.

The Life of a Landscape Idea

In 1937, promotion efforts paid off, and the reservoir was finally inaugurated. This event also marked the opening of a new chapter in the history of this landscape, namely the chapter on its actual material existence. The imaginary that had evolved over 40 years was suddenly confronted with the infrastructure's concrete manifestation. At the time of that encounter, discourses celebrated the new infrastructure and its landscape as a major achievement. However, it gradually became apparent over the following years that the actual landscape of Lake Sihl differed from what had been promised. The warning words of Schmid and Roth published four decades ago, portraying an unsightly empty lakebed, became a stark reality for several months every year during the late winter and early spring. During those times, the reservoir could not provide the aesthetics of a natural lake as had been promised (Figure 1.3). In other words, the natural reference had reached its limit.

As the years passed, these limits became clearly apparent, compelling people to take action. Three years after the inauguration, voices from both within and beyond the district called for large-scale plantation works along the shoreline, resulting in a commission granted to a landscape gardening firm.[39] The designers of this plantation scheme explicitly sought to mitigate the visual impact of the dry shoreline. They used vegetation to provide a foreground for passers-by plying the new road along the shoreline to frame features in the distance and direct attention away from the empty lakebed.[40] This action marked the first of several efforts to actively steer the development of Lake Sihl's landscape since the reservoir's inauguration. One decade later, the district authorities decided to construct some paths along the shore and designated several portions of the shore as nature reserves.[41] Nature protection frameworks were thus used to develop a new, anthropogenically established body of water. Despite these efforts, however, the handling of new constructions around the lake remained challenging. Even the hydropower operator raised this matter with the district authorities and requested more regulation.[42] To address the issue, in 1962, the canton put the entire reservoir under protection within the ambit of regulations on the preservation of nature, heritage, and art works. The objective

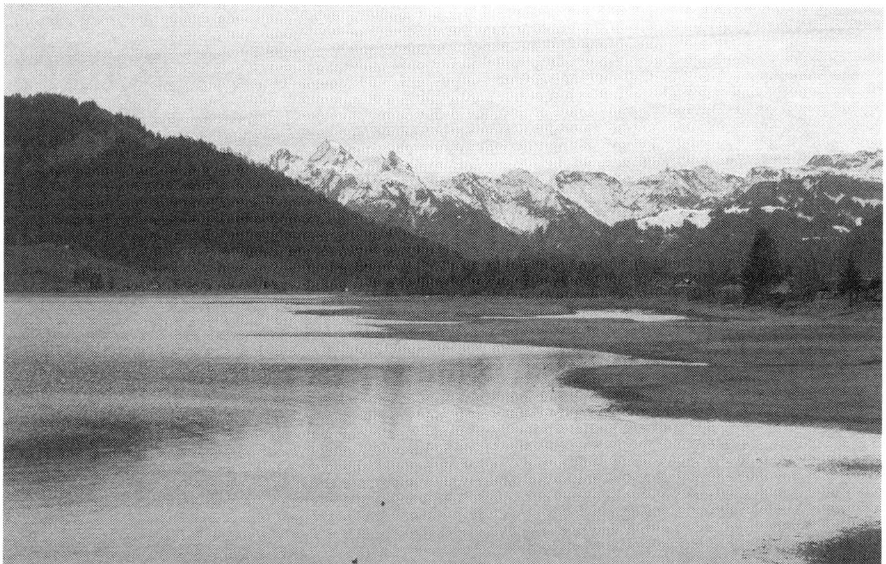

Figure 1.3 The exposed shoreline during drawdown. Even with variations of only a few metres in the water level, large portions of the lakebed become visible in shallow areas, such as in Gross, depicted here. Photograph by the author, April 2021.

of this measure was 'the maximum possible preservation of the view from the lakeside roads to the lake, the preservation of unused shore, and the protection of the natural landscape's appearance.'[43] More recently, the canton has worked on an integrated development concept, involving actors who use the lake, which was published in 2014.[44] These efforts reveal how the landscape of Lake Sihl was shaped by conceptualizations of this landscape before the reservoir's impoundment as well as by material practices afterwards. In some cases, they supported new development projects and in others, they sought to prevent certain developments.

Lake Sihl has been an acclaimed recreational area since the late twentieth century. Leisure activities can be undertaken in the summer (hiking, cycling, bathing, fishing, and water sports) and in the winter (hiking, cross-country skiing, snowshoeing, and ice skating). These activities attract locals as well as visitors from further away, particularly the Zurich metropolitan area. In parallel, the lake has frequently been featured in visual representations: paintings of the landscape by local or external professional and amateur artists, postcards, and posters. The composition logic established by Wydler in 1900 has become a recurrent feature in published photographs, drawings, and paintings. This indicates that little has changed in the way Lake Sihl is viewed, from its conceptualisation in a proposal that was yet to materialise in the early twentieth century to its establishment as a landmark site up to the early twenty-first century.

This success, however, has a dark side. Even if recent images illustrate the enduring popularity of the first drawing of the reservoir created in 1900, the period of the year where the shores become exposed has not disappeared over time. Year after year, the scene appears and disappears while remaining absent from the reservoir's iconography. Just as topographical maps show rivers at a designated 'normal' maximum water level rather than at other levels, the reservoir was depicted at selected moments in the year when the water level was high because the empty lakebed was deemed unsightly.

Recently, however, the situation has begun to change. In the last few decades, perspectives on reservoirs have shifted throughout Europe. Instead of repelling people, the empty lakebeds of reservoirs are now attracting them because of the unusual views that they provide, while also rendering visible the remnants of previous constructions, such as buildings, roads, and agricultural landforms. This attraction and interest in reservoirs can be observed in and beyond the Alps: in Lake Serre-Ponçon in France, the Molveno and Valvestino Lakes in Italy, the Eder and Forggen Lakes in Germany, the la Breña reservoir in Spain, and the Alto Lindoso reservoir in Portugal to cite some well-known examples.[45] Lake Sihl too can be counted among these attractions. Drawdown has become a moment of interest for many, transforming a 'recurrent steppe-like landscape' into 'a public magnet for visitors from all over Switzerland,'[46] as observed by a local resident in 2020. However, the same resident complained that this recurrent phenomenon may be less attractive to those who see it every day, with little being done to change things. Thus, this new appreciation of the empty lakebed only speaks to some inhabitants and visitors.

The relatively new appreciation of the margins of reservoirs has transformed their shadow sides and ambiguities into opportunities. What was previously considered accidental and unsightly is becoming an object of fascination. Apart from its association with leisure and attracting visitors, this shift has generated an opportunity for local communities to reappropriate their daily environment and relate to their past by viewing traces of former buildings, paths, agricultural patterns, and even trees. It has also become a topic of exploration for artists and especially photographers, with the work of the professional photographer Martin Linsi being notable in this regard. Linsi has explored the reservoir margins of Lake Sihl over a period of years, revealing facets of human activity along this temporary shoreline, ranging from its use as free ground for leisure activities to the landforms and remnants that usually remain underwater. His images explore the interaction of well-established visual codes for a sublime and romantic alpine setting with the less common element of reservoir margins and their oddity, reminiscent of wastelands. Linsi's photograph taken along the valley axis (Figure 1.4) reflects

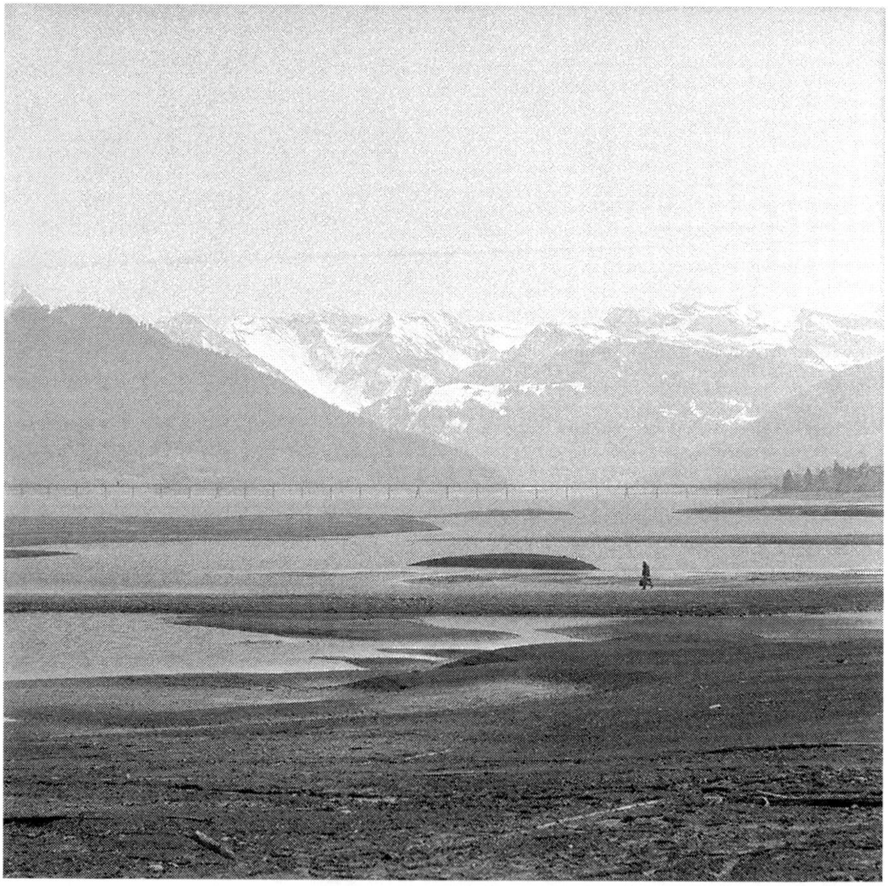

Figure 1.4 The valley during drawdown. Photograph by Martin Linsi.

this new direction, portraying a scene that taps into deeply rooted artistic traditions relating to natural alpine lakes, while also integrating the darker sides of this landscape that differ from that tradition and exploring their qualities to produce something new and specific to reservoirs.

Conclusion

The story of Lake Sihl, considered from the time of its inception in the late nineteenth century rather than since its construction in the 1930s, provides insight into the landscape culture of hydroelectric reservoirs. It invalidates the frequently made assumption that reservoir landscapes are accidental or that they are intrinsically beautiful, revealing the considerable extent to which this landscape may be a product of active practices. In the case of the Etzelwerk, these practices included the construction of an aesthetic and its wider propagation. Moreover, this aesthetic construction often unfolded at the sidelines of the infrastructural project. Even if the Etzelwerk project was not intrinsically motivated by the idea of a landscape, and little was said about this matter, a variety of actors appropriated and imagined the idea of the reservoir and sought optimal ways of integrating it into the Upper Sihl Valley. They communicated their views through words and images and contested the views of others. The reservoir's landscape thus became a highly political matter, instrumentalised either to promote or reject the infrastructure.

It is also important to consider the limits of this outcome. In the broader context of the infrastructural project, the landscape was one among several issues, which notably included financial compensation or the loss of land and homes. One can hardly quantify the extent to which landscape mobilisation influenced the district residents' opinion and the approval of the Etzelwerk proposal. Other factors may have held more weight. For instance, among the individuals called to vote, the number of people who benefitted from the project clearly exceeded that of those impacted negatively. Resistance to the idea of a beautiful landscape was also much lower than resistance relating to other issues. For instance, an attempt to expand the Etzelwerk project in the 1920s entailing the construction of a larger lake met with fierce opposition, as this posed a much larger threat to settlements in the valley. As a result, the proposal was eventually abandoned. When it comes to landscape, framing the scenery as a benefit of the project contributed to the greater weight of benefits than that of losses. It steered opinions on the infrastructural project, resulting in a general appreciation of the reservoir and the implementation of concrete measures, such as regulation of the water level in the summer and the planting of trees along the shoreline. In retrospect, these efforts were successful: the idea of a beautiful Lake Sihl remains as strong as ever. Such cultural constructions may vary considerably from case to case, but they are too important to ignore when it comes to landscape.

The first part of this study confirms the outcomes of previous infrastructural histories, wherein the idea of the infrastructure leads to an 'enthusiasm of the imagination,'[47] carried by notions of progress and resulting in mobilisations of visibility and invisibility.[48] It adds depth to an understanding of the aesthetic foundations at stake and their implications. The 'politics of (in)visibility'[49] observed by Ferrario and Castiglioni along the Italian Piave differ from the case of Lake Sihl. In the Etzelwerk, the reservoir was clearly apparent and lauded from the beginning, being hidden only when the water level was low. This altogether more positive relation to the reservoir's scenery reflects a time when landscape transformation was a less sensitive matter and thus easier to promote, in contrast with the later decades of the Piave projects and the current situation. Furthermore, many of the dynamics revealed in the Lake Sihl project, such as the naturalisation of the reservoir and material efforts to sustain this natural image, are still clearly apparent. The case of Lake Sihl highlights some of the limits of these structuring logics. Even in the early days of the project, some individuals pointed out issues that persist 120 years later. The empty lakebed and its temporary shore continue to be unsettling for many observers, as they do not fit an established aesthetic ideal. What is noteworthy is that some people are now gradually finding ways to overcome these logics, such as the 'natural' reference, and have begun to explore the qualities of reservoirs considered as specific landscapes rather than as natural lakes with evident shortcomings.

These outcomes show that a nuanced approach to landscape impacts (negative and positive) is essential for understanding what actually happens around the construction of a reservoir. This consideration applies not only to historical perspectives but also to current practices. At present, planning and design procedures for reservoirs are stuck in the outdated and limited logic of negative impacts. Projects for new reservoirs or for the transformation of existing ones remain exclusively focused on mitigation and often encounter similar impediments to projects of a century ago. For hydropower to play a constructive role in the energy transition, these planning and design procedures would greatly benefit from not only mitigating the negative impacts of hydroelectric infrastructures but also from cultivating their qualities, such as the specific aesthetic of reservoirs, the activities they allow (in all seasons), and the traces of the past they reveal at low water. This is important for metropolitan visitors escaping their urban environment but even more critical for local communities that have sacrificed their land and homes for the sake of large-scale energy politics.

Notes

1 For a critique of this situation. which is based on the French context but widely applicable throughout Europe, see Folléa, 'Pour une politique paysagère.'
2 Examples include the contributions in Pavia, *Paesaggi elettrici*; Jakob and Stahel, *Berge versetzen*; Kaika, "Dams as Symbols"; Mittmann, "Architecture de l'industrie électrique"; and Zorgno, "Elettricità e territorio."
3 Cosgrove, Roscoe, and Rycroft, "Landscape and Identity."
4 Ferrario and Castiglioni, "Visibility/Invisibility."
5 Ferrario and Castiglioni, "Visibility/Invisibility," 833.
6 Another relevant work here is Briffaud et al., "The Nature of Resources."
7 Schwager, "Technischer Fortschritt," 54.
8 The legal divisions differ from canton to canton. Both cantons of Zurich and Schwyz are divided in districts, each of which is subdivided into municipalities. However, districts

play a more important role in Schwyz, whereas municipalities hold greater weight in Zurich.

9 These numbers are taken from the survey of the organisation tasked with the resettlement scheme, which remains the most precise survey to date. Bernhard, *Die Randbewohnerfrage*, 5, 22, 57.

10 For an extensive account of the interests and negotiations involved in the Etzelwerk project, see Schwager, "Technischer Fortschritt."

11 Maschinenfabrik Oerlikon. "Konzessionsgesuch," 1.

12 Maschinenfabrik Oerlikon, "Konzessionsgesuch," 1–2.

13 See, for instance, Kürsteiner's descriptions of his other projects, starting with Kürsteiner, "Das Elektrizitätswerk Kubel [I]'; and Kürsteiner, "Das neue Elektrizitätswerk [I]."

14 *Einsiedler Anzeiger*, September 15, 1897, section Kt. Schwyz, 2.

15 *Einsiedler Anzeiger*, September 29, 1897, section Kt. Schwyz, 2.

16 *Einsiedler Anzeiger*, October 23, 1897, section. Kt. Schwyz, 2.

17 Wyss, Oskar. "Ueber den Einfluss," 2.

18 Wyss, "Ueber den Einfluss," 2–3.

19 See, for instance, Schmid, and Roth, "An den Tit," 1.

20 Schmid and Roth, "An den Tit," 2.

21 Maschinenfabrik Oerlikon, "Konzessionsgesuch," September 6, 1897, 2–3.

22 Schmid and Roth, "An den Tit," 16, 18.

23 Schmid and Roth, "An den Tit," 18.

24 Bezirksrat Einsiedeln, *Elektrizitätswerk am Etzel*, 26, 29, 37.

25 Bezirksrat Einsiedeln, *Elektrizitätswerk am Etzel*, 41.

26 Einsiedler Kalender, "Ein neuer See," 52.

27 Several works have explored both of these changes. Corbin, *Le territoire du vide* is a seminal work that explores shifting perceptions of water. There is a rich literature on changing perceptions of mountains. Notable works are Nicolson, *Mountain Gloomy*; Groh and Groh, "Von den schrecklichen"; and Reichler, *La découverte des Alpes,* 8. On the parallels between the rediscovery of mountains and of waters, see Walter, *Les figures paysagères*, 83–84 and Briffaud, "Alain Corbin," 694.

28 Jakob, "L'invention du Léman," 14.

29 "Unser Turnfest," April 5, 1905. A miniature of this mural can be found in the private collection of Albert Bingisser, Einsiedeln.

30 Information on the painting's story was provided by Albert Bingisser and the *Kulturverein Chaernehus Einsiedeln*.

31 *Zürcher Illustrierte*, "Sihlsee und Etzelwerk," 2; Eggenberger, "Das projektierte Etzelwerk," 300; Eggenberger, "Das Etzelwerk," 31; Krause, *Das Etzelwerk: erbaut 1932–1937*.

32 Frohmut, *Ein Sihl-Alps-See*, 7; "Eine Fata Morgana," 1.

33 Bezirksrat Einsiedeln, *Bericht und Antrag 1926*, 47–48. The fine was collected when the operator did not comply with these requirements.

34 Bezirksrat Einsiedeln, *Bericht und Antrag*, 25.

35 Gelpke, *Entartungserscheinungen*, 8.

36 Gelpke, *Entartungserscheinungen*, 7.

37 Gelpke, *Entartungserscheinungen*, 6. Italics in the original.

38 Leisi, "Delegierten- und Hauptversammlung," 79.

39 A, (as signed by the author, who is assumed to be Gustav Ammann)), "Uferbepflanzung des Sihlsees erwünscht," 2. [*Neue Einsiedler Zeitung*, July 23, 1940; Ammann, "Naturschutz und Landschaftsgestaltung," 244.

40 Nussbaumer, "Der Sihlsee," 33.
41 "78 D V 1." Bezirk Einsiedeln, 1952. G II 7.2. Bezirksarchiv Einsiedeln.
42 Etzelwerk A.G. "An den Bezirksrat Einsiedeln."
43 Regierungsrat Kanton Schwyz, "Verordnung zum Schutz des Sihlsees," 1
44 Tsp raumplanung, "Entwicklungskonzept Sihlsee (EKS): Bericht."
45 See, for instance, Göbel, "Wegen der Trockenheit"; Il Dolomiti, "Nel lago di Valvestino"; The Guardian, "Ghost Village Emerges."
46 Letter from Hans Kälin (Euthal) in: "Verlandung des Sihlsees."
47 Mrázek, *Engineers of Happy Land*, 166.
48 Larkin, "The Politics and Poetics," 336. Larkin's paper contains a rich review of key precedents.
49 Ferrario and Castiglioni, "Visibility/Invisibility," 833.

Bibliography

A. "Uferbepflanzung des Sihlsees erwünscht" ["Planting of Lake Sihl Requested"]. *Neue Einsiedler Zeitung*, 23 July 1940.

Ammann, Gustav. "Naturschutz und Landschaftsgestaltung" ["Nature Protection and Landscape Design"]. *Schweizerische Bauzeitung*, May 25, 1940. https://doi.org/10.5169/SEALS-51181.

Bernhard, Hans. *Die Randbewohnerfrage beim Stauwerk am Etzel: Nach Aufnahmen der Schweizerischen Vereinigung für Innenkolonisation und industrielle Landwirtschaft* [*The Question of Nearby Residents at the Etzel Hydroelectric Complex: According to Surveys by the Swiss Association for Internal Colonization and Industrial Agriculture*]. *Schriften der Schweizerischen Vereinigung für Innenkolonisation und industrielle Landwirtschaft* 42. Bern-Bümpliz: Benteli, 1931.

Bezirksrat Einsiedeln. *Bericht und Antrag des Bezirksrates Einsiedeln über die Etzelwerkkonzession* [*Report and Proposal of the District Council of Einsiedeln on the Etzelwerk Concession*]. Einsiedeln: Meinrad Ochsner, Buch und Akzidenzdruckerei, 1926.

Bezirksrat Einsiedeln. *Elektrizitätswerk am Etzel mit der Seeanlage im Hinterthal: Bericht und Anträge des Bezirksrates Einsiedeln* [*Electrical Power Plant at the Etzel with the Lake Facility in Hinterthal: Report and Proposals of the District Council of Einsiedeln*]. Einsiedeln: Eberle & Rickenbach, 1900.

Briffaud, Serge. "Alain Corbin, Le Territoire du vide. L'Occident et le désir du rivage, 1740-1840," (book review). *Revue d'Histoire Moderne & Contemporaine* 40, no. 4 (1993): 691–695. www.persee.fr/doc/rhmc_0048-8003_1993_num_40_4_1699_t1_0691_0000_2.

Briffaud, Serge, Emmanuelle Heaulmé, Véronique André-Lamat, Bernard Davasse, and Isabelle Sacareau. "The Nature of Resources: Conflicts of Landscape in the Pyrenees during the Rise of Hydroelectric Power." In *Renewable Energies and European Landscapes*, edited by Marina Frolova, María-José Prados, and Alain Nadaï, 135–153. Dordrecht: Springer Netherlands, 2015. https://doi.org/10.1007/978-94-017-9843-3.

Corbin, Alain. *Le territoire du vide: l'Occident et le désir de rivage* [*The Lure of the Sea: The Discovery of the Seaside in the Western World, 1750-1840*]. Champs – Histoire. Paris: Flammarion, 1990.

Cosgrove, Denis, Barbara Roscoe, and Simon Rycroft. "Landscape and Identity at Ladybower Reservoir and Rutland Water." *Transactions of the Institute of British Geographers* 21, no. 3 (1996): 534–551. https://doi.org/10.2307/622595.

Eggenberger, H. “Das Etzelwerk” [“The Etzelwerk”]. *Schweizerische Wasser- und Energiewirtschaft* 25, no. 3 (1933): 27–33. https://doi.org/10.5169/SEALS-922405.

Eggenberger, H. “Das projektierte Etzelwerk” [“The Planned Etzelwerk”]. *Schweizerische Bauzeitung*, December 14, 1929, 299–304. https://doi.org/10.5169/SEALS-43472.

“Eine Fata Morgana: Luftspiegelung oder eitles Truggebilde über Einsiedeln” [“A Fata Morgana: Loom or Conceited Delusion over Einsiedeln]. *Einsiedler Anzeiger* April 16, 1921.

Einsiedler Anzeiger. September 15, 1897, section Kt. Schwyz.

Einsiedler Anzeiger. September 29, 1897.

Einsiedler Anzeiger. October 23, 1897, section Kt. Schwyz.

Einsiedler Kalender. “Ein neuer See bei Einsiedeln” [“A New Lake near Einsiedeln”]. 1901. Einsiedeln.

Etzelwerk A. G. “An den Bezirksrat Einsiedeln” [“To the District Council of Einsiedeln”]. April 4, 1959. G II 7.2. Bezirksarchiv Einsiedeln.

Ferrario, Viviana, and Benedetta Castiglioni. “Visibility/Invisibility in the ‘Making’ of Energy Landscape. Strategies and Policies in the Hydropower Development of the Piave River (Italian Eastern Alps).” *Energy Policy* 108 (September 2017): 829–835. https://doi.org/10.1016/j.enpol.2017.05.012.

Folléa, Bertrand. “Pour une politique paysagère de la transition” [“For a Landscape Policy of the Transition”]. *Les Carnets du Paysage* 36 (2019): 97–115.

Frohmut. *Ein Sihl-Alps-See und der Hochsihlsee* [*A Lake Sihl-Alp and the High Lake Sihl*]. Einsiedeln: M. Ochsner, 1921.

Gelpke, Rudolf. *Entartungserscheinungen auf dem Gebiete der schweizerischen Wasserwirtschaft: Vortrag gehalten an der 14. Naturschutzlandsgemeinde in St. Gallen, am 15. Mai 1927* [*Signs of Degeneration in the Field of Swiss Water Management: Lecture Given at the 14th National Nature Protection Assembly in St. Gallen on May 15, 1927*]. Basel: Schweizerischer Bund für Naturschutz, 1927.

Göbel, Phil. “Wegen der Trockenheit: In Hessen ist dieses versunkene Dorf aufgetaucht” [“Due to the Drought: This Sunken Village Has Emerged in Hesse”]. *Der Spiegel*, August 20, 2018, section Panorama. https://www.spiegel.de/panorama/edersee-in-hessen-duerre-legt-dorf-ruinen-im-see-wieder-frei-a-00000000-0003-0001-0000-000002761459.

Groh, Dieter, and Ruth Groh. “Von den schrecklichen zu den erhabenen Bergen. Zur Entstehung ästhetischer Naturerfahrung” [“From the Terrifying to the Sublime Mountains: On the Origin of the Aesthetic Experience of Nature”]. In *Vom Wandel des neuzeitlichen Naturbegriffs* [*On the Change of the Modern Notion of Nature*], edited by Heinz-Dieter Weber, 53–95. Konstanz: Universitätsverlag, 1989.

Il Dolomiti. “Nel lago di Valvestino il livello dell’acqua si abbassa e riemerge l’antica dogana” [“In Lake Valvestino, the Water Level Drops and the Old Customs House Reappears”]. February 28, 2020. www.ildolomiti.it/ambiente/2020/nel-lago-di-valvestino-il-livello-dellacqua-si-abbassa-e-riemerge-lantica-dogana.

Jakob, Michael. “L’invention du Léman” [“The Invention of Lake Geneva”]. In *Autour du Léman: Histoire et Esthétique d’un Espace Lacustre* [*Around Lake Geneva: History and Aesthetic of a Lacustrine Space*], edited by Michael Jakob, 8–25. Geneva: MétisPresses, 2018.

Jakob, Michael, and Urs Stahel, eds. *Berge Versetzen: Girolas Kraftwerke in den Alpen* [*Moving Mountains: Girola’s Power Plants in the Alps*]. Winterthur: Scheidegger & Spiess, 1997.

Kaika, Maria. "Dams as Symbols of Modernization: The Urbanization of Nature between Geographical Imagination and Materiality." *Annals of the Association of American Geographers* 96, no. 2 (2006): 276–301. https://doi.org/10.1111/j.1467-8306.2006.00478.x.

Krause, Oskar. *Das Etzelwerk: erbaut 1932-1937* [*The Etzelwerk: Built 1932–1937*]. Wollerau: M. Theiler-Helbling, 1937.

Ksr. "Die Schweiz: das Land der blauen Seen" ["Switzerland: The Land of Blue Lakes"]. *Die Schweiz: Offizielle Reisezeitschrift der Schweiz* 29, no. 6 (June 1, 1956). https://doi.org/10.5169/SEALS-779566.

Kürsteiner, Louis. "Das Elektrizitätswerk Kubel bei St. Gallen [I]" ["The Kubel Electrical Power Plant near St. Gallen [I]"]. *Schweizerische Bauzeitung* 43, no. 14 (April 2, 1904): 161–165. https://doi.org/10.5169/SEALS-24703.

Kürsteiner, Louis. "Das neue Elektrizitätswerk der Stadt Chur an der Plessur bei Lüen [I]" ["The New Electrical Power Plant of the City of Chur on the Plessur near Lüen (I)"]. *Schweizerische Bauzeitung* 69, no. 1 (January 6, 1917): 4–8. https://doi.org/10.5169/SEALS-33815.

Larkin, Brian. "The Politics and Poetics of Infrastructure." *Annual Review of Anthropology* 42, no. 1 (2013): 327–343. https://doi.org/10.1146/annurev-anthro-092412-155522.

Leisi, Ernst. "Delegierten- und Hauptversammlung in Einsiedeln" ["Delegates Assembly and General Assembly in Einsiedeln"]. *Heimatschutz* 24, no. 5 (1929): 78–80.

Maschinenfabrik Oerlikon. "Konzessionsgesuch" ["Application for a Concession"]. September 6, 1897. G II 1.1. Bezirksarchiv Einsiedeln.

Mittmann, Elke. "Architecture de l'industrie électrique et cadre urbain. La confrontation entre la France et l'Allemagne" ["Architecture of Electric Industry and Urban Environment: The Confrontation between France and Germany"]. *Bulletin d'histoire de l'électricité* 24, no. 1 (1994): 43–64. https://doi.org/10.3406/helec.1994.1251.

Mrázek, Rudolf. *Engineers of Happy Land: Technology and Nationalism in a Colony*. Princeton, NJ: Princeton University Press, 2002. https://doi.org/10.1515/9780691186931.

Nauer, Heinz. *Fromme Industrie: Der Benziger Verlag Einsiedeln 1750–1970* [*Pious Industry: Benziger Verlag Einsiedeln 1750-1970*)]. Baden, Switzerland: Hier und Jetzt, 2017.

Nicolson, Marjorie H. *Mountain Gloom and Mountain Glory: The Development of the Aesthetics of the Infinite*. Seattle: University of Washington Press, 2014.

Nussbaumer, Hans. "Der Sihlsee" ["Lake Sihl"]. *Anthos: Zeitschrift für Landschaftsarchitektur* 3, no. 1 (1964): 32–37. https://doi.org/10.5169/SEALS-131888.

Pavia, Rosario, ed. *Paesaggi Elettrici: Territori, Architetture, Culture* [*Electrical Landscapes: Territory, Architecture, Culture*]. Venice: ENEL, 1998.

Regierungsrat Kanton Schwyz. "Verordnung zum Schutz des Sihlsees" ["Ordinance for the Protection of Lake Sihl"]. Schwyz, July 12, 1967. J II 27. Bezirksarchiv Einsiedeln.

Reichler, Claude. *La découverte des Alpes et la question du paysage* [*The Discovery of the Alps and the Question of the Landscape*]. Chêne-Bourg/Geneva: Georg, 2002.

Salathé, O. *Wägital und Etzelwerk: Führer in das Etzel- und Wägital-Werk* [*Wägital and Etzelwerk: Guide to the Etzelwerk and Wägital-Werk*]. Einsiedeln: Karl Zehnder, 1933.

Schmid and Roth. "An den Tit. Bezirksrat von Einsiedeln" ["To the Titular District Council of Einsiedeln"]. December 6, 1898. G II 1.2. Bezirksarchiv Einsiedeln.

Schwager, Gerhard. "Technischer Fortschritt in der bäuerlichen Schweiz: Die Realisierung des Etzelwerkprojekts 1897–1937." ["Technological Progress in Rural Switzerland: The Realization of the Etzelwerk Project 1897–1937"]. PhD Diss., University of Zürich, 1992.

The Guardian. "Ghost Village Emerges in Spain as Drought Empties Reservoir." *The Guardian*, February 11, 2022, section World News. www.theguardian.com/world/2022/feb/11/ghost-village-emerges-in-spain-as-drought-empties-reservoir-aceredo.

Tsp raumplanung. "Entwicklungskonzept Sihlsee (EKS): Bericht" ["Development Concept Lake Sihl (EKS): Report"]. Umweltdepartement, Kanton Schwyz, 2014.

"Unser Turnfest" ["Our Gymnastics Festival"]. *Einsiedler Anzeiger* April 5, 1905.

"Verlandung des Sihlsees: Leserbriefe" ["Silting of Lake Sihl: Letters to the Editor"]. *Einsiedler Anzeiger*, May 8, 2020. www.einsiedleranzeiger.ch/2020/05/08/verlandung-des-sihlsees.

Walter, François. *Les Figures Paysagères de la Nation: Territoire et Paysage en Europe (16e-20e siècle)*. [*Landscape Figures of the Nation: Territory and Landscape in Europe (16th-20th Century)*]. Paris: Éditions de l'École des hautes Études en Sciences Sociales, 2004.

Wyss, Oskar. "Ueber den Einfluss eines künstlichen Sees im Sihlthal bei Einsiedeln auf die Gesundheitsverhältnisse der Bewohner des Thales" ["On the Influence of an Artificial Lake in the Sihl Valley near Einsiedeln on the Health Conditions of the Valley's Inhabitants"]. Zurich, May 28, 1898. G.4.5.1.4. Historisches Archiv ABB Schweiz.

Zorgno, Anna Maria. "Elettricità e territorio" ["Electricity and Territory"]. *Rassegna* 63, no. III (1995): 70–80.

Zürcher Illustrierte. "Sihlsee und Etzelwerk" ["Lake Sihl and Etzelwerk"]. Conzett & Cie., August 23, 1926. https://doi.org/10.5169/seals-833808.

2 Land of Fire, Temples of Extraction

Azerbaijan's Geo-Architectural Assemblage of Oil and Nation[1]

Leyla Sayfutdinova

On a crisp December night, I visited the National Seaside Park in Baku, popularly known as 'the Boulevard,' to take a photograph of the Flame Towers. This complex of high-rise buildings in the shape of three flames was constructed in 2013 and has since become a symbol of the 'new Baku,' with its spectacular architecture and windfall oil revenues. I knew that at night the buildings' windows were transformed into LED screens lit up in different colours, including the colours of the national flag. Thinking that this would be a fitting image for my research on the symbolism of oil in Baku, I found a spot in the upper alley of the park with a good view of the Towers. They were diligently lit up in the colours of the flag, and I snapped a few shots. As the flag faded, however, the Towers lit up again in red and orange. This view of huge burning flames of fire in the dark winter sky floating over the city was surreal. On my left, the Christmas tree decorations on an exploration drilling rig twinkled softly in Baku Bay. I recalled a poem that I learnt in school many years ago:

> Азербайджан!
> Купель огня,
> Гряда несокрушимых башен.
> Я твой навеки,
> Ибо я
> Из праха твоего заквашен.
>
> Azerbaijan!
> the font of fire,
> the ridge of towers that never break.
> I am yours forever
> for I am made of your dust.
> Mansur Vekilov, *Aziz Diyar* (1970). Author's translation.

The high-rise buildings, the drilling rig, and the images of fire in front of me bore a striking resemblance to the imagery of the poem published in 1970 in Soviet-ruled Azerbaijan. This articulation of national and local belonging conveyed through the

DOI: 10.4324/9781003473534-4

geological and architectural metaphors of fire, land, and buildings is the central focus of this chapter. Specifically, I explore how fire, which for centuries denoted the presence of hydrocarbon resources in the subsurface of the Abşeron Peninsula, became linked to the built environment of Baku over the last century and a half and was subsequently extended symbolically to the entire nation. Fire, thus, serves as a metaphor that articulates the conjuncture between hydrocarbon resources and belonging at both local and national scales. This chapter contributes to a growing scholarship on resource nationalism encompassing both political-economic and cultural-symbolic forms, which can be 'interwoven and mutually reinforcing.'[2]

Baku is an especially interesting place in which to explore this interweaving. In the mid-nineteenth century, it was one of the pioneering sites of industrial oil production, but unlike many other coeval production sites, it remained an oil city well into the twenty-first century. However, as a national capital, Baku is a highly unusual site within the global landscape of the oil industry, wherein extraction sites are usually located in the resource-rich peripheries, far removed from the centres of power. To some extent, this used to be the case in Baku, which had been a resource periphery within the Russian Empire and the Soviet Union. However, the cumulative concentration of economic, human, and cultural resources in Baku, initially after it became the administrative centre of the Baku Governorate and subsequently the capital of the Azerbaijan Soviet Socialist Republic, consolidated its unique role in the construction of Azerbaijan's resource nationalism.

The mechanisms behind this conjuncture between resources, territory, and nation in Azerbaijan remain surprisingly underexplored, perhaps because of the unusually strong normalisation, and even naturalisation, of the links between them. Thus, while scholars are working to reconstruct the relationships between what appear to be separate spheres of geology and politics in other extraction sites,[3] in Baku, and in Azerbaijan more broadly, the task is conversely to disentangle the natural and the social and the geological and architectural, which often appear inseparable.

Many observers have noted the geopolitical significance of spectacle and architecture in post-Soviet Baku and their connection to revenues from resource extraction.[4] Some of these accounts briefly mention the cultural importance of fire worship and Zoroastrianism. Yet, these analyses treat both the architecture and symbols of fire as detached from the landscape, emphasising processes of signification and surplus accumulation as opposed to the history of resource extraction. Yet, in other contexts, the relationship between subterranean resources and the urban built environment has been explored in greater depth. Thus, for example, Gray Brechin described the skyscrapers of San Francisco metaphorically as 'inverted mines,' referring to the massive wealth generated from extractive industries that financed the construction of high-rise buildings in the city.[5] Natalie Koch explored the relationship between resource wealth and the spectacularisation of cities in non-democratic states of Asia,[6] while Carola Hein introduced the concept of a 'petroleumscape' to describe the network of the built environment 'carrying physical and financial flows of petroleum.'[7] What these studies have in common is a focus on the workings of capital accumulated from resource extraction and on the trade in petroleum as a commodity.

In this chapter, I extend this research further by examining what I term the 'geo-architectural assemblage,' which links Baku's built environment to the hydrocarbon reserves beneath the surface. I explore how architecture became a site of material and symbolic articulation of the conjuncture between the state, territory, and subterranean resources. Bridging the literature on urban verticality and the territorialisation of the subterranean, I show how fire, an element of the physical landscape of the Abşeron Peninsula originating in subterranean hydrocarbon reserves, has been re-mapped onto the urban landscape via key iconic buildings in Baku. In doing so, I draw on the assemblage thinking which highlights how more-than-human open-ended wholes are constituted from heterogeneous elements.[8] To this end, I adopt a resource materialities perspective according to which resources are constituted relationally across the nature/culture divide through extractive and distributive infrastructures and technologies, valuation devices, and cultural discourses.[9]

From this perspective, fire's importance can be seen to stem from its role in the physical landscape, where it has historically been a marker of the hydrocarbon resources that lie beneath the surface. The use of fire symbolism in the built environment serves to re-create the local energy landscape, reinforcing both the symbolic links and the material connections between land, fire, and the oil that feeds it. While I focus specifically on several landmarks, they are not unique; rather, they are representative of the entanglements of nature, architecture, and extractive infrastructure within Baku's dense urban fabric.

In the first section of this chapter, I trace the intersections between the literatures on energy landscapes, urban verticality, and volumetric territoriality and suggest how assemblage theory can be used to examine these intersections. Next, I briefly describe the geographic background of the Abşeron Peninsula, where Baku is located and from where oil has been historically extracted. I subsequently examine the linked process of adopting fire as a symbol of the city and nation and the mechanisms of its inscription and embedding within iconic architecture over three historical periods: the beginning of industrial oil extraction (from the mid-nineteenth century to 1920), the Soviet period (1920–1991) and the post-Soviet period (from 1992 to the present). In conclusion, I discuss how discursive, visual, and material practices co-produced the imaginaries and material installations of resource nationalism and probe the possibilities of change in the future.

Resource Nationalism in an Urban Energy Landscape

Within urban energy landscapes, energy infrastructures are located close to and are enmeshed with the built environment of cities.[10] Most of the existing research on urban energy landscapes has focused on the entanglements of dwellings and other urban forms with infrastructures for energy distribution and consumption, such as electricity networks and transportation.[11] This scholarship generally differs from that on landscapes of energy extraction, notably those of coal and petroleum, the two main fossil extractive industries, which are usually found in rural or remote areas.[12] Both of these industries are characterised by the appropriation of large swathes of land, bringing about large-scale transformations of the physical landscape, especially when they operate onshore. However, coal mining has historically been located closer to people's dwellings, given difficulties in transporting coal. By contrast, oil's fluidity and lightness facilitated its transportation across long distances, thus allowing for a different, less concentrated spatial organisation.[13] Landscapes of oil extraction thus tend to be more spatially dispersed and removed from sites of processing and consumption compared with coal mining landscapes. Hannah Appel takes this idea even further. In light of her fieldwork conducted in Equatorial Guinea, she argues that the contemporary oil industry is organised within a transnational network of 'bundled and repeating' modular projects that are as detached from local conditions as possible.[14]

Baku's urban energy landscape differs, however, from the landscapes of oil extraction described in many scholarly accounts. Far removed from the maritime transportation routes, Baku has one of the highest concentrations of extraction sites in the world. Beginning in the nineteenth century, Baku's urban development was closely linked to the growth of the oil industry, which dictated the city's spatial organisation and the location of its industrial and residential districts.[15] Baku thus exhibits features of both direct and indirect energy landscapes co-produced by extractive activities and supporting industries alike.[16] Because of the socialist experience of over-industrialisation, the historically contingent concentration of extractive activities did not unravel into dispersed capitalist modular organisation. Instead, it led to increased concentration, adding refining, processing, and manufacturing of oilfield equipment to the industrial mix. Even the city's transportation system, including its motorways, railroads, and public transport, was organised to serve the oil industry and to connect the city with the surrounding oilfields.

Within this enmeshment of the built environment with energy extraction, oil and its derivatives—natural gas and fire—saturate, fill, and mingle with both nature and culture. Oil spills from the extractive infrastructure, seeping back into the ground and polluting soil and water, while natural gas oozes out from the cracks in the earth's surface, sometimes self-combusting into fires. Oil-related infrastructure is an integral part of the urban landscape: derricks and pumpjacks punctuate the landscape to the north and west of the city, while the flare tip of the refinery burns like a beacon in the eastern part of Baku. Beyond this functional infrastructure, oil is inscribed in the built environment of the city as well through monuments and memorials, bas reliefs and billboards, and, above all, through the image of fire that is part of both urban and national heraldry.

The concept of landscape is inherently topographical, focusing on observable and visual objects.[17] For Rose and Wylie, this topographical richness enables the landscape concept to animate the geographical scholarship that foregrounds the study of topological connections between different elements. However, in studies on energy landscapes, the focus on the visible and observable can be considered a limitation because these landscapes are co-produced by invisible matter, which is often hidden underground. This is especially true for petroleum, which has been described as ubiquitous but invisible.[18] This invisibility of oil is to some extent a product of intentional action, such as fencing, restricted access to extractive sites, and the separation of extraction sites from places of residence and consumption, but it is inherently due to the physical location of most petroleum reserves underground. The production of petroleum landscapes entails interfering with matter that lies below the surface that is invisible to the naked eye. Writing about energy extraction, Bridge described these landscapes as 'portals, worm-holes between two worlds.'[19] Yet on the Abşeron Peninsula, the self-igniting fires denote the presence of hydrocarbons in the subsurface, thus bringing petroleum into the visible landscape. As I argue in this chapter, the symbols of fire perform much the same function of signalling the presence of oil and gas, which may otherwise be invisible. Here, the volumetric approach that draws attention to the vertical plane in relations between the surface, the underground, and the space above the ground is of value in informing my analysis of the enmeshment of oil, fire, and built form in Baku.[20]

The volumetric turn has been influential within the scholarship on cities as well as natural resources, albeit in different ways. Within urban studies, the focus has been on the relationship between power and height, or above-the-ground aspects of verticality.[21] This approach has expanded and evolved from an initial focus on political technologies of aerial military control to one that includes 'more ordinary vertical urbanisms'[22] and 'subtle workings of power in and through crooked vertical and horizontal planes.'[23] Andrew Harris identified four modes of urban verticality within landscapes: religious buildings connecting this world with heaven above and the subterranean realm of the dead below it, the high-rise iconography of corporate capitalism in fin-de-siecle American cities, modernist high-rise building blocks, and the rise of urban verticality of Asian cities, linked to their changing role in the global economy. These modes are not exclusionary; rather, they are interconnected and overlapping, with the vertical rise of Asian cities being a prime example of the juncture between the power of capital accumulation, commodification of space, and 'the spectacle of self-assured sovereignty.'[24] By contrast, research on natural resources, particularly fossil fuels, has focused on depth and the subterranean component of the vertical. Scholars in this field have examined the extension of sovereignty to areas underground and the various technologies used to produce knowledge about the subterrane for military and extractivist purposes.[25] Imaginaries of the subsurface also play a role in cultural production. Thus, for example, Douglas Rogers examined how corporate public relations in the Perm region of Russia mobilised the imaginaries of depth to link oil extraction to local culture.[26]

A multitude of objects are linked both materially and symbolically to oil and fire within Baku's dense urban energy landscape, ranging from kiosks to still-functioning derricks. Although a geo-architectural assemblage, as I conceive it, comprises all of these structures, here I focus on several iconic buildings, given their prominence in the national imaginary. Two of the buildings are historical monuments, representing the heritagisation of oil as part of the assemblage. I also consider the volumetric space of this energy landscape as it intersects with the horizontal plane.[27] In considering the horizontal, however, my focus is on the symbolic rather than the material aspects of assemblage. Here, I explore how an element of the local physical landscape is symbolically extended to the nation.

My analysis of the geo-architectural assemblage of fire in Baku and, more broadly, Azerbaijan, contributes to the growing literature on resource nationalism as 'the political, economic, and discursive articulation of natural resources with national interests and identity.'[28] Departing from the political-economic definition of resource nationalism and state ownership of resources, scholars drawing on this perspective have shown how resources are imbricated in the production of belonging at both local and national scales.[29] I extend this approach from a focus on the geopolitical discourse to the symbolic and material assemblage within the landscape of Abşeron Peninsula, where Baku is situated.

Fire Symbolism and Early Industrialisation on the Abşeron Peninsula

The Abşeron Peninsula is a strip of land, approximately 60 km long and 30 km wide, which protrudes from the Caucasus Mountains into the Caspian Sea. Although geologically linked to the Caucasus, the peninsula's landscape is mostly flat, with some hills. The lowest parts are located 27 m below sea level. This is a semi-desert zone, with an arid climate, no natural tree cover, and few sources of fresh water. The oil-bearing strata—productive series—are concentrated mostly in the southern part of the peninsula and extend further into the Caspian Sea along the South Caspian depression.[30]

The hydrocarbon deposits located along the Abşeron Peninsula vary in depth, but in many places, they lie fairly close to the surface. It is this upper, shallow layer from which oil has been seeping onto the surface and from which natural gas, which feeds the 'everlasting' fires, has been escaping. The everlasting fires are caused by the spontaneous combustion of hydrocarbon gases that either seep through the subsurface or are ejected into the atmosphere during the eruption of mud volcanoes. The Western Caspian geological depression, which stretches from the Abşeron Peninsula into the Caspian Sea, has the densest concentration of mud volcanoes in the world.[31] In the past, before the hydrocarbon reserves of the Abşeron Peninsula were depleted by industrial oil extraction, many such flames could be seen both on land and in the sea, with columns of fire sometimes reaching 500 m.[32] Nowadays, only one 'everlasting fire,' known as Yanardağ (Burning Mountain; see Figure 2.1), which is fed by gas microseepage, remains in the vicinity of Baku. However, temporary spontaneous eruptions of fire still occur, the most recent one being in 2021, when a mud volcano erupted on a small Caspian island, Daşlı Ada.[33]

The established narrative is that the natural fires attracted Zoroastrians, who value fire above all other elements of nature, to this area. The name of the country, Azerbaijan, is also often linked to fire. According to one popular theory, 'Azerbaijan' is an Arabicised form of the Persian word, 'Adar-Baykan,' which means the 'guardian of fire' and refers to the Zoroastrian practice of fire worship.[34] Historically, the name 'Azerbaijan' referred to a larger area, most of which is now part of Iran. Takht-e Soleyman, one of the three great fire temples in the Zoroastrian tradition, is located in the province of West Azerbaijan in Iran. However, the

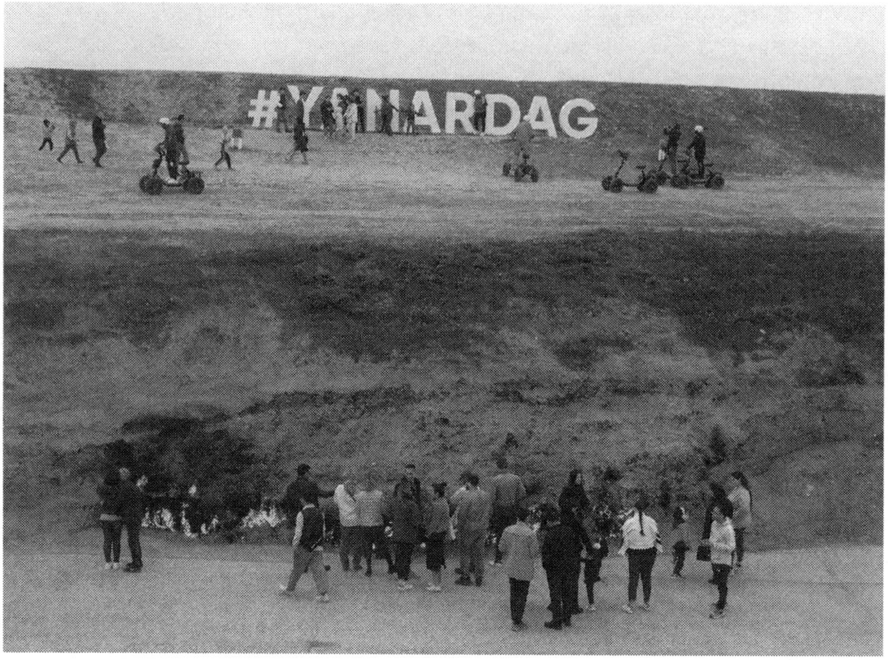

Figure 2.1 Yanardağ near Baku. Photograph by the author, April 2023.

moniker 'land of fire' became more prevalent in the northern part of historical Azerbaijan, which was annexed by the Russian Empire in 1828.[35]

In Baku, oil had been used for lighting, heating, cooking, lubrication of cartwheels, and in medicine for many centuries. Traditionally, it was collected from microseepages and hand-dug wells. Industrial extraction only began after Azerbaijan's annexation by Russia. In 1806, the Russian authorities confiscated almost all oil-bearing plots and made an initial attempt to develop the industry as a Crown monopoly. Russian engineers and geologists worked in the area to explore more reserves and improve extractive technologies, as the traditional methods of extraction were economically unproductive.[36] This model proved unprofitable for the Crown, and commencing from 1821, a lease system was implemented, whereby entrepreneurs could lease oil-bearing plots for a period of four years. However, the lease system provided little incentive for entrepreneurs, did not protect their property rights, and was mired in bureaucracy, which posed many obstacles. As a result, despite early technological innovations, such as drilling, and the abundance of reserves, the output of oil lagged far behind that of the main global producer, the USA. In 1872, an auction system was introduced, which allowed individuals and private companies to buy oil-bearing plots.[37] Within the new system, oil production rose rapidly: in just one year, production almost tripled. By 1901, half of the world's oil output was produced in Baku, which became a boomtown, with its population rising from 12,191 in 1870 to 214,672 in 1913.[38]

Along with changes in the population, the landscape and built environment were transformed. Oil fields covered with derricks constituted an iconic oil landscape in Baku. A large portion of the city's lands was designated for industrial activities, creating an industrial zone known as the 'Black City.' New housing and social infrastructure were built to accommodate large-scale migration to the city. Villages in Abşeron, which were previously connected to Baku via nomadic pastoralist and trade routes, became industrial suburbs, and a network of roads connected extraction sites in the villages around Baku to the transportation and refining facilities in the city.[39] A new extractive technology, drilling, transformed the relationship between the subsurface, surface, and aerial environments: oil gushers flooded the surroundings with oil, some of which was collected in oil lakes. However, a lot of the spillage was uncontrolled. This led to long-term contamination of the land, making much of it unusable for traditional agriculture and animal husbandry. To this day, oil extraction, agriculture, and animal husbandry are often implemented side by side on the Abşeron Peninsula.[40] The gushers also changed the fire emanating from below the ground: instead of spontaneous combustion, the burning gushers were now human-made, burning for days and throwing out smoke, ashes, and soot.

Despite fires being a feature of the natural landscape for centuries, they did not become a visual symbol of Baku until the onset of Russian colonisation. Prior to this time, Baku's medieval symbols included the heads of a bull and a lion as well as the sun and moon, but not fire. One of the first visual representations of the links between the land and fire was the introduction of fire into heraldry.[41] In the city's coat of arms adopted by the Russian administration in 1843, three flames of fire emerging from the ground constituted one of the four heraldic symbols, along with a lion—the old symbol of power, a camel as a symbol of trade, and an anchor as a symbol of a seaport (Figure 2.2, first row, left).

In 1877, four years after the initiation of the oil boom in 1873, the three flames of fire became the main symbol of Baku. In the new city emblem, they were depicted on a black heraldic shield, topped with what looked like a crown composed of a sandstone citadel wall (Figure 2.2, first row, right). A similar image was also adopted for the coat of arms of the Baku Governorate, the territory comprising the eastern part of the present-day Republic of Azerbaijan. This was the first of many efforts to extend the reach of fire symbolism horizontally beyond the Abşeron Peninsula, where the physical fires were burning. In the beginning of the twentieth century, the symbol began to be linked to the built environment: stone bas reliefs of flames decorated the Baku City Duma building (constructed between 1900 and 1904) and the medieval citadel walls of the Içəri Şəhər (the Inner City, added in 1906).

During this period of colonial extractivism, fire was also adopted as a symbol by foreign oil entrepreneurs in order to ground their economic ventures within the locality. A case in point is the Nobel Brothers Company, Branobel, the largest oil company in Baku until 1917. Branobel was established in 1873 by Robert

Figure 2.2 First row (left to right): Baku Governorate Coat of Arms, 1843 and 1877; second row (left to right): coats of arms of Azerbaijan SSR, 1941–1991, and the Republic of Azerbaijan, 1993; third row (left to right): Baku city emblems, 1967 and 2001.

and Ludvig Nobel, who hailed from the famous Swedish dynasty but were based at the time in Saint Petersburg.[42] Branobel soon became the leading oil company throughout the Russian Empire, benefiting from Ludvig Nobel's technical vision and the financial prowess of his Paris-based brother, Alfred. The company is credited with several innovations of global significance, including refining kerosene and inventing the world's first oil tanker. This tanker, built in Stockholm in 1878, was named 'Zoroaster,' and marked the first instance of the Nobel Brothers drawing on the local Zoroastrian heritage. In 1904, Branobel issued a memorial medal with an image of the Atəşgah fire temple[43] to commemorate the twenty-fifth anniversary of the company. Atəşgah is an eighteenth-century structure located in Suraxanı, one of the main sites of onshore oil production on the Abşeron Peninsula, where the company owned many wells. Although the current building was most likely built in the eighteenth century by fire-worshipping Hindu and Parsee pilgrims from India, it is believed to stand on the site of an older, abandoned Zoroastrian fire temple.[44] The complex includes a *chahartag*, a four-cornered altar with a fire outlet, and a *caravanserai*, where pilgrims and other travellers used to stay. Self-ignited fire burned in the altar space, in the torches of the walls, and in the area around the temple until the beginning of the twentieth century, which is how the temple is depicted on the medal. However, the fires themselves went out around the same time because the subterranean reserves that fed them were depleted as a result of oil extraction. For the Nobels, the meaning of these symbols went beyond local concerns and had to do with their civilising, messianic self-representations. Branobel's corporate narratives were full of colonial tropes, which revealed that they saw oil as the fuel of the future and themselves as torchbearers of progress and civilisation, somewhat akin to Prometheus who, according to legend, gifted fire to the people on Earth.[45]

Early visual representations of Baku's industrial landscape often emphasised the scale of both the extractive infrastructure and the fires that escaped from it, juxtaposing small human figures against the huge structures of derricks and even larger gushers. Derricks became visually linked to fire at the turn of the century. Wooden derricks covered Abşeron's oilfields in the nineteenth century and soon became one of the paradigmatic representations of the early oil industry. Popular images of the derricks, on postcards, for example, often also included oil gushers and gusher fires. Indeed, Baku's shallow reservoirs frequently produced tall gushers of oil mixed with water and sand in the early days of the industry.[46] When gushers caught fire, they produced tall pillars of flames and smoke. Today, such images are usually perceived by Western observers as eco-disasters, but at the time, those fires were perceived as both business-as-usual and a spectacle.[47] This perspective remained dominant in Azerbaijan until the end of the twentieth century, with Azerbaijani intellectuals still describing gusher fires as a fascinating spectacle as opposed to an ecological disaster in the late 1990s.[48]

Soviet Geo-architectural Assemblage

Oil and fire featured in the national symbolism of the short-lived Azerbaijan Democratic Republic (ADR), which was formed in 1918 and dissolved following the country's conquest by the Bolshevik Red Army in 1920. ADR's official postage stamp, now considered the first oil postage stamp in the world, depicted Atəşgah surrounded by oil derricks.[49] The proposed national coat of arms, which was never formally adopted, also included flames as its central element. After Azerbaijan was reoccupied by the Red Army, the Bolsheviks immediately embarked on the reconstruction of the newly nationalised oil industry.[50] Under Soviet rule, Baku's landscape continued to develop as a combination of urban form with extractive activities. Oilfields were connected to Baku by road networks and light rail systems, while settlements were built near the oilfields to accommodate the workforce. In the 1950s, the extractive installations extended into the Caspian Sea with the construction of the first Soviet offshore oil production town, the Oil Rocks (Neft Daşları/*Нефтяные Камни*).[51] Baku retained its key role within the Soviet oil industry until after World War II, when it was superseded by Volga-Urals and subsequently by the Siberian basins. However, oil and related industries, such as refining and manufacturing of oilfield equipment, continued to dominate Azerbaijan's economy, and extractive infrastructure, both active and decommissioned, continued to dominate its landscape.

This dominance is reflected in the choice of symbols during the Soviet regime. One part of this infrastructure, the derrick, became the heraldic symbol of the Azerbaijan Soviet Socialist Republic in the 1930s and remained so until the dismantling of the Soviet Union in 1991 (Figure 2.2, second row, left). The new coat of arms also consolidated the extension of oil symbolism to the territory of Soviet Azerbaijan, including to areas where oil had never been produced or even discovered. This symbolism reflected the centrality of oil in Azerbaijan's economy throughout the Soviet period.[52] City coats of arms were abandoned during the early Soviet period but were reinstated in some cities in the 1960s.[53] The Soviet emblem for Baku, reinstated in 1967, retained the three flames along with the black background in the Tsarist coat of arms, but it also included sea waves. The new symbol reflected a more voluminous territorial imaginary, incorporating columns of water and air (Figure 2.2, third row, left).

In Suraxanı, after years of abandonment, Atəşgah was reconstructed as a tourist site as part of a wider programme of restoring historical monuments and developing tourism in the Republic,[54] and fires were once again lit in the temple in 1969. In the restored temple, gas was delivered to the altar through the pipelines of the city's gas network, thus physically linking the sacral architecture of the past with the energy infrastructure of the present.

The most thorough reconfiguration of relations between petroleum, fire, and architecture took place in another medieval building, Qız Qalası (the Maiden Tower). Prior to the construction of the Flame Towers in 2013, the Maiden Tower was Baku's flagship architectural symbol. Situated within Baku's walled medieval İçəri Şəhər (Inner City), the history of this 30-metre-tall tower is shrouded in mystery, but it remains a repository of local legends. Historians and archaeologists disagree on when the different parts of this structure were constructed, and there is also no agreement on its purpose. It differs from all other defence towers on the Abşeron Peninsula, leading many to reject the most obvious explanation that its purpose was military. Currently, most scholars agree that the Tower was probably part of a larger complex, which included underground chambers and tunnels as well as above-ground sections, probably built at different times between the sixth and twelfth centuries. However, the reasons for these constructions remain a subject of debate,[55] and this uncertainty provides a canvas on which to project an imagined history of fire.

Like Atəşgah, the Maiden Tower was reconstructed and converted into a heritage and tourism site in the 1960s. The archaeological excavations and reconstruction prompted a proliferation of new interpretations of the structure's origins. Drawing on local legends and the medieval literature, some scholars have attempted to link the Maiden Tower to Zoroastrianism, although there is little archaeological evidence to support this. According to one of these theories, the Maiden Tower served as a burial tower or *dakhma*.[56] However, the Tower's architecture differs from that of all known *dakhmas*, which are usually much lower. According to another theory posited by Davud Akhundov, a historian of architecture, the Tower was a seven-level temple dedicated to the supreme god of Zoroastrians, Ahura Mazda, and six lower deities, with fires lit at each of the seven levels of the Tower.[57] Akhundov bases his theory on references to the Baku fires in various legends and literary sources but provides no direct evidence that these references specifically relate to the Maiden's Tower. Furthermore, according to Seyran Veliyev, there is no trace of fire or soot in the structure that could support this theory.[58]

The Maiden Tower was also connected to hydrocarbon reserves, Zoroastrianism, and the national identity through performative acts, notably those occurring during the Novruz festival celebrations. This festival, which is celebrated on the day of the spring equinox, has been observed for centuries across parts of the Middle East, the Caucasus, and Central Asia. Within Azerbaijan, this festival is a national holiday, and its roots in Zoroastrianism are widely acknowledged.[59] Fire plays a central role in folk traditions, entailing the lighting of candles and bonfires. Children, and occasionally adults, jump over these bonfires, intended to purify people and bring about renewal, on four consecutive Tuesdays leading up to the festival. In Soviet Azerbaijan, Novruz was banned from 1928 up to the mid-1960s but was restored and reinvented in 1967 in a highly secularised form purged of religious content. The restoration was part of the establishment of new socialist rituals during the Khruschevian 'Thaw' when many folk festivals were reinterpreted in a secular way. The new socialist festivities were usually organised by state officials who officiated the ceremonies.[60]

The script of the first official Soviet Novruz ceremony[61] held at the Maiden Tower provides a good illustration of this process, with details of the assemblage work entailed in reimagining Novruz as a national holiday. In this reimagined version, the Baku City Gas Administration (BakGaz) assumed a leading role in the festival instead of a Zoroastrian priest:

> 11:30 am: seven trumpeters and seven drum players stand on top of the Maiden Tower. Old communists, the participants of the October Revolution, [and] young pioneers gather around [the] gas flare tip.

> 11:35 am: the host announces (in Azerbaijani and Russian) that according to the astronomical data, the equinox arrives in two minutes at 11:37 am. This marks the end of winter and the beginning of spring.
>
> 11:36 am: trumpeters signal, "attention, everyone." Then the drums play. A timer counts down the last seconds of winter.
>
> 11:37 am: BakGaz employee opens the gas tap. M. Dadashev, [a] member of the Communist Party since 1917, lights the gas torch and [so] the fire in the Maiden Tower is lit. Everyone congratulates each other, and applause follows. The loudspeaker [is] turned on and the music, the overture to the Koroğlu Opera composed by Uzeyir Hajibekov, plays. The host congratulates Bakuvians and guests of the city with the [onset of the] first minute of spring and invites everyone to the public celebration in the evening.
>
> 11:40 am: The ceremony ends, and participants go down [the stairs of Maiden Tower] and leave.

This ceremony brings together several elements of the geo-architectural national assemblage. It takes place in the medieval Maiden Tower, connecting an ancient tradition with an 'old' structure of uncertain age. Although Novruz originates in the Persian culture, the ceremony is linked to the Turkic ethnic roots of modern Azerbaijanis through the incorporation of music from the Koroğlu Opera, the national Soviet–Azerbaijani opera composed by Uzeyir Hajibeyov in 1936 and based on an eleventh-century Turkic epic. This epic, *Kitabi Dede Qorqud*, is widely considered to be a repository of Azerbaijani national narratives.[62] The material connection to fire is made by connecting the building to the distributive energy infrastructure—the city gas network—and the fire is lit on a specially installed gas torch. By lighting the fire with natural gas, the festival is connected to the sub-surface petroleum reserves, and it also departs from tradition, whereby fires are generated from wood and other combustibles. In the reinvented Soviet tradition, an employee of the gas company, who lights the fire, replaces the Zoroastrian *magus*, recreating the relationship between fire, subsurface, architecture, and infrastructure in new ways.

The Post-Soviet Urban Landscape

After Azerbaijan gained independence following the Soviet Union's dissolution in 1991, it did not sever its links with the oil industry. One of the priorities of the leaders of independent Azerbaijan was to attract investments and revitalise the oil industry. In 1994, the first contract with a foreign-led consortium for the development of deep-water oilfields in the Caspian, collectively known as Azeri-Chirag-Guneshli (ACG), was signed.[63] This was followed by several new contracts, mostly offshore.[64] Onshore oilfields, as well as the first offshore town, Neft Daşları, remained under the operation of the State Oil Company of Azerbaijan (SOCAR).

The flow of revenue from the new oil deals that began during the early 2000s propelled a massive transformation of Baku's urban space. The new constructions were usually high-rise and spectacular, creating, as one of my interlocutors put it, 'a vertical ring' around Baku. Many have compared this spectacular urban transformation to that of the Persian Gulf cities, particularly Dubai, and have highlighted the links between oil revenues and real estate development.[65] Similar to cities in the Persian Gulf and in other parts of Asia, the vertical rise of Baku was also linked to the 'worldling' efforts of the national leadership, as architecture became a synecdoche of modernisation and wealth.[66] Beyond changing the vertical dimensions of the city, urban renovation began to transform the relationship between the infrastructure and urban forms. The Eurovision Contest, held in Baku in 2012, was the first mega-event to trigger these transformations.[67] In preparation for the contest, a new ultra-modern concert venue, Crystal Hall,[68] was built in Baku off the coast of the Caspian Sea. The waterfront just behind the Crystal Hall and adjoining the existing seaside park was beautified, and some of the old industrial enterprises and debris were removed from this area. Some of the old oil wells were decommissioned, whereas others were heritagised, including a replica of a nineteenth-century derrick and several Soviet pumpjacks.[69] The oil fields around Bibi-Heybət, one of the oldest oil-producing districts in Baku, now constitute a peculiar mix of active oil installations, open views of the sea, and post-industrial art and consumption spaces. On the other side of the city, the oil fields around the main airport road were fenced off, thus slotting Baku into a capitalist regime in which oil production was rendered invisible.[70]

The incorporation of fire into the urban and national symbols and the reconfiguration of the geo-architectural assemblage continued alongside changes in the landscape. In 1993, the national coat of arms of the Azerbaijan Republic was adopted. It draws on proposals for the coat of arms of the short-lived ADR emblem and features fire at its centre. Strikingly, this fire is even further detached from the supposed Zoroastrian origins and is instead linked to Islam by way of a calligraphic inscription of Allah's name in the shape of flames (Figure 2.2, second row, right).[71] Thus, the elevation of a local geological process to the status of a national symbol applied to the entire territory of the Azerbaijani state was accomplished.

In the Baku city emblem, the territorial imaginary was further extended into the vertical plane. The new emblem was adopted in 2001 and includes, in addition to the three already existing traditional golden flames from earlier emblems, a blue sky, the green waves of the sea, and a black line running along the seabed, representing the subsea oil reserves (see Figure 2.2, third row).

Returning to Baku's two iconic buildings, Atəşgah and the Maiden's Tower, they are respectively embedded in the imaginaries of nation and resources. Following another round of renovation, Atəşgah was reopened, now featuring a larger square, a museum of exhibits, and tourist shops. The building is still situated at the centre of the extractive landscape, with oil derricks operating beyond its walls. Although the site is now mainly a tourist attraction, it continues to be a place of worship for Hindu and Parsee pilgrims, subject to their obtaining special permission from Azerbaijani state authorities to close the site to tourists to perform their rituals. However, it is also a site for performing national rituals, which include—but are not limited to—the celebration of Novruz. When I visited the site in the spring of 2023, a television programme on national music was being filmed there. The dances that were performed and filmed there hailed from different parts of Azerbaijan and the wider Caucasus region. Although they have no specific relation to Zoroastrianism or to this specific place, they have become spatially linked to Atəşgah as the literal hearth of the nation.

The Maiden Tower remained the site of the annual Novruz celebrations, now led by the President of Azerbaijan rather than a BakGaz employee, until the Second Karabakh War in 2020.[72] Differing from the Soviet celebration, the ceremony was moved from the top of the Tower to the square in front of it, and the fire was lit more traditionally, with wood. This is an outcome of the most recent restoration of the monument in 2012–2013, when fears of a destructive explosion prompted the disconnection of the Tower from the gas piping and the removal of the gas torch from its rooftop.

Notwithstanding the disconnection from the material gas infrastructure, the symbolic link with subsurface resources endures. Beyond Novruz celebrations, a new folk theory on the connection between the Maiden Tower and oil emerged in 2020. Amateur historian Veyis Ibrahimov published an article on his blog in which he hypothesised that the Maiden Tower is the distillation column of a medieval refinery.[73] According to this theory, oil was pumped from a well onto the first floor of the building. The well does indeed exist, even now, but it only contains water. According to this theory, the oil was heated using steam from the baths, the ruins of which still surround the Tower. The oil would then be separated into different fractions at different levels of the Tower. I have not come across any professional evaluation of the theory, but it is currently being circulated in social media and appears to be attracting supporters as well as critics among social media users. In the following years, several online media outlets reprinted the article. Like the other narratives presented here, this theory connects the subsurface reserves with the verticality of the built environment. It also transforms what was previously considered sacral architecture into an object of extractive infrastructure. Moreover, its retrospective mapping of refining technology onto the distant past serves to reinstate Baku and Azerbaijan as the historical centre of the oil industry.

These elements—fire, extractive infrastructure, and vertical sacred architecture—are brought together within a single imaginary combining different historical periods in the painting of an acclaimed Azerbaijani artist, Tahir Salahov (1928–2021), a leading artist in the Soviet Union and post-Soviet Russia (Figure 2.3). Salahov began his career as a socialist realist painter and is considered to be one of the founders of the 'severe style' in Soviet painting.[74] He gained acclaim in the 1960s for his depictions of oil workers in Neft Daşları. In this early work, which combines portraiture with landscape, oil workers are depicted against a background of industrial infrastructure and the sea. After moving to Moscow in 1968, Salahov shifted more towards portraiture, but as he claimed in an interview conducted in 2012, he remained fascinated with the themes of oil and the Caspian throughout his life.[75] In his monumental 'Land of Fire' (2007) triptych dedicated to Azerbaijan, Salahov returned to portrayals of the built environment, including Atəşgah, the Maiden Tower, and a modern offshore oil platform. Here, his style is more magical-realist, combining realism with fantastical elements. On the left side of the triptych,

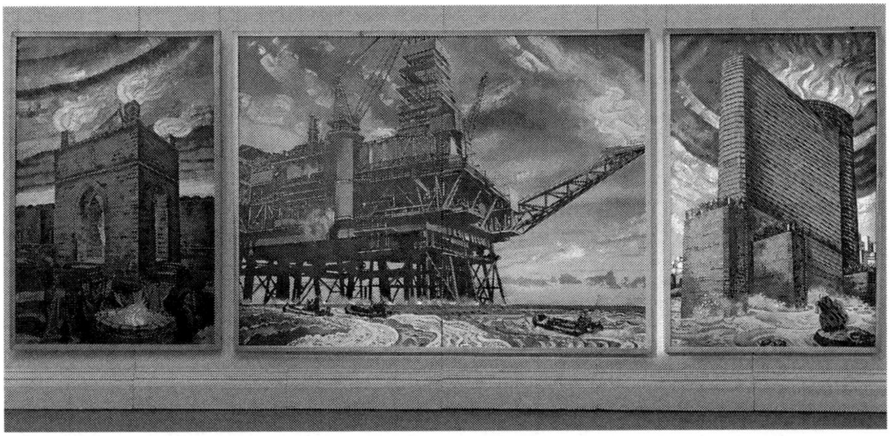

Figure 2.3 'The Land of Fire' painted by Tair Salakhov (2007) at Sotheby's exhibition, 2010. Tony Kyriacou/Shutterstock.com.

several fire worshippers can be seen sitting around a fire erupting from a well in Atəşgah, while larger flames burn within it and atop the altar. The right pane of the triptych depicts the Maiden Tower, surrounded by the waves of the rising Caspian Sea, while worshippers, standing on a rectangular attachment, cling to the structure. In a scene taken straight out of the pages of Davud Akhundov's book about a seven-level fire temple, an enormous fire burns atop the Tower.

The middle pane of the triptych pulls the assemblage together temporally as well as spatially. At its centre is a modern offshore platform, presumably the same one that was used in the first post-Soviet oil extraction project in the ACG oilfield in the Caspian Sea. A fiery flame burns atop the platform's flare tip, with the flaring associated gas produced during the extraction of oil from the subsurface. People are present in this painting too, in the small lifeboat floating on the water. The composition of this painting differs starkly from Salahov's early work on the Caspian oilfields, which foregrounded the workers. Here, similar to the depictions of people in nineteenth-century images of derricks and gush fires, workers are dwarfed by the enormous platform. Fire is a central element in all three parts of the triptych. The paintings also reproduce the temporal continuity between the ancient past and the present moment, and while architectural forms change, the relations between the elements of the assemblage remain stable: fire, fuelled by natural gas from the subsurface, burns on all three structures, the Atəşgah temple, the Maiden Tower, and the oil platform. The gas—and the subsurface—remain invisible and unseen, but fire indicates the presence of hydrocarbons under the surface.

Salakhov's painting may appear as a culmination of Azerbaijan's geo-architectural assemblage, but it was surpassed in 2012 with the construction of the Flame Towers mentioned at the beginning of this chapter.[76] This high-rise complex comprising three tall towers in the shape of flames was designed by an American firm, HOK Architects, with the Dubai-based DIA Holding as the main contractor. The cost of the project was estimated at USD 300–350 million.[77] Its design explicitly drew on the history of fire worship in Azerbaijan and the country's natural gas resources.[78] The structure, standing at 180 m tall, was imagined and implemented as a new architectural 'dominanta,'[79] and this position of ascendance is amplified by the placement of the Towers atop the hill overlooking Baku Bay. The national flag and the fire transmitted by the LED screens further spectacularise the connections between fire and the nation. The building can also be read as an expression of post-colonial defiance of Soviet urban planning: the spot was previously occupied by the 'Moscow' Hotel, a symbol of Russian domination that was torn down to make way for the construction of the Towers.[80]

The hill where the Flame Towers stand also hosts several important and symbolic institutions. On one side of the square is the National Parliament (Milli Məclis*)*; further down the street is the Ministry of National Security, a 'descendant' of the KGB, and the Honorary Cemetery, where Azerbaijan's key Soviet and post-Soviet political and cultural figures are buried. The long list includes Heydar Aliyev, generally regarded as the founder of the post-Soviet Azerbaijan state. However, the most important landmark in the vicinity of the Flame Towers, located just across the street, is the Alley of Martyrs—a cemetery and memorial for civilians killed during 'Black January' in 1990 when the Soviets launched an attack on Baku and for soldiers who fell during the First Karabakh War (1992–1994). The complex is located on a hill overlooking the city and was built in the place of an earlier cemetery, where Muslims killed in the Armenian–Tatar massacres of 1905 and 1918 were buried.[81] During the Soviet era, in an attempt to erase the memory of intercommunal strife in the city, the cemetery was converted into a park, and a statue of the Bolshevik leader, Sergei Kirov, who oversaw the early Soviet reconstruction of Baku's oil industry, was installed in its centre.[82] The statue was removed in 1994 as part of efforts to decommunise Baku's public space. Currently, the Alley of Martyrs is the most important national commemoration site in Azerbaijan, drawing together the memory of Azerbaijani victims of early twentieth-century strife, Soviet oppression, and the Karabakh Wars.[83] The memorial complex was constructed in 1998 and includes a mosque and an 'eternal flame' embedded in an octagonal structure similar to the medieval tombs of Muslim saints. The Alley is thus an assemblage in itself, drawing on various traditions of memorialisation, including secularised Soviet memorial landscape architecture, religious symbolism, local architectural traditions, and fire.

It is the fire aspect that interests me the most here. In Azerbaijan, commemoration with fire in modern times stems from the Soviet tradition of eternal fires, which were maintained in the memorials of World War II and the October Revolution. Interestingly, this Soviet tradition itself is linked to the gas industry. The first fire commemorating World War II was lit in Tula by the director of a gas plant in 1966.[84] Soviet ideological literature on commemoration links this tradition to both the Christian rites of lighting candles and to fire worship and the temple in Suraxanı.[85] Technically, these commemorative fires were not 'everlasting' in the sense applied to the natural fires of Abşeron, which burned indefinitely; they were always connected to gas networks, and many, especially those dedicated to the October Revolution, were put out after the Soviet Union was dismantled. In this instance, 'everlasting' referred to the eternal memory of those who gave up their lives for Communism or victory in World War II. In contemporary Azerbaijan, the continued use of fire for commemoration constitutes a form of postcolonial reclaiming and reinterpretation of what is also perceived as a local tradition. In the Alley of Martyrs, framed with two fires—live flames at one end and the gigantic Flame Towers at the other—Abşeron's fire comes full circle: from a natural, geological, volatile phenomenon of everlasting fire in *Yanardağ* to the human-made, monumental, solid, and spectacular symbol of the nation.

Conclusion

This chapter has examined the material and discursive processes through which the geo-architectural assemblage of petroleum reserves, fire, and land in Baku and Azerbaijan has been constructed over the last 150 years. I have argued that this assemblage, which began with the natural landscape in one locality, was subsequently extended to the entire nation through the construction of a national imaginary, which is inextricably linked to petroleum resources. In the early days of petromodernity, the self-combusting fires of the Abşeron Peninsula were part of the natural landscape, while simultaneously serving as signs denoting the presence of hydrocarbons beneath the surface. As a result of oil extraction, the gas was vented, putting out most of the natural flames, but fire became part of the urban built environment within what I have termed a geo-architectural assemblage. The construction of this assemblage required the building of material connections—the infrastructure that linked the subsurface with the area above ground—as well as symbolic, discursive, and performative work. As subterranean hydrocarbon resources assumed growing economic and political importance, fire became an increasingly dominant visual symbol of the place. Attaching these imaginaries to the built environment enables the resource to be spatially fixed in Baku and extended to the territory of Azerbaijan, making the buildings into secular temples of both oil and the nation. The assemblage also has a temporal dimension, with the connections between the subsurface and the aerial area imagined as being stable over time in addition to being fixed in space. The self-combusting flames of Yanardağ connect humanly measured time to deep geological time, while the medieval Maiden Tower and Atəşgah structures link the petromodern present to an imagined national past. The recently constructed Flame Towers further project this connection between petroleum and the nation into the future. The geo-architectural assemblage constructed around fire is an example of resource nationalism constructed as a socio-material imaginary. In this chapter, I have thus extended the critical resource nationalism framework from the discursive to the socio-material plane. This work can help to advance understanding of how petromodernity and extractivism become culturally embedded, locally, at different scales.

Azerbaijan's geo-architectural assemblage of oil developed over a long duration, spanning the entire period of petromodernity from the mid-nineteenth century to the present and transcending three different forms of statehood, from the Russian Empire to the USSR and, ultimately, an independent Azerbaijani state. Yet it is in the last period from 1991 to the present that the assemblage was consolidated and linked to processes and symbols of national statehood, ranging from heraldry to nation-building and territorial sovereignty. As such, it has been entangled with the consolidation of authoritarian rule and the Nagorno–Karabakh conflict. Both oil and Karabakh have featured prominently in Azerbaijan's post-Soviet development. In the early post-Soviet period, these were the two issues that challenged Azerbaijan's sovereignty. While Armenia contested Azerbaijan's territorial sovereignty in Karabakh and surrounding regions, Russia challenged Azerbaijan's control over the offshore oil fields in the Caspian Sea and sought to exert influence on the export routes out of the country.[86] The resolution of both of these issues was also interlinked. In the 1990s, Azerbaijan's ability to attract foreign investment in its oil industry was contingent on the cessation of violence in Karabakh. In May 1994, Heydar Aliyev signed the Bishkek Protocols, which established a ceasefire with Armenia and froze the conflict for over two decades. The ceasefire in Karabakh paved the way for the signing of the contract to develop the ACG field in September of the same year.

The revenues from this and subsequent contracts were used to finance Baku's urban renovation programmes, through which the geo-architectural assemblage was consolidated. These revenues also financed increases in military spending, the restructuring of the army, and the purchase of modern weapons, which made the Second Karabakh War in 2020 possible. The role of Karabakh in Azerbaijan's national imaginary has been steadily growing since it was fully recaptured in 2023, often competing with that of oil. In Baku's urban landscape, Karabakh and war-related toponymy and memorials coexist or compete with oil-derived street names and monuments. Besides the Alley of Martyrs discussed in the previous section, examples include the controversial Military Trophies Park in the redeveloped section of the former 'Black City' industrial district and the renaming of Nobel Avenue as November 8 Avenue (marking the date of the recapture of Şuşa in Karabakh).[87] The state programme of rebuilding and reconstructing recaptured territories, known as the 'Great Return,' which envisages large-scale transformation of the region, is also financed primarily through revenue generated from oil and gas, indicating that the material entanglement with petroleum will continue into the foreseeable future.

At the same time, the elevation of the Karabakh issue in the national imaginary indicates the possibility of change. This chapter has focused on processes of fixing and stabilising, but it is also important to consider how the assemblage can be reconfigured; how petroleum can be decoupled from the national imaginary and rescaled back to the local scale.[88] Assemblage thinking necessitates the rejection of linear causality, making predictions impossible. However, it is possible to indicate some pathways of potential reconfiguration. In addition to the possibility of oil being supplanted from the national imaginary by the Karabakh theme, another potential area of change is linked to the tension between Azerbaijan's resource nationalism and the demands for an energy transition. Azerbaijan has recently embraced the discourse of energy transition, sustainability, and green energy. At the time of writing, preparations for the UN Climate Change Conference, COP29, scheduled to be held in Baku in November 2024, are underway. As part of this process, Azerbaijan has also been active in promoting alternative sources of energy generation, particularly solar and wind energy. All of this is happening alongside increasing production of gas and continued, if decreasing, oil production. Can Azerbaijan's geo-architectural assemblage be reconfigured in ways that are more sustainable and do not celebrate extractivism? At present, natural gas, for example, is being touted in Azerbaijan and elsewhere as a greener transitional fuel. This framing is fully compatible with the existing assemblage, but other possibilities exist as well. Several of my interlocutors in Azerbaijan, while speaking about the Zoroastrian legacies of fire worship, mentioned the cosmology of balance between the earth–fire–water–air elements, emphasising regeneration, renewal, and rebirth.[89] Taking these beliefs seriously could reveal a path towards reconstructing Azerbaijan's geo-architectural assemblage in less extractivist and more sustainable ways.

Notes

1 The research on which this chapter is based received funding under the European Union's Horizon 2020 research and innovation programme through a Marie Skłodowska-Curie grant (agreement no. 897155). The chapter benefited from discussions during the 'Cultural Landscapes of Energy' workshop held in Leipzig in 2022 and from comments provided by several other colleagues. I am especially grateful to Mette High, Pauline Destrée, Sean Field, Itay Noy, Sarah O'Brien, and Anna Rauter at the Centre for Energy Ethics at the University of St Andrews, and David Smith at the University of Glasgow for their comments on earlier drafts. Special thanks go to the editors of this volume, Corinne Geering and Torsten Meyer, for their attentive editing and suggestions that helped to improve this chapter.

2 Koch and Perreault, "Resource Nationalism," 612.

3 See, for example, Marston, "Strata of the State"; Oguz, "Cavernous Politics."

4 Grant, "The Edifice Complex"; Koch, *The Geopolitics of Spectacle*; Koch and Valiyev, "Urban Boosterism."

5 Brechin, *Imperial San Francisco*, 66.

6 Koch, *The Geopolitics of Spectacle.*

7 Hein, "Oil Spaces"; Hein, "Space, Time, and Oil," 3–18.

8 Anderson et al., "On Assemblages and Geography"; McFarlane and Anderson, "Thinking with Assemblage"; Li, "What is Land?"; Buchanan, "Assemblage Theory"; Baker and McGuirk, "Assemblage Thinking as Methodology"; Anderson et al., "On Assemblages and Geography"; Bueger, "Thinking Assemblages Methodologically."

9 Bakker and Bridge, "Material Worlds?"; Richardson and Weszkalnys, "Introduction: Resource Materialities."

10 Castán Broto and Sanzana Calvet, "Sacrifice Zones."

11 Castán Broto, "Energy Landscapes and Urban Trajectories."

12 Pasqualetti and Stremke, "Energy Landscapes in a Crowded World."

13 This argument was first put forward by Timothy Mitchell in Mitchell, *Carbon Democracy*, 33–34. According to Mitchell, the spatial organisation of Baku during the pre-Soviet

period resembled the concentrated model of coal mining He did not, however, consider Baku's development after 1920.

14 Appel, "Offshore Work."

15 Blau and Rupnik, *Oil and Urbanism.*

16 Pasqualetti and Stremke, "Energy Landscapes."

17 Rose and Wylie, "Animating Landscape"; Pasqualetti and Stremke, "Energy Landscapes."

18 Szeman and Whiteman, "Oil Imag(e)inaries."

19 Bridge, "The Hole World."

20 Elden, "Secure the Volume." Billé and Battaglia, "Voluminous States."

21 Harris, "Vertical Urbanisms"; McNeill, "The Volumetric City"; Graham and Hewitt, "Getting off the Ground."

22 Harris, "Vertical Urbanisms," 602.

23 Ebbensgaard et al., "Introduction: Verticality, Radicalism, Resistance," 622.

24 Ong, "Hyperbuilding," 206; see also Donald McNeill, "Volumetric Urbanism."

25 Marston, "Strata of the State"; Marston and Himley, "Earth Politics"; Oguz, "Cavernous Politics."

26 Rogers, "Deep Oil," 61–71.

27 Graham and Hewitt, "Getting off the Ground"; Ebbensgaard et al., "Introduction."

28 Perreault, "Materializing Space," 126.]

29 Marston, "Strata of the State"; Koch and Perreault, "Resource Nationalism."

30 Alizade et al., *Antropogennaia geomorfologiia*; Veliev, "Tainy Baku."

31 Rakhmanov, *Griazevye vulkany.*

32 Aliev, Guliev, and Rakhmanov, "Katalog Izverzhenii."

33 Ghaderi and Rahbani, "Mud Volcano."

34 Aliev, "Ocherk Istorii."

35 The territory of the present-day Azerbaijan Republic was incorporated into the Russian Empire between 1806 and 1828 with the ratification of the Turkmanchay Treaty between Iran and the Russian Empire. See Altstadt, *The Azerbaijani Turks.*

36 Martellaro, "The Acquisition and Leasing"; McKay, "Baku Oil."

37 Martellaro, "The Acquisition and Leasing."

38 Hastings, "Oil Capital," 52.

39 Blau and Rupnik, *Oil and Urbanism*, 34.

40 The proximity of industrial, horticultural, and husbandry uses of land is a consistent feature of the Abşeron landscape, along with the urban built environment. In my fieldwork, I encountered sheep grazing near the oil fields in Balaxanı, instances of people utilising decommissioned pieces of industrial infrastructure in their *dachas* (Az: *bağ,* literally 'garden'), as well as stories of accidentally hitting subsurface oil during gardening. See also Rumiantsev, "Neft' i Ovtsy."

41 This section draws on my blog post at https://energyethics.st-andrews.ac.uk/blog/the-flame-towers-of-baku-new-building-old-symbol/.

42 For the history of Branobel, see Tolf, "The Russian Rockfellers."

43 Seits, "From Ignis Mundi."

44 Mamed-zade, "Stroitel'noie Iskusstvo," 61.

45 Seits, 'From Ignis Mundi," 66–68.

46 Some of the gushers reached 250–350 feet in height. See, for example, Marvin, *The Region of the Eternal Fire;* Vasilieff and Anderson, "The Oil-Wells of Baku."

47 Murray and Heumann, "The First Eco-Disaster Film?" 44–51.

48 Badalov, "Oil, Revolution, and Cinema"; Kazimzade, "The Pre-Soviet Era."

49 I am grateful to Dr Laura Hildelang for providing this information.

50 Sicotte, "Alexandr Serebrovskiĭ"; Sicotte, "Baku and Its Oil Industry."
51 Blau and Rupnik, *Oil and Urbanism*, 162–168.
52 Two other industries that were developed during the Soviet period were petrochemicals and manufacturing of oilfield equipment. Beyond that, Azerbaijan remained largely agricultural and produced cotton, nuts, and fruit.
53 Naumov, "Ikonografiia i Semantika."
54 Gasanov, "Istoriia Razvitiia Turizma," Kerimli, "Restavratsiia Khramovogo."
55 For a summary of these debates, see Veliev, "Tainy Baku," 221–328. For the dating of the structure in the most recent archaeological work, see Döring-Williams and Albrecht, "Sasanian Construction Technology"; Döring-Williams and Albrecht, "Summary of Findings."
56 Nabiev, "Taina Bakinskoi Gyz Galasy," cited in Veliev, "Tainy Baku," 240.
57 Akhundov, *Architektura Drevnego.*
58 Veliev, "Tainy Baku."
59 See Militz, *Affective Nationalism,* 77–85, for a discussion of Novruz enactment in Azerbaijan and its differences from Persian traditions of celebration.
60 Zhidkova, "Sovetskaia Grazhdanskaia Obriadnost'."
61 In 2007, Xalida Alizade, the widow of the Communist Party functionary, Vəli Axundov, who organised the celebration in 1967, published the script of the official ceremony held at the Maiden's Tower, which she extracted from her late husband's archive. Əlizadə, *Xalqın Vəlisi.* I am grateful to Yusif Ismayil for bringing this document to my attention.
62 Rumiantsev, "Geroicheskii Epos."
63 For a discussion of the contract deal, see Ciarreta and Nasirov, "Development Trends." Interestingly, the second of the oilfields, Çıraq, means "lamp" in Azerbaijani, referring specifically to the medieval oil lamp.
64 Ciarreta and Nasirov, "Development Trends; Nasibov, "Energy Governance."
65 Koch and Valiyev, "Urban Boosterism"; Valiyev, "The Post-Communist Growth Machine"; Valiyev and Wallwork, "Post-Soviet Urban Renewal"; Grant, "The Edifice Complex."
66 Koch, "Introduction."
67 Kaczmarska, "Contemporary Symbols."
68 Some of my interlocutors compared the shape of the Crystal Hall to a gas burner in yet another example of the symbolic inscription of petroleum onto the built environment.
69 I describe some of these heritagisation efforts here: https://bakuresearchinstitute.org/en/the-oil-industry-as-heritage-bakus-untapped-resource/.
70 Szeman and Whiteman, "Oil Imag(e)inaries."
71 This is a popular interpretation of the emblem, drawing on the words of a graphic designer who worked on the coat of arms. Other symbols were also suggested, but fire was eventually chosen. For the official explanation of state symbols, see: https://azerbaijan.az/en/information/106. For a discussion of the adoption of the emblem in 1993, see: https://teleqraf.com/news/art/379903.html.
72 Since 2021, the President of Azerbaijan has led the Novruz celebrations in various places in the recaptured Karabakh territory. Karabakh and several districts around it were occupied by Armenia following the First Karabakh War (1991–1994). The districts around the former Nagorno-Karabakh Autonomous Region (NKAO) were recaptured in 2020, and the self-proclaimed 'Nagorno-Karabakh Republic' was disbanded in 2023. On the origins of the conflict and the history of the First Karabakh War, see Laurence Broers, *Armenia and Azerbaijan;* on the Second Karabakh War, see De Waal, "The Nagorny

Karabakh Conflict." On memory politics relating to Karabakh in Azerbaijan, see for example: Huseynova, "A Tool of Propaganda."

73 See Ibrahimov's blog: https://veyis-ibrahimov.livejournal.com/899.html.

74 McCallum, *The Fate of the New Man.*

75 Mikeladze, "Formula uspekha."

76 I first made this argument in a blog post: https://energyethics.st-andrews.ac.uk/blog/the-flame-towers-of-baku-new-building-old-symbol/.

77 Dome+Partners, "Flame Towers."

78 HOK, "Baku Flame Towers."

79 'Dominanta' is a term from socialist urban planning, describing prominent buildings that dominate both space and social life in a city. See Murawski, *The Palace Complex.*

80 Soviet planning legacies are a controversial topic in Baku. While some lament the loss of Soviet ensemble planning, others praise the lifting of restrictions on verticality and celebrate high rise construction as an indication of worlding and modernisation of Baku's architecture. Significantly, the push for high-rise constructions in Baku dates back to at least the 1970s, although most of these projects were rejected by Moscow. The '"Moscow" Hotel, built in 1977, used to be the tallest building in Baku. Its demolition was part of both the worldling and erasure of the Soviet past from the urban landscape. See Grant, "The Edifice Complex"; see also Murawski, *The Palace Complex*, 136–172 for a discussion on the post-communist future of the Palace of Culture in Warsaw.

81 Shafiyev and Huseynova, "Violence and Memory," 33.

82 Sergei Kirov (1886–1934) first arrived to Baku as a member of the 11th Red Army that conquered the city in 1920. From 1921 to 1926, he served as the First Secretary of the Communist Party of Azerbaijan. He subsequently moved on to assume the position of the First Secretary of the Leningrad Committee of the Communist Party, a post that he held until his assassination in 1934.

83 Sayfutdinova, "The Alley of Martyrs," 14–16; Shafiyev and Huseynova, "Violence and Memory."

84 Makhotina and Bürger, "Making (Monumental) Sense of War."

85 Adonyeva and Warner, *We Love, We Grieve*, 227

86 On Russia's efforts to control the production and transit of oil from the Caspian, see Stulberg, "Moving Beyond the Great Game."

87 On the commemoration of the Second Karabakh War, see Huseynova. *A Tool of Propaganda,* 10–12; On the renaming of Nobel Avenue, see: https://abc.az/en/news/71027.

88 For a conceptualisation of changes within assemblages, see Haarstad and Wanvik, "Carbonscapes and Beyond."

89 Foltz and Saadi-nejad, "Is Zoroastrianism an Ecological Religion?"

Bibliography

Adonyeva, Svetlana, and Elizabeth Warner. *We Remember, We Love, We Grieve. Mortuary and Memorial Practice in Contemporary Russia.* Madison: University of Wisconsin Press, 2021.

Akhundov, Davud. *Arkhitektura Drevnego i Rannesrednevekovogo Azerbaijana.* [*The Architecture of Ancient Azerbaijan*]. Baku: Azerneshr, 1986.

Aliev, A. A., Guliev, I. S., and Rakhmanov, R. R. *Katalog Izverzhenii Griazevykh Vulkanov Azerbaidzhana (1810–2007).* [*The Catalogue of Mud Volcano Eruptions in Azerbaijan (1810–2007*]. Baku: Nafta Press, 2009.

Aliev, Igrar. *Ocherk istorii Atropateny*. [*A Review of the History of Atropatene*]. Baku: Azerbaidzhanskoe gosudarstvennoe izdatel'stvo, 1989.

Alizade, E. K., S. A. Tarikhazer, S. G. Mamedov, and Z. A. Gamidova. *Antropogennaia geomorfologiia bol'shogo Baku*. [*Anthropogenic Geomorphology of Larger Baku*]. Baku: AFpoligrAF, 2017. https://ebooks.az/ru/book/hsylvxa.

Altstadt, Audrey L. *The Azerbaijani Turks: Power and Identity under Russian Rule*. Stanford: Stanford University Press, 1992.

Anderson, Ben, Matthew Kearnes, Colin McFarlane, and Dan Swanton. "On Assemblages and Geography." *Dialogues in Human Geography* 2, no. 2 (2012): 171–189. https://doi.org/10.1177/2043820612449261.

Appel, Hannah. "Offshore Work: Oil, Modularity, and the How of Capitalism in Equatorial Guinea." *American Ethnologist* 39, no. 4 (November 2012): 692–709. https://doi.org/10.1111/j.1548-1425.2012.01389.x.

Badalov, Rahman. "Oil, Revolution, and Cinema." *Azerbaijan International* 5.3 (Autumn 1997): 57–64.

Baker, Tom, and Pauline McGuirk. "Assemblage Thinking as Methodology: Commitments and Practices for Critical Policy Research." *Territory, Politics, Governance* 5, no. 4 (October 2017): 425–442. https://doi.org/10.1080/21622671.2016.1231631.

Bakker, Karen, and Gavin Bridge. "Material Worlds? Resource Geographies and the 'Matter of Nature.'" *Progress in Human Geography* 30, no. 1 (2006): 5–27. https://doi.org/10.1191/0309132506ph588oa.

Billé, Franck, and Debbora Battaglia. *Voluminous States: Sovereignty, Materiality, and the Territorial Imagination*. Durham: Duke University Press, 2020.

Blau, Eve, and Ivan Rupnik. *Oil and Urbanism*. Zurich: Park Books, 2018.

Brechin, Gray Allan. *Imperial San Francisco: The Environmental Impact of Urban Elites upon the Pacific Basin*. Berkeley: University of California Press, 1998.

Bridge, Gavin. "The Hole World: Scales and Spaces of Extraction." *Scenario Journal* 8 (October 2015). https://scenariojournal.com/article/the-hole-world/.

Broers, Laurence. *Armenia and Azerbaijan: Anatomy of a Rivalry*. Edinburgh: Edinburgh University Press, 2019.

Buchanan, Ian. "Assemblage Theory, or, the Future of an Illusion." *Deleuze Studies* 11, no. 3 (August 2017): 457–474. https://doi.org/10.3366/dls.2017.0276.

Bueger, Christian. "Thinking Assemblages Methodologically: Some Rules of Thumb." In *Reassembling International Theory: Assemblage Thinking and International Relations*, edited by Michele Acuto and Simon Curtis, 58–66. London: Palgrave Macmillan, 2014. https://doi.org/10.1057/9781137383969_7.

Castán Broto, Vanesa. "Energy Landscapes and Urban Trajectories towards Sustainability." *Energy Policy* 108 (2017): 755–764. https://doi.org/10.1016/j.enpol.2017.01.009.

Castán Broto, Vanesa, and Martin Sanzana Calvet. "Sacrifice Zones and the Construction of Urban Energy Landscapes in Concepción, Chile." *Journal of Political Ecology* 27, no. 1 (2020): 279–299. https://doi.org/10.2458/v27i1.23059.

Childs, John. "Geography and Resource Nationalism: A Critical Review and Reframing." *The Extractive Industries and Society* 3, no. 2 (April 2016): 539–546. https://doi.org/10.1016/j.exis.2016.02.006.

Ciarreta, Aitor, and Shahriyar Nasirov. "Development Trends in the Azerbaijan Oil and Gas Sector: Achievements and Challenges." *Energy Policy* 40 (2012): 282–292.

de Waal, Thomas. "The Nagorny Karabakh Conflict in its Fourth Decade." Centre for European Policy Studies Working Document No. 2021-02, September 2021. https://cdn.ceps.eu/wp-content/uploads/2021/09/WD2021-02_The-Nagorny-Karabakh-Conflict-in-its-Fourth-Decade.pdf.

Dome+Partners. "Flame Towers." https://dome.ws/project/flame-towers/.

Döring-Williams, Marina, and Luise Albrecht. *Summary of Findings from the Historical Analysis of the Architecture of the Maiden Tower in Baku*, 2016. https://repositum.tuwien.at/handle/20.500.12708/39013.

Döring-Williams, Marina, and Luise Albrecht. "Sasanian Construction Technology in the Maiden Tower Complex as Evidence of Late Antiquity Building Activities in Baku (Azerbaijan)." In *6th International Congress on Construction History (6ICCH 2018), July 9–13, 2018, Brussels, Belgium*, 581–589, 2018.

Ebbensgaard, Casper Laing, Michał Murawski, Saffron Woodcraft, and Katherine Zubovich. "Introduction: Verticality, Radicalism, Resistance." *Urban Studies* 61, no. 4 (2024): 619–635. https://doi.org/10.1177/00420980231216884.

Elden, Stuart. "Secure the Volume: Vertical Geopolitics and the Depth of Power." *Political Geography* 34 (2013): 35–51. https://doi.org/10.1016/j.polgeo.2012.12.009.

Əlizadə, Xalidə. *Xalqının Vəlisi, millətin səsi: Vəli Məmmədovun siyasi və ictimai həyatından ştrixlərXalqın Vəlisi.* [*Veli of his People, the Voice of His Nation: Sketches from the Political and Public Life of Veli Mamedov*]. Baku: Adiloğlu, 2007.

Foltz, Richard, and Manya Saadi-nejad. "Is Zoroastrianism an Ecological Religion?" *Journal for the Study of Religion, Nature & Culture* 1, no. 4 (2007): 413–430. https://doi.org/10.1558/jsrnc.v1i4.413.

Gasanov, A. "Istoriia Razvitiia Turizma v Azerbaidzhane do Nachala 1990ykh Godov." ["The History of tourism in Azerbaijan until the 1990s"] *Servis Plus*, no. 3 (2013): 3–10.

Ghaderi, Danial, and Maryam Rahbani. "Mud Volcano as a Feature of Emergence in Caspian Sea." *Oceanologia* 64, no. 3 (July 2022): 503–513. https://doi.org/10.1016/j.oceano.2022.03.006.

Graham, Stephen, and Lucy Hewitt. 'Getting off the Ground: On the Politics of Urban Verticality." *Progress in Human Geography* 37, no. 1 (February 2013): 72–92. https://doi.org/10.1177/0309132512443147.

Grant, Bruce. "The Edifice Complex: Architecture and the Political Life of Surplus in the New Baku." *Public Culture* 26, no. 3 (2014): 501–528. https://doi.org/10.1215/08992363-2683648.

Haarstad, Håvard, and Tarje I. Wanvik. "Carbonscapes and Beyond: Conceptualizing the Instability of Oil Landscapes." *Progress in Human Geography* 41, no. 4 (2017): 432–450. https://doi.org/10.1177/0309132516648007.

Harris, Andrew. "Vertical Urbanisms: Opening up Geographies of the Three-Dimensional City." *Progress in Human Geography* 39, no. 5 (2015): 601–620. https://doi.org/10.1177/0309132514554323.

Hastings, Rebecca Lindsay. "Oil Capital: Industry and Society in Baku, Azerbaijan, 1870–Present." PhD diss., University of Oregon, 2020.

Hein, Carola. "Oil Spaces: The Global Petroleumscape in the Rotterdam/The Hague Area." *Journal of Urban History* 44, no. 5 (September 2018): 887–929. https://doi.org/10.1177/0096144217752460.

Hein, Carola. "Space, Time, and Oil: The Global Petroleumscape." In *Oil Spaces*, edited by Carola Hein, 3–18. New York: Routledge, 2021.

HOK. "Baku Flame Towers." Accessed July 18, 2024. www.hok.com/projects/view/baku-flame-towers/.

Huseynova, Sevil. "A Tool of Propaganda: Thirty Years of Memory Politics in Independent Azerbaijan." Accessed March 29, 2024. https://ge.boell.org/sites/default/files/2023-05/sevil-huseynova_en.pdf.

Kaczmarska, Elżbieta. "Contemporary Symbols in the Space of Baku." *Budownictwo i Architektura* 19, no. 2 (2020): 121–136.

Kazimzade, Aydin. "The Pre-Soviet Era: Celebrating 100 Years in Film, not 80." *Azerbaijan International* 5.3 (Autumn 1997): 30–35.

Kerimli, Telman. "Restavratsiia Khramovogo Kompleksa 'Ateshgiakh' v Poselke Surakhany Goroda Baku." ["The Restoration of the Atəşgah Temple Complex in Suraxanı Settlement of Baku City"]. *Architecture and Modern Information Technologies* 4 (29) (2014): 2.

Koch, Natalie. *The Geopolitics of Spectacle: Space, Synecdoche, and the New Capitals of Asia*. Ithaca: Cornell University Press, 2018.

Koch, Natalie, and Tom Perreault. "Resource Nationalism." *Progress in Human Geography* 43, no. 4 (2019): 611–631.

Koch, Natalie, and Anar Valiyev. "Urban Boosterism in Closed Contexts: Spectacular Urbanization and Second-Tier Mega-Events in Three Caspian Capitals." *Eurasian Geography and Economics* 56, no. 5 (2015): 575–598.

Li, Tania Murray. "What Is Land? Assembling a Resource for Global Investment." *Transactions of the Institute of British Geographers* 39, no. 4 (2014): 589–602.

Makhotina, Ekaterina, and Philipp Bürger. "Making (Monumental) Sense of War: Memorials of the 'Great Patriotic War' in the Soviet Union and in Post-Soviet Russia." In *The Political Cult of the Dead in Ukraine: Traditions and Dimensions from the First World War to Today*, edited by Guido Hausmann and Iryna Sklokina, 197–222. Göttingen: Vandenhoeck and Ruprecht Unipress, 2021.

Mamed-zade, K. M. *Stroitel'noie Iskusstvo Azerbaidzhana s Drevneishikh vremen do 19 v.* [*The Art of Architecture of Azerbaijan from Ancient Times to the 19th Century*]. Baku: Elm, 1983.

Marston, Andrea. "Strata of the State: Resource Nationalism and Vertical Territory in Bolivia." *Political Geography* 74 (2019). https://doi.org/10.1016/j.polgeo.2019.102040.

Marston, Andrea, and Matthew Himley. "Earth Politics: Territory and the Subterranean–Introduction to the Special Issue." *Political Geography* 88 (2021). https://doi.org/10.1016/j.polgeo.2021.102407.

Martellaro, Joseph A. "The Acquisition and Leasing of the Baku Oilfields by the Russian Crown." *Middle Eastern Studies* 21, no. 1 (1985): 80–88.

Marvin, Charles. *The Region of the Eternal Fire: An Account of a Journey to the Petroleum Region of the Caspian in 1883*. London: WH Allen & Company, 1891.

McCallum, Claire. *The Fate of the New Man: Representing and Reconstructing Masculinity in Soviet Visual Culture, 1945–1965*. Ithaca: Cornell University Press, 2018.

McFarlane, Colin, and Ben Anderson. "Thinking with Assemblage." *Area* 43, no. 2 (2011): 162–164. https://doi.org/10.1111/j.1475-4762.2011.01012.x.

McKay, John P. "Baku Oil and Transcaucasian Pipelines, 1883–1891: A Study in Tsarist Economic Policy." *Slavic Review* 43, no. 4 (1984): 604–623.

McNeill, Donald. "Volumetric Urbanism: The Production and Extraction of Singaporean Territory." *Environment and Planning A: Economy and Space* 51, no. 4 (2019): 849–868. https://doi.org/10.1177/0308518X19830699.

McNeill, Donald. "The Volumetric City." *Progress in Human Geography* 44, no. 5 (2020): 815–831. https://doi.org/10.1177/0309132519863486.

Mikeladze, Galina. "Formula Uspekha. Tair Salakhov: 'Ia nikogda ne iskal legkikh putei.'" ["The Formula of Success. Tair Salakhov: 'I Never Took the Easy Route']. *1news.az*, November 14, 2012. https://1news.az/news/20121114091928590-Formula-uspekha-Tair-Salakhov-YA-nikogda-ne-iskal-legkikh-putei-FOTO.

Militz, Elisabeth. *Affective Nationalism: Bodies, Materials and Encounters with the Nation in Azerbaijan*. Vol. 15. Zürich: LIT Verlag, 2019.

Mitchell, Timothy. *Carbon Democracy*. London: Verso, 2011.

Murawski, Michał. *The Palace Complex: A Stalinist Skyscraper, Capitalist Warsaw, and a City Transfixed*. Bloomington: Indiana University Press, 2019.

Murray, Robin L., and Joseph K. Heumann. 'The First Eco-Disaster Film?' *Film Quarterly* 59, no. 3 (March 1, 2006): 44–51. https://doi.org/10.1525/fq.2006.59.3.44.

Nabiev, M. "Taina Bakinskoi Gyz-Galasy." ["The Mystery of the Maiden Tower of Baku"]. In *Doklad AN Az.SSR, [The Reports of the Academy of Sciences of the AzSSR]*. No. 6, 98. Baku, 1974.

Nasibov, Murad. "Energy Governance in Azerbaijan." In *Handbook of Energy Governance in Europe*, edited by Michèle Knodt and Jörg Kemmerzell, 483–509. Cham: Springer International Publishing, 2022. https://doi.org/10.1007/978-3-030-43250-8_3.

Naumov, O. N. "Ikonografiia i Semantika Sovetskikh Gorodskikh Gerbov." ["The Iconography and Semantics of the Soviet City Emblems"] In *Aktual'nye Problemy Istorii Rossii XX veka, [Topical Problems in Russian History of the Twentieth Century]*, 108–128. Moscow: Moscow State Regional University Publishing 2017. https://gerboved.ru/pdf/Naumov-2017-Iconography-pp108%E2%80%93128.pdf.

Oguz, Zeynep. "Cavernous Politics: Geopower, Territory, and the Kurdish Question in Turkey." *Political Geography* 85 (2021). https://doi.org/10.1016/j.polgeo.2020.102331.

Ong, Aihwa. "Hyperbuilding: Spectacle, Speculation, and the Hyperspace of Sovereignty." In *Worlding Cities*, edited by Ananya Roy and Aihwa Ong, 1st ed., 205–226. Chichester: Blackwell Publishing Ltd., 2011. https://doi.org/10.1002/9781444346800.ch8.

Pasqualetti, Martin, and Sven Stremke. "Energy Landscapes in a Crowded World: A First Typology of Origins and Expressions." *Energy Research & Social Science* S36 (2018): 94–105. https://doi.org/10.1016/j.erss.2017.09.030.

Perreault, Tom. "Materializing Space, Constructing Belonging." In *Routledge Handbook of Critical Resource Geography*, edited by Matthew Himley, Elizabeth Havice, Gabriela Valdivia, 126–139. London: Routledge, 2021.

Rakhmanov, R. R. *Griazevye vulkany i ikh znachenie v prognozirovanii gazoneftenosnosti nedr*. [*Mud Volcanos and Their Significance for the Forecasting of Oil and Gas Presence in the Subsurface*]. Moscow: Nedra, 1987. https://ebooks.az/ru/book/UqePrG2N.

Richardson, Tanya, and Gisa Weszkalnys. "Introduction: Resource Materialities." *Anthropological Quarterly* 87, no. 1 (2014): 5–30. https://www.jstor.org/stable/43652719.

Rogers, Douglas. "Deep Oil and Deep Culture in the Russian Urals." In *Subterraneous Estates: Life Worlds of Oil and Gas*, edited by Hannah Appel, Arthur Mason, and Michael Watts, 61–71. Ithaca: Cornell University Press, 2015.

Rose, Mitch, and John Wylie. "Animating Landscape." *Environment and Planning D: Society and Space* 24, no. 4 (2006): 475–479. https://doi.org/10.1068/d2404ed.

Rumiantsev, S. "Geroicheskii Ėpos i Konstruirovanie Obraza Istoricheskogo Vraga." ["Heroic Epic and the Construction of the Historical Enemy Image"]. *Ab Imperio* 2005, no. 2 (2005): 441–468.

Rumiantsev, S. "Neft' i Ovtsy: iz Istorii Transformatsii Goroda Baku iz Stolitsy v Stolitsu." ["Oil and Sheep: from the History of the Transformations of Baku City from One Capital to Another"]. In *PS Landshafty: Optiki Gorodskikh Issledovanii [PS Landscapes: In the Optic of Urban Studies]*, edited by N. Mileryus and B. Cope, 228–266. Vilnius: EHU, 2008.

Sanzana Calvet, Martín, and Vanesa Castán Broto. "Sacrifice Zones and the Construction of Urban Energy Landscapes in Concepción, Chile." *Journal of Political Ecology* 27, no. 1 (2020). https://doi.org/10.2458/v27i1.23059.

Sayfutdinova, Leyla. "The Alley of Martyrs: Deaths, Memory and the Nation." *Caucasus Analytical Digest* 66 (2014): 14–16.

Seits, Irina. "From Ignis Mundi to the World's First Oil-Tanker: The Legacy of the Nobels' Oil Empire in Baku." *Approaching Religion* 13, no. 2 (2023): 57–76. https://doi.org/10.30664/ar.131878.

Shafiyev, Farid, and Ilaha Huseynova. "Violence and Memory in Urban Park Space: The Alley of Martyrs in Baku." *Eurasian Turkic Studies* 4 (2020): 33–53.

Sicotte, Jonathan. "Alexandr Serebrovskiĭ and the Modernization of the Soviet Oil Industry—1924 to 1929." *The Soviet and Post-Soviet Review* 48, no. 3 (2021): 315–344. https://doi.org/10.30965/18763324-BJA10032.

Sicotte, Jonathan. "Baku and Its Oil Industry through War and Revolution: 1914–1920." *The Extractive Industries and Society* 5, no. 3 (2018): 384–392. https://doi.org/10.1016/j.exis.2018.04.001.

Szeman, Imre, and Maria Whiteman. "Oil Imag(e)inaries: Critical Realism and the Oil Sands." *Imaginations: Journal of Cross-Cultural Image Studies* 3, no. 2 (2012): 46–67. https://doi.org/10.17742/IMAGE.sightoil.3-2.5.

Tolf, Robert. *The Russian Rockfellers: The Saga of the Nobel Family and the Russian Oil Industry*. Stanford: Hoover Institution Press, 1976.

Valiyev, Anar. "The Post-Communist Growth Machine: The Case of Baku, Azerbaijan." *Cities* 41 (2014): S45–S53. https://doi.org/10.1016/j.cities.2014.06.008.

Valiyev, Anar, and Lucy Wallwork. "Post-Soviet Urban Renewal and its Discontents: Gentrification by Demolition in Baku." *Urban Geography* 40, no. 10 (2019): 1506–1526. https://doi.org/10.1080/02723638.2019.1627147.

Vasilieff, F., and W. Anderson. "The Oil-Wells of Baku." *Minutes of the Proceedings of the Institution of Civil Engineers* 83 (1886): 405–414. https://doi.org/10.1680/imotp.1886.21445.

Veliev, S. S. *Tainy Baku: S Drevneishikh Vremen do Nashikh Dnei*. [*The Mysteries of Baku: from Ancient Times to Our Days*]. Baku: JekoPrint, 2019.

Zhidkova, E. "Sovetskaia Grazhdanskaia Obriadnost' kak Al'ternativa Obriadnosti Religioznoi." ["Soviet Civil Alternative to Religious Rituality."] *Gosudarstvo, Religiia, Tserkov' v Rossii i za Rubezhom*, [*State, Religion, and the Church in Russia and Abroad*], no. 3–4 (30) (2012): 408–429.

3 Energy Transition as Cultural Trauma

The Making and Unmaking of the Finnish Peat Industry

Hanna Lempinen

Finland, a Nordic country located on the northern outskirts of the European Union (EU), ranks globally among countries with the highest prevalence of peatland, which has had profound and unparalleled impacts on biodiversity and the conditions of human life.[1] Owing to their omnipresence, peatland landscapes have also played an instrumental role as a material 'backdrop' to Finnish everyday life, experience, and economic activity, and as building blocks of national culture and identity construction.[2] Finland is also one of the few nations that have made use of their peatlands to generate energy on an industrial scale. Moreover, it is one of an even smaller number of countries that have continued to include peat, a solid fossil fuel consisting of partially decomposed organic matter, in its energy mix.[3] For Finland, which is dependent on energy imports, peat has been a valuable source of domestic energy, which has contributed to approximately 5% of the country's annual primary energy consumption. It continues to play a vital role in providing heat to Finnish homes through the nation's vast district heating networks.[4]

In addition to meeting the energy needs of households and industries alike, peat extraction has been an important source of employment and income in rural Finland during the twentieth century. Traditionally, it has provided employment in the Ostrobothnian provinces, central, and, historically, eastern Finland (Figure 3.1).[5]

These socio-economic benefits of peat have, however, been undermined by its adverse environmental and climatic impacts. Not only has the extraction of peat been detrimental for peatland ecosystems, peatland landscapes and adjacent waterways but also its consumption has accounted for over 10% of Finland's annual carbon dioxide emissions.[6] Specifically, the climatic impacts of peat consumption as an energy source led to the Finnish government's decision to phase out its use in 2019. The goal of halving peat use by 2030 was to be achieved in a 'socially and regionally just' manner.[7] However, studies have shown that lived experiences of this transition have been anything but just. Those working with peat have endured heavy financial losses, and the forced severance of ties with a personally meaningful and fulfilling livelihood has caused distress at individual and community levels.[8] Moreover, concerns about the availability and affordability of energy in the aftermath of Russia's invasion of Ukraine have led to the phase-out

DOI: 10.4324/9781003473534-5

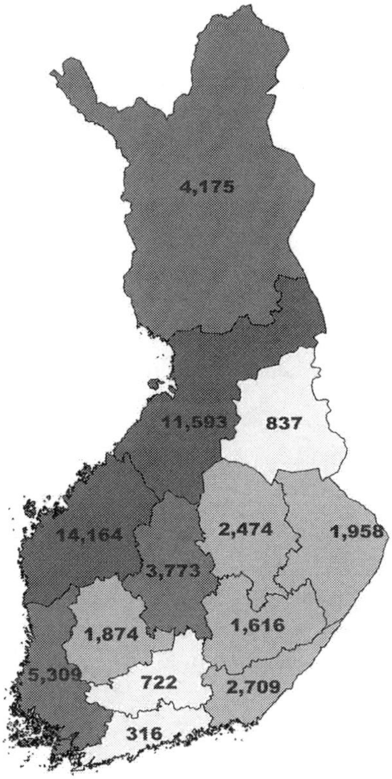

Figure 3.1 Map of Finland showing the numbers of hectares reserved for peat extraction by region in 2019. Darker shades of blue correspond to increasing numbers of hectares reserved for peat extraction within the region.

Source: Arto Vitikka, Arctic Centre, University of Lapland.

decision being contested from another perspective. Such concerns, which have prompted questioning of the rationale for giving up a readily available domestic energy resource, have added to the societal debate.

The Finnish state was responsible not only for the decision to phase out the use of peat as an energy source but also for the initial decision to initiate and develop the peat-based energy industry. Its formal inception can be traced to the establishment of the state-owned energy company Valtion polttoainetoimisto (State Fuel Office), or VAPO, in 1941, ultimately culminating in a mass-scale, multi-million effort to develop the peat industry during the 1970s. Peat extraction served to strengthen the energy security of the rapidly modernising and urbanising nation, while also advancing employment prospects in the country's relatively underdeveloped rural regions.[9] In this process, peat and peatlands were constructed as a 'national treasure'[10] synonymous with the prosperity, modernity, and independence of a remote, energy-poor nation. As such, it was conceived as essential for the nation's survival. The intimate intertwining of peat as an energy and livelihood source with rural and national identities that were once woven into Finland's socio-economic and cultural fabric continues to linger in the contemporary political and societal setting, which demands rapid decarbonisation.

Despite its centrality in energy and regional development policies, peat has remained an understudied topic in the social sciences and humanities. Until recently, its prominent role in the Finnish energy system and society was taken for granted by scientists, policymakers, and practitioners. Since its emergence as a social scientific research topic during the last decade, the literature on energy peat has largely focused on its political history and significance,[11] the environmental and discursive conflicts around peat and its extraction,[12] and, since the transition was announced, on various aspects of its implementation.[13] In this chapter, I focus on the entwinements of peat as an energy source, peat extraction as a livelihood, and the wider Finnish society to elucidate the symbolic aspects of a given resource that make its phasing out much more painful than is belied by employment and energy statistics that convey its relatively insignificant role.

In this chapter, I adopt the notion of *cultural trauma* as framed by Robert J. Brulle and Kari Marie Norgaard in their analysis of the social dynamics of climate change *in*action. They define cultural trauma as 'a social process that involves the systematic disruption of the cultural basis of a social order … resulting in profound challenges to routine, taken-for-granted ways of interacting.'[14] They frame the culturally traumatic consequences of climate change either in terms of the immediate disruption and damage that it causes or the challenges that it poses to familiar ways of living and being. I argue that climate change constitutes a potential cultural trauma in a third, as yet unexplored sense: through the impacts of the policies designed for its mitigation. It is not the immediate or long-term impacts of climate change that have upended the lives of those engaged in peat extraction. Rather, it is the national climate policies whose impacts on the peat workers and on the Finnish nation—both of which are symbolically interwoven with the destiny of peat as an energy source—that constitute the cultural trauma, which is the focus of this chapter.

I frame the emergence of the Finnish peat industry as a conscious state-led initiative intended as a response to national energy security concerns as well as the acute concern regarding rural depopulation in the rapidly urbanising Finland of the 1960s–1970s. In the first section of the chapter, I trace the societal and political contexts of the emergence of peat as an energy and livelihood source. I show how within these contexts, peat became synonymous with (energy) independence, self-sufficiency, and regional prosperity in ways that have continued to shape how Finland relates to its peat reserves and to itself as a nation. Next, I examine two lobbying campaigns of the peat industry during the previous decade to illustrate how these associations continued to be reinforced in the public sphere only a few years before the national phase-out was announced. In the final section, drawing on my fieldwork and interview data, I highlight how the announced transition severed cultural ties between peat and Finland's (rural) self-understandings and transformed the everyday lives of those whose incomes, identities, and communities were still entwined with peat extraction. Common threads running through the chapter are the basis for advocacy of peat as an energy source; the ways in which these framings entwine with personal, rural, and national identities; and how the dismantling of these ties come to be perceived as a threat to personal and national meanings, cultures, and destinies.

The Making of a 'Traditional' Energy Source

Of Finland's entire surface area, which is close to 34 million hectares, an estimated 10.4 million hectares were originally peatlands.[15] These omnipresent peatlands have occupied a central role in the national imagination, inspiring mythology, fears, and desires that constitute the country's living (bio)cultural heritage and its contemporary cultural landscapes.[16] Moreover, the functions of Finland's peatlands are not limited to the symbolic domain; they have also been deliberately harnessed for economic gain by private landowners and within state-coordinated campaigns alike. Today, the remaining 1.2 million hectares of peatland are protected to varying (and contested) degrees.[17]

Finnish peatlands have been subjected to anthropogenic use for as long as the areas surrounding them have been inhabited. Their large-scale industrial extraction for energy is situated along an extended cultural continuum of sourcing energy from local landscapes and moulding peatlands in the service of human comfort and gain. Peatlands have traditionally been a source of nutrition and warmth, providing berries, game, hay, and bedding for thousands of years, and peat extraction for agricultural and horticultural purposes persists even now.[18] Finland's industrial use of peatlands emulates that of other European countries, including Russia/the Soviet Union, although its trajectory unfolded somewhat later.[19] Practices of clearing and draining peatlands for agriculture were introduced during the early seventeenth century in a process that transformed peatlands, which till then had been perceived as wastelands, into 'productive' land.[20] The transformation of peatlands intensified during the nineteenth century, mainly driven by agricultural and, later, by forestry demands,[21] with peat extraction only following towards the end of the century.[22] In these socio-economic and socio-cultural processes, the vast majority of the nation's natural state peatland landscapes were transformed into cultural landscapes of production,[23] equally in material, symbolic and visual terms. Finland's abundant peatlands came to be predominantly characterised by 'straight ditches, clean streams and totally changed peatland landscape'[24] and, as such, visible entanglements of the interactions between the human and nonhuman.[25]

Compared to its European counterparts, Finland's energy history has been characterised as "peculiar": the country had no oil, gas, or coal deposits. However, owing to its late industrialisation, poor standard of living and abundant forest and hydropower resources, Finland remained largely independent from energy imports until the early twentieth century. As the processes of industrialisation and modernisation continued to unfold, the resulting increase in energy demand began to pose growing concerns about the overuse of timber for energy supply and increasing dependence on energy imports.[26] Combined with delivery disruptions during the First World War, these developments motivated the diversification of the nation's energy supply, including the establishment of a national peat energy industry.

While the first attempts to achieve this goal failed, largely because of a lack of knowledge about Finland's peat reserves and the continued abundance of wood available for energy use,[27] energy shortages in the 1940s gave new impetus to state-led visions of extracting and using peat as an energy source.[28] Increasing energy consumption across the nation also led to increased dependence on energy imports, entailing risks, which manifested as delivery disruptions during World War II and its aftermath. These experiences contributed to the first coordinated state-led efforts to explore national peat reserves and develop the technologies and expertise required for their industrial use. VAPO was established as a state-owned company and tasked with developing the peat industry as part of their mandate to secure national energy security.[29] Despite these efforts, however, peat continued to have minimal importance in the national energy mix during subsequent decades until the advent of the oil crisis of the 1970s, which fuelled efforts to decrease dependence on energy imports and led to the establishment of the national energy peat industry.[30]. Alongside the institutionalisation of the industry itself, the arguments emphasising the importance of the peat industry and its development were also 'institutionalised' into the national political arenas and societal debates. These arguments, used in the 1970s in defence of peat, continue to justify its use.[31]

However, a view of the emergence of the peat industry solely as a response to growing energy needs or national energy shortages obscures the complex ways in which the craft of and demand for peat extraction were intertwined with other societal challenges and developments at the time. This embeddedness is in no way unique to the resource-making endeavour in the Finnish peatlands: parts of the biophysical world are always constructed and reframed as 'resources' in any given social, cultural, and political setting, and energy policies seldom focus as much on energy itself as they do on achieving and securing other societal functions and goals.[32] In the decades following World War II, Finland returned to the path of rapid industrialisation and urbanisation.[33] The mechanisation of both agriculture and forestry that reduced the labour demand in rural regions coincided with the period when the so-called 'baby boomers'—the populous generation born after the world wars in the 1940s and 1950s—were entering the workforce. Shrinking numbers of rural jobs combined with the increase in the working-age population forced many rural residents to seek employment opportunities in Finland's expanding cities as well as abroad. Between 1969 and 1970 alone, an estimated 80,000 people out of a population of approximately 4.5 million moved abroad, mainly to neighbouring Sweden. The extent of the emigration wave was so massive that it caused a visible dip in the national population curve.[34]

The Finnish state responded to these unfavourable developments with interventions on multiple fronts, encompassing multiple industries. The period between the late 1960s and the end of the 1970s was characterised by regional development and settlement policies that were aimed at reversing these disturbing trends, improving living prospects in the countryside, and diversifying rural economies, especially in the eastern and northern regions of the sparsely populated country.[35] An extensive campaign was initiated in which peatlands were drained to allow for the expansion of forestry and farmlands[36] and the state-led establishment of the Finnish peat industry.[37] Peat extraction became an integral element in the puzzle of seasonally changing and overlapping livelihoods for rural residents, with the new industries resonating culturally with the existing 'ethos of hard work.'[38] Peatlands thus became a source of income, identity, and belonging for thousands of peat workers, their families, and rural communities and cultures.

From 'National Treasure' to Climate Villain

For most of the early decades following its introduction, industrial peat extraction encountered little societal resistance: its economic and employment benefits were sufficient to counter the arguments that had begun to surface regarding the impacts of the peat industry on the nation's lakes and rivers.[39] However, the emergence of the environmental movement in Finland during the 1970s also transformed peatlands into sites of contestation given their conflicting values. Over the following decades, peat-related conflicts unfolded predominantly in the form of localised environmental conflicts over the environmental impacts of peat extraction.[40] At the same time, the role of peat in the national energy mix remained uncontroversial: peat was such a natural part of the national energy and economic landscapes that its extraction did not warrant discussion. Peat had become interwoven within the fabrics of rural economies, national mindsets, and energy infrastructures.[41]

The share of peat in Finland's total annual energy consumption continued to increase well into the 1990s, peaking at 7.5% and stabilising at this figure over the following decades. However, the international political landscape in which Finland was embedded was undergoing a gradual transformation which would have profound implications for how the nation would relate to its peat reserves. Growing awareness about the impacts of climate change and the importance of reducing carbon dioxide emissions had begun to shape both regional and international political agendas.[42] Another nationally definitive event was Finland's accession to the EU in 1995, which would make it subject to the common climate and energy politics that would be introduced a decade later.[43] EU membership would also later integrate Finland into the Union's emissions trading system, within which the carbon dioxide emitted by an energy source would have a defining role in dictating the economic feasibility of its use.

International developments also forced Finland to explicate its stance on peat. Whereas peat was domestically categorised as a non-renewable energy source as a result of a tight committee vote,[44] internationally Finnish politicians and industry continued to lobby actively to define peat as a renewable energy resource within the frameworks of the EU and the Intergovernmental Panel on Climate Change (IPCC). These attempts remained unsuccessful,[45] although IPCC did change the classification of peat, assigning it as a separate category.[46] Despite peat's extremely slow renewal rate and CO_2-intensive emission profile—both from its combustion as well as from the emissions of drained peatlands—that distinguish it from renewable energy sources, an understanding of peat as a 'slowly renewable biomass fuel'[47] was deeply rooted in the parlance of many actors in the Finnish (energy) political arenas. These discursive framings continue to be reflected in societal debates over peat, even in a situation in which the dominant view is that of peat no longer conceived as a local environmental issue but rather as a global climate-related concern.[48]

The battle over definitions of peat and its future unfolded not just on the stages of domestic and international politics. Around the turn of the millennium, societal acceptance of peat as an energy source peaked, with three out of four Finns viewing peat favourably and more than 50% strongly supporting its increased share in the Finnish energy mix. However, towards the beginning of the second decade of the twenty-first century, this support began to decline slowly to a point at which less than 50% of Finns favoured peat as an energy source.[49] The extraction and use of peat had also begun to decline slightly compared with extraction and use during the peak years in the 2000s. No single reason for this downward trend can be pinpointed; changes in energy taxation, environmental licensing practices at production sites, and rapidly declining societal acceptance of peat as an energy source all played a role.[50] At the same time, the impacts of peat as a source of employment gradually declined to around 10,000 person-years annually.[51] It was in this societal context that two massive nation-wide lobbying campaigns were initiated by the peat industry to influence public perceptions of this energy source. The first, titled *2 prosenttia* (2 percent), was implemented by VAPO in 2010, and the second, *Turveinfo* (peat info), was launched as a collective effort of peat-producing and using industries under the umbrella of the Finnish Bioenergy Association in 2016.

Neither of these campaigns entailed energy advertising as understood in the traditional sense because in Finland, energy peat is not a product that is directly purchased by consumers. Instead, the campaigns were aimed at influencing public perceptions of peat and encouraging their audiences to seek to influence peat-related decision making to improve the prospects of its use for energy generation. The *2 prosenttia* campaign was launched online, supported by short television and radio advertisements, while *Turveinfo* ran as a series of four posters placed at bus stops in key locations and as whole-page advertisements placed in prominent national newspapers. Each of these campaigns stirred up considerable public debate, with both campaigns being accused of distorting scientific facts related to peat as an energy source and exaggerating the importance of peat as an energy source in Finland.[52] This widespread criticism led to their dismantling earlier than planned and, before that, to modification of their content. The campaigns shared a common storyline in which the historically significant role of peat as the symbolic saviour of rural (and energy-dependent) Finland was framed by interlinking the fate of peat as an energy source and the destiny of Finland as a nation in ways that will be highlighted below. Accordingly, I view the campaigns as presenting a single pro-peat discourse, distinguishing between citations sourced from individual campaigns only when this is necessary.

Indeed, both campaigns portrayed peat as an incomparable resource in several ways. Together, they framed it as a 'clean,' 'renewable,' and 'locally produced' energy source and as Finland's own and 'most important energy source' that has a 'positive impact on the everyday life of every Finn.' Peat was also framed as an integral part of the Finnish landscape and of Finnishness: 'the land of thousands of lakes' was also 'a land of thousands of peatlands.' As such, peat was Finland's very own 'national treasure.' The images featured in the campaigns visually reinforced the idea(l)s of Finnishness invoked in the textual content: sunny forests, aerial views of lakes and trees, green fields in summer, dirt roads, and hay barns. These images do not stem from or depict contemporary Finland, which is highly industrialised, urbanised, and energy-intensive; rather, they convey rural nostalgia for times gone by. Similar themes are depicted in illustrations in the *Turveinfo* campaign posters. One is a caricature of the national symbol of a Finnish maiden with the head of a donkey and a flag of Finland in one hand. Another is an abandoned elderly man sitting alone in his wheelchair in darkness, representing the generation who fought for Finland's independence, now left to their own devices with the crumbling of the welfare state.

Indeed, as 'the most foolish nation in the world,' the Finland that is not availing of its peat reserves is heading straight towards its own doom, according to this narrative. However, the two lobbying campaigns differ slightly in their interpretations of what specifically this doom entails. Both effectively convey responses to ongoing societal concerns of the time. In the *2 prosenttia* campaign, launched in the aftermath of the 2006–2008 European gas delivery disruptions, looming threats mainly relate to the consequences of energy shortages and delivery disruptions. They focus on 'what happens when a cold day arrives' in an energy-deficient world where 'what is scarce will be fought for.' Images of warships, overpopulated cities, energy-intensive machines, and trains conveying oil tankers passing the viewer not only emphasise but also exaggerate the textual message conveyed.

Concerns about energy dependence are also reflected in the subsequent campaign, which presents energy scarcity as a 'ticking time bomb' as a result of which 'the pipes in Tampere will freeze, the lights in Turku will go out, or Helsinki will not have heat anymore … with concrete and potentially fatal consequences.' In this campaign, the urgency of addressing societal challenges is considerably amplified. The Finland of 2016 that it depicts is undergoing rapid change following strict austerity policies and government budget cuts. It is 'sinking faster than ever,' and the 'welfare state is constantly being stepped on' as a result of Finland being 'in debt to our ears from buying expensive energy from abroad.' Both sides of this miserable situation are also depicted visually: energy dependence takes the shape and form of the hungry Russian bear, while the old man abandoned in his wheelchair becomes the symbolic embodiment of what is envisioned as the wider collapse of the welfare state (Figure 3.2).

The impending doom is not, however, portrayed as inevitable: by including more peat consumption for energy, Finns could 'save money,' generate more heat, become 'twice as rich as the Norwegians,' and 'lift the national economy, employment rates, and Finland out of the [proverbial] bog.' Peat is advocated as *the* one-size-fits-all solution serving national interests, while simultaneously addressing the country's continuing struggle to improve its national economy and reduce its

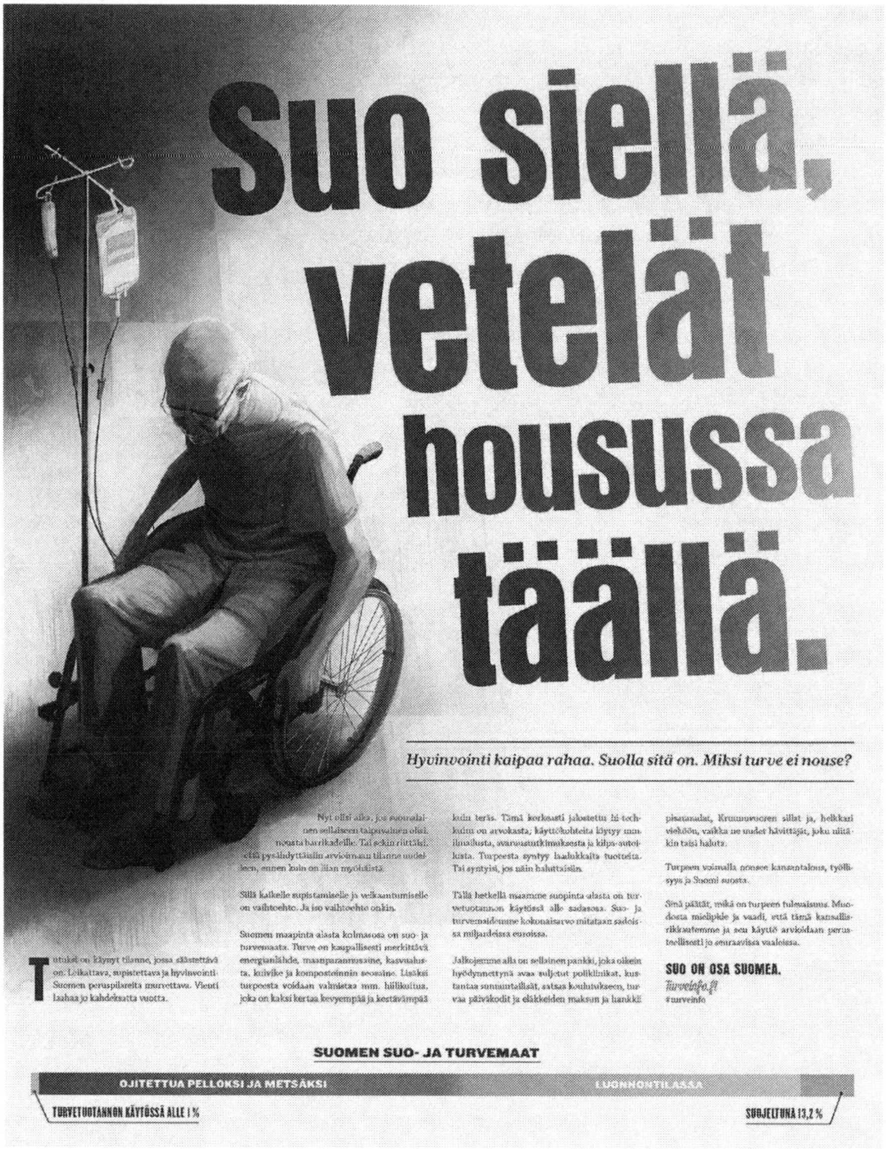

Figure 3.2 Poster from the 'Peat info' campaign depicting the collapse of the Finnish welfare state. Its text argues that the collapse could be avoided through the increased use of peat.

Source: Finnish Bioenergy Association (2016), reprinted with permission.

energy dependence, much like it did at the turn of 1970s. Overall, the campaign draws strong analogies between peat's role in turning around the rural decline in the 1960s and its potential role in reverting the current decline of the Finnish economy and society.

In retrospect, the intense and at times aggressive campaigning had little observable effect on national attitudes towards peat as an energy source. During the 2010s, societal acceptance of peat as an energy source declined to the point at which approximately two-thirds of Finnish survey respondents indicated that peat use should be decreased in favour of renewable and low-carbon energy alternatives.[53] However, the pro-peat lobbying campaigns did succeed in strengthening the views of those involved with or strongly in favour of peat use. Irrespective of their 'truth value,' many of the campaign arguments continue, even now, to circulate among livelihood interest groups, lobbyists, policymakers, and practitioners across industries.

The Downward Spiral

In the Finnish government's original timeline for phasing out peat, the target was to reduce, by half, the use of peat as an energy source by 2030 and to phase out its use shortly thereafter. The rationale for including a transition period was to enable the transition process to unfold in a manner that would be 'socially and regionally just.'[54] A broad-based working group with representatives from different ministries, the peat industry, and other key stakeholder groups, ranging from nature conservation associations to regional authorities, was established to support the planning and implementation of the transition.[55] However, the timeline set for its implementation was soon overridden by the drastic increase of carbon emission prices under the EU's emissions trading scheme, which nearly quadrupled within just one year.[56] Combined with modest changes in the national peat taxation regime, these developments pushed the use of peat into a rapid downward spiral, with the amount of energy extracted annually from peat falling from 21.6 terawatt hours in 2018 to just 2.85 terawatt hours in 2021.[57] The target of halving peat use for energy generation was thus surpassed in just a few years as opposed to the planned timeframe of 2030.

When Finland's peat phase-out was announced, peat accounted for approximately 2,500 person-years of employment in the national economy. However, less than 1,200 of these hours were directly connected to peat extraction.[58] This number indicates a dramatic decline from the industry's peak years to a nationally insignificant level of 0.56% of the national workforce. However, the geographically concentrated and entrepreneurial nature of the Finnish peat industry complicated the transition in several ways. What remained of Finland's peat extraction activity was geographically concentrated in the Ostrobothnian provinces and the province of Central Finland.[59] The approximately 500 small- and medium-sized enterprises at the time had incurred significant debts from their investments in machinery, administrative permits, and extraction sites.[60] Moreover, it was not only the financial capital and professional expertise invested in the industry that were making its phase-out painful but also the symbolic and cultural capital of peat as an industry and a livelihood, which would continue to play a prominent role in complicating the announced phase-out.

The experiences of transition of those working with peat and their preferences for support mechanisms have been described in studies based on regional[61] and national[62] surveys and reports as well as by expert working groups involved in planning and monitoring the transition process.[63] This research paints a picture of a precarious livelihood in a context of an uncontrolled and unmanaged transition characterised by delayed or dysfunctional support measures and multidimensional injustices associated with the transition. The timeframe of the transition was perceived as inadequate, national political stances as unpredictable, and available support mechanisms as ill-suited and inappropriately timed in their implementation. There was also insufficient inclusion of those most affected and the undervaluing or lack of recognition of losses experienced.[64] Whereas the impacts of transitions tend to be framed mainly in terms of lost employment and income, the results of the national survey of peat workers hint at a much more all-encompassing role of peat extraction as a livelihood for those who worked with it daily. Peat workers framed their occupation in terms that signalled deep personal meaning and attachment; rather than referring to their work as a livelihood, they used words such as 'lifestyle' and a 'family member.'[65] These are the kinds of remarks that propelled me to perform further empirical research: a short stint of fieldwork conducted at a northern Finnish peat bog in June 2022 and semi-structured interviews conducted with northern Finnish peat entrepreneurs during the winter of 2022–2023. This research was the source of the observations and citations presented in this section.

In light of the findings of the above-mentioned research, it is apparent that the original framings of peat continue to resonate among those who have spent their lives extracting peat, which is perceived as a 'steady domestic fuel,' a 'local [source of] energy,' and a significant source of employment in rural communities. The entrepreneurs spoke with great pride about the summer jobs that they have been able to offer the youth in their communities, about the role of peat in 'keeping cities warm,' and the positive impacts of peat extraction on local economies. They spoke equally fondly and passionately about the livelihood itself. Perceived attributes of peat work that they mentioned were 'freedom,' long days, unpredictability, challenges, and 'engaging with the forces of nature,' all of which they considered personally meaningful and fulfilling. Moreover, this work fostered a sense of community. At the bog, the work is done collectively: 'everyone is equal,' and 'boys are not boys anymore' as they are transformed into men. Stories from the bog are told and retold at coffee tables, during family dinners, and in school classrooms, making this 'extremely difficult sport' a part of the shared cultural heritage of peat-extracting communities.[66] For peat workers, their work is not only a source of income; it is also a means of self-realisation and achievement, fostering a sense of belonging to a community and of giving back to society. Conversely, its loss constitutes a personal and cultural trauma, posing an existential threat of abrupt severance of these intimate ties and the upending of familiar values and worldviews with little or nothing gained in return (Figure 3.3).[67]

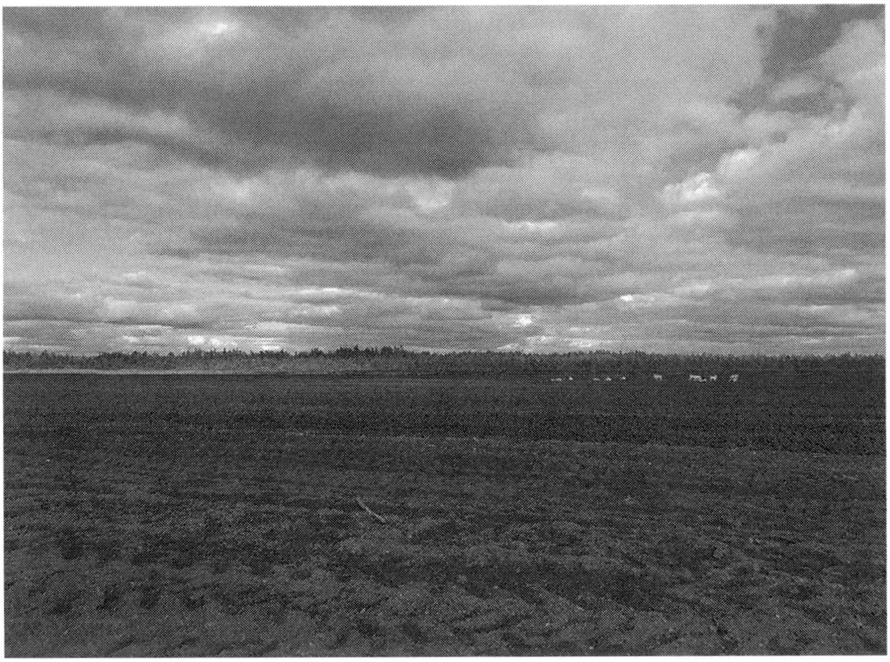

Figure 3.3 An active peat bog in northern Finland, photographed by the author during the summer of 2022.

Juxtaposed with the emotional ways in which peat workers relate to their livelihood, the harshness of the impacts of the transition comes as no surprise. Work that was previously a source of great enjoyment and pride has lost a great deal of its meaning and inspiration, and workers who were once respected members of society now experience themselves as 'outcasts' and 'environmental criminals.' Both the injustices experienced and the painful everyday personal losses are reflected in the peat workers' feelings of 'disappointment and bitterness' towards the Finnish political system and society from which they have become alienated. In addition to their personal losses, peat workers have deep concerns about the impacts that the phasing out of peat will have for Finland as a welfare society and an independent state. The interviews conveyed a picture of 'extremely worried' citizens who care deeply about the future of the Finnish countryside now grappling with the 'snowball effect' of disappearing livelihoods, declining services, closing schools, and depopulation, which serve as a painful reminder of the challenges that peat extraction as a livelihood was originally intended to address. Concerns about the availability and affordability of energy in the interplay of exploding state debt and unrest within international politics and markets are equally grave. Questions raised include: 'what will we be rubbing [generating] heat with?' and 'How can we make sure Finland will not be the first one falling to its knees?'

Much of the failure in implementing a just transition for the peat industry can be attributed to the accelerated and unmanageable time frame within which the still ongoing transition has occurred. However, injustices also stem from transition policies that have failed to grasp the meanings attached to the livelihood of peat extraction, with its diverse embodied, sensory, and situated practices. Environments considered as integral parts of concrete livelihood practices carry manifold meanings and functions. Through livelihood practices entailing continued engagements with these environments, they are permeated by individual, rural, and community heritage; private and shared memories; and social encounters and practices.[68] The meanings associated with livelihoods thus extend beyond their economic importance: work can provide a source of community building, appreciation, personal identity and cultural attachment.[69] Individual experiences, practices, and work environments convey a dimension of intergenerational belonging that exceeds the economic functions of this work.[70] At the same time, policies on the transition from peat-based energy continue to reduce work to simply a source of employment and income, thereby contributing to policy responses that are deeply misaligned with the needs of those who urgently require support. Finland's transition from peat-based energy offers an exemplary illustration of the ways in which livelihoods may penetrate all aspects of human experience and community life and of policies that fail to recognise this plethora of meanings and functions, with detrimental consequences for individuals and communities. At the same time, this case study critically reveals how 'the risk of cultural trauma is met with resistance and attempts to restore and maintain the status quo'[71] by those affected at the expense of climate action.

Reimagining Peat(lands)

This chapter has focused on the emergence and decline of Finland's energy peat industry set against the broader societal context of (rural) Finland. While it has framed peat as an important domestic energy resource, its explicit focus has been on the manifold and intimate ways in which the state-established peat industry has been woven into the fabrics of national, rural, and community identities. The tightly knit symbolic entwining of peat, prosperity, identity, and national independence was established with the introduction of the peat industry, conceived as the solution to energy security concerns and rural depopulation. During the 2010s, it was emphatically reconstructed when the industry was fighting for its own future. These entwinements were abruptly severed with the announcement of a national transition and the subsequent years of the unanticipated decline of the peat industry. Russia's invasion of Ukraine and its political and energy market impacts complicated the situation further, both in terms of the national energy mix and how the nation relates to its peat and peatlands as a resource.

Finnish environmental historian, Esa Ruuskanen, noted the 'cyclical' nature of the emphasis on peat as an energy resource in Finnish society. According to Ruuskanen, its importance is accentuated during times of crisis of different kinds, such as in the aftermath of World War II in the 1940s and in response to the oil crisis of the 1970s.[72] A similar pattern can be observed now in relation to the Russian invasion of Ukraine, which prompted a reassessment of the role of peat in the Finnish energy mix. The demand for peat as an energy source, which had plummeted following the announcement of the transition, stabilised to the 2021 level before the Ukraine invasion,[73] and for the first time in over a decade, an upward trend in the long-declining societal support for energy peat could be observed.[74] However, the national decision to phase out peat as an energy source remains in force.

Throughout its existence, the share of the peat industry in the national energy mix never even amounted to 10%, and its impacts on employment have gradually been reduced to numbers that are nationally irrelevant. Therefore, the economic value addition of peat extraction has become insignificant.[75] Yet, for decades, the role of peat in the national energy palette, which has been and continues to be actively debated within political and societal arenas, and its importance in local and regional contexts cannot be refuted. Those most affected by the peat transition are experiencing not only economic distress but also hopelessness, anger, sorrow, alienation, and immeasurable loss. The everyday and symbolic significance of peatlands, both for those whose livelihood depends on it and for the wider society, bear observable parallels with how other livelihoods were crafted into being by 'the leaders of economic and political life'[76] during the era of regional development campaigns in the 1960s and 1970s continue to be framed.[77]

At the beginning of this chapter, I framed the process and impacts of the Finnish peat energy transition as constitutive of a cultural trauma.[78] For an event or a developmental process to be constitutive of an actual cultural trauma, or of the threat of one, it must cause a profound and systematic disruption of essentialised ways of thinking, acting, and being for individual lives and communities as well as for the functioning of a society and its institutions. The narratives that fostered the birth of the Finnish peat industry, the discursive battles waged over peat as an energy source, and the experiences of the transition of those who engaged in peat extraction all underline the profound ways in which peat was woven into the fabric of rural everyday life and into Finland's national storylines about itself. The loss of a symbolically important livelihood and energy source and the failure of transition policies to support those acutely impacted by this loss and to construct alternative visions for regional and national futures have served to 'dislocate the social reality that anchors individual identities and social interactions,'[79] with detrimental consequences for peat workers. However, the (failed) transition has also posed a threat to the shared stories that Finland as a nation has relied on in narrating its own history, self-understanding, and its place on the international (energy) map. As a small nation dependent on (Russian) energy imports, Finland has taken great pride in its peat endowments, considered as its own precious energy resource whose loss has yet again become synonymous with the risk of loss of energy security and independence. The continued constructions of peat as a politically divisive topic among citizens and decision makers alike adds yet another element to the debate, endangering the shared understanding of a collective Finnish 'we.'[80]

During the last 50 years, the Finnish peat debate has gradually shifted from viewing peat as a natural resource to be utilised for maximum economic gain to one where peatlands are depicted as a resource for global climate change mitigation. Whereas for a long period, the economic motives were more or less 'the norm against which conservation needs to be justified,'[81] the waning importance of peat as a source of energy and employment prompted a shift in policies as well as societal attitudes. Internationally, perceptions of peatlands have shifted. Initially considered wasteland, they were later viewed as a resource to be harnessed for local and national economic gain. However, a gradually formed consensus on the integral role of peatlands as carbon sinks, biodiversity hotspots, and sources of recreation, aesthetic experience, and cultural belonging has been evident in recent decades.[82] The effects of these societal developments are again having a concrete impact on Finland's peatland landscapes, as an increasing number of drained peatlands are undergoing rewetting and restoration to return (closer) to their natural state, in an attempt to erase the signs and impacts of a century of intensive human intervention. The challenge—and opportunity—for Finland lies in reimagining its peatlands as a resource in a manner that is compatible with its own cultural and symbolic relations with them as well as the demands of a rapidly warming world.

Notes

1 Heikkilä and Lindholm, "Soidensuojelun vaikeat vuosikymmenet," 95.
2 Tanskanen, *Näkyvän takana*, 13–28; Raivo, "The Finnish Landscape."
3 World Energy Council, *World Energy Resources.*
4 Korhonen et al., *Turvetyöryhmän loppuraportti*, 19; AFRY, *Selvitys turpeen energiakäytön*, 5–10.
5 Korhonen et al., *Turvetyöryhmän loppuraportti*, 37–38; Pellervo Economic Research (PTT), *Turvetoimialan vaikutukset*; Ruuskanen, *Suosta voimaa ja lämpöä*, 141.
6 Soimakallio et al., *Turpeen rooli*, 3; Konttinen et al., "Turvetuotannon sosiaalisen toimiluvan menetys" 63; Albrecht, "Peatland Politics," 27–30.
7 Government of Finland, *Programme of Prime Minister Sanna Marin*, 36.
8 Lempinen and Vainio, "Lost in Transition," 6–8; Lempinen, "Pohjoissuomalaisen turvesiirtymän," 7–10
9 Ministry of Economic Affairs and Employment, "Working Group on Peat"; AFRY, *Selvitys turpeen energiakäytön*, 37–41; Pellervo Economic Research (PTT), *Turvetoimialan vaikutukset*, 2–4.
10 Lempinen, "Barely Surviving," 5.
11 Ruuskanen, *Suosta voimaa ja lämpöä*; Laihanen et al., "Future Role of Domestic Biomass," 155; Kortelainen and Rytteri, "EU Policy on the Move," 4–5.
12 Albrecht, "Peatland Politics," 13–16; Pettersson et al., "Ympäristökonfliktit Suomessa," 90; Konttinen et al., "Turvetuotannon sosiaalisen toimiluvan menetys," 62–63; Möttönen et al., "Ympäristönsuojelulaki vuotaa," 83–84.
13 Laasasenaho et al., "A Just Transition," 5–9; Lempinen and Vainio, "Lost in Transition," 5–8.
14 Brulle and Norgaad, "Avoiding Cultural Trauma," 886–888. These authors build on the writings of Alexander, "Toward a Theory of Cultural Trauma", and Sztompka, "The Trauma of Social Change."

15 Heikkilä and Lindholm, "Soidensuojelun vaikeat vuosikymmenet," 95.
16 Laurén et al., "Soiden muuttuvan kulttuuriperinnön jäljillä," 99–100; 110; Latvala-Harvilahti, "Kohtaamispaikkana suo," 83–85.
17 Ministry of Agriculture and Forestry, "Suot ja turvemaat."
18 Laurén, "Suo karnevalistisena tilana," 100; Heikkilä and Lindholm, "Saarijärven Paavon henki Suomen soilla," 186–188; Soimakallio et al., *Turpeen rooli*, 44–50.
19 cf. Bruisch, "Nature Mistaken," 362–364; Heikkilä and Lindholm, "Saarijärven Paavon henki Suomen soilla," 181–185.
20 Heikkilä and Lindholm, "Saarijärven Paavon henki Suomen soilla," 181–185; Laurén, "Suo karnevalistisena tilana," 100; Ruuskanen, *Suosta voimaa ja lämpöä,* 53.
21 Heikkilä and Lindholm, "Soidensuojelun vaikeat vuosikymmenet," 95–96; Heikkilä and Lindholm, "Saarijärven Paavon henki Suomen soilla," 181–185.
22 Albrecht, "Peatland Politics," 21; Ruuskanen, *Suosta voimaa ja lämpöä,* 18.
23 Tanskanen, *Näkyvän takana*, 55.
24 Tanskanen, *Näkyvän takana*, 1.
25 Atha et al., "Introduction."
26 Myllyntaus, "Farewell to Self-sufficiency," 31–35.
27 Ruuskanen, *Suosta voimaa ja lämpöä*, 18.
28 Heikkilä and Lindholm, "Saarijärven Paavon henki Suomen soilla," 183.
29 Ruuskanen, *Suosta voimaa ja lämpöä*, 18; Albrecht, "Peatland Politics," 23.
30 Albrecht, "Peatland Politics," 23; Heikkilä and Lindholm, Saarijärven Paavon henki Suomen soilla" 183.
31 Ruuskanen, *Suosta voimaa ja lämpöä,* 45–47.
32 Lähde, "Politics in a World of Scarcity," 60–62; Scrase and Ockwell, "The Role of Discourse," 2226–2228; Prontera, "Energy Policy," 8–12.
33 Orasmaa, *Maaseudun tulevaisuus*, 76–77.
34 Korkiasaari, *Suomalaisten Ruotsiin suuntautuneen siirtolaisuudenm*, 14–15; Statistics Finland, *Väestötieteen perusteet*; Sverigefinländarnas Arkiv, *Joukkomuutto Suomesta Ruotsiin*, 2.
35 Sippola, "Kiinnekohtia aluepolitiikan," 89–90; Karkinen, "Sitkeä sissi," 67–68, 74–75; Orasmaa, *Maaseudun tulevaisuus*, 76–79.
36 Heikkilä and Lindholm, "Saarijärven Paavon henki Suomen soilla," 184–185.
37 Ruuskanen, *Suosta voimaa ja lämpöä*, 43–47; Albrecht, "Peatland Politics," 21.
38 Laurén, "Suo karnevalistisena tilana," 100.
39 Möttönen et al., "Ympäristönsuojelulaki vuotaa," 83–85; Konttinen et al., "Turvetuotannon sosiaalisen toimiluvan menetys," 62–63.
40 Albrecht, "Peatland Politics," 45–47; Pettersson et al., "Ympäristökonfliktit Suomessa," 89–90.
41 Laihanen et al., "Future Role of Domestic Biomass," 150.
42 UNFCCC 1992; UNFCCC 1997; Ruuskanen, *Suosta voimaa ja lämpöä*, 138–139.
43 Ruuskanen, *Suosta voimaa ja lämpöä*, 139–146; Kortelainen and Rytteri, "EU Policy on the Move" 5–6.
44 cf. Ruuskanen, *Suosta voimaa ja lämpöä*, 141.
45 Ruuskanen, *Suosta voimaa ja lämpöä*, 139–141.
46 IPCC, *IPCC Guidelines, 2006.*
47 Government of Finland, *Government Programme of Prime Minister Matti Vanhanen.*
48 Lempinen, "Suomalainen turvesuo keskellä maailman muutosta," 2000.
49 Finnish Energy, *Finnish Energy Attitudes 2022*, 16.
50 Albrecht, "Peatland Politics," 23.

51 Pellervo Economic Research (PTT), *Bioenergia työllistää maakunnissa*, 1.
52 Lempinen, "'Jos se ei riitä, siitä riidellään,'" 56–57; Lempinen, "'Suo siellä, vetelät housussa täällä,'" 40.
53 Finnish Energy, *Finnish Energy Attitudes 2022*, 16.
54 Government of Finland, *Government Programme of Prime Minister Sanna Marin*, 36.
55 cf. Korhonen et al., *Turvetyöryhmän loppuraportti,* 11–13.
56 Laasasenaho et al., "A Just Transition," 1.
57 Laasasenaho et al., "A Just Transition," 1–2.
58 Soimakallio et al., *Turpeen rooli,* 3.
59 Soimakallio et al., *Turpeen rooli*, 25–27; Korhonen et al., *Turvetyöryhmän loppuraportti*, 37–38.
60 cf. Korhonen et al., *Turvetyöryhmän loppuraportti*, 38–39.
61 Laasasenaho et al., "A Just Transition," 45–46.
62 Lempinen and Vainio, "Lost in Transition," 3–4.
63 Korhonen et al., *Turvetyöryhmän loppuraportti*; Pellervo Economic Research (PTT), *Turvetoimialan vaikutukset.*
64 Lempinen and Vainio, "Lost in Transition," 4–7; Lempinen, "Pohjoissuomalaisen turvesiirtymän," 7–12; for discussions on justice in transitions see, for example, Williams and Doyon, "Justice in Energy Transitions"; McCauley and Heffron, "Integrating Climate."
65 cf. Lempinen and Vainio, "Lost in Transition," 6.
66 See also Lempinen and Vainio, "Lost in Transition," 5–6; Lempinen, "Pohjoissuomalaisen turvesiirtymän," 11–12.
67 cf. Brulle and Norgaard, "Avoiding Cultural Trauma," 896–897.
68 Flood et al., "Assigning Value," 7–11; Latvala-Harvilahti, "Kohtaamispaikkana suo," 84–85.
69 Whalley Hammell, "Belonging," 40–41; McGrath and McGonagle, "Exploring 'Wicked Problems,'" 312–316.
70 Bell and York, "Community Economic Identity," 117–118; Della Bosca Gillespie, "The Coal Story," 739–740; Crowe and Li, "Is the Just Transition Socially Accepted?" 2–3.
71 Brulle and Norgaard, "Avoiding Cultural Trauma," 3.
72 Lempinen, "Barely Surviving," 3
73 Statistics Finland, *Energy in Finland.*
74 Finnish Energy, *Finnish Energy Attitudes 2022*, 16.
75 Statistics Finland, *Energy in Finland*; Laasasenaho et al., "A Just Transition," 1–2.
76 Tanskanen, *Näkyvän takana*, 1.
77 For discussions on fur farming, see Karkinen, "Sitkeä sissi," and for studies on agricultural work, see Orasmaa, *Maaseudun tulevaisuus,*
78 See Brulle and Norgaard, "Avoiding Cultural Trauma," 894–896.
79 Brulle and Norgaard, "Avoiding Cultural Trauma," 897.
80 Vainio et al., "Citizens' Images," 612; Ruostetsaari, "Stealth Democracy," 98.
81 Ratamäki et al., "Framing the Peat," 9.
82 Heikkilä and Lindholm, "Saarijärven Paavon henki Suomen soilla," 286–187; Ratamäki et al., "Framing the Peat," 9–10. Latvala-Harvilahti, "Kohtaamispaikkana suo," 84–85.

Bibliography

AFRY. *Selvitys turpeen energiakäytön kehityksestä Suomessa.* [*Investigation of Energy Peat Use Trends in Finland*]. 2020. https://afry.com/sites/default/files/202008/tem_turpeen_kayton_analyysi_loppuraportti_0.pdf

Albrecht, Eerika. "Peatland Politics in Finland: Local Movements, Deliberative Environmental Governance and Arguments." PhD diss., University of Eastern Finland, 2018.

Alexander, Jeffery C. "Toward a Theory of Cultural Trauma." In *Cultural Trauma and Collective Identity*, edited by Jeffery C. Alexander, 1–30. Berkeley: University of California Press, 2004.

Arjonranta, Jonne, Irma Hirsjärvi, Urpo Kovala, Maria Ruotsalainen, and Tuija Saresma. "Turvetta tupaan: faktat, valheet ja affektiivinen vastaanotto Turveinfo-mainoskampanjassa." ["Facts, Lies and Affective Reception in Peat Info Advertisement Campaign"]. *Media & Viestintä* 40, no. 3–4 (2017): 76–99. https://journal.fi/mediaviestinta/article/view/67795.

Atha, Mick, Peter Howard, Ian Thompson, and Emma Waterton. "Introduction: Ways of Knowing and Being with Landscapes: A Beginning." In *The Routledge Companion to Landscape Studies*, edited by Peter Howard, Ian Thompson, Emma Waterton, and Mick Atha, xix–xxviii. London: Routledge, 2019.

Banerjee, Aparajita, and Geertje Schuitema. "How Just are Just Transition Plans? Perceptions of Decarbonisation and Low–Carbon Energy Transitions among Peat Workers in Ireland." *Energy Research & Social Science* 88 (June 2022). https://doi.org/10.1016/j.erss.2022.102616.

Bell, Shannon E., and Richard York. "Community Economic Identity: The Coal Industry and Ideology Construction in West Virginia." *Rural Sociology* 75, no. 1 (March 2010): 113–143. https://doi.org/10.1111/j.1549-0831.2009.00004.x.

Bruisch, Katja. "Nature Mistaken. Resource–Making, Emotions and the Transformation of Peatlands in the Russian Empire and the Soviet Union." *Environmental History* 26, no. 3 (August 2020): 359–382. https://doi.org/10.3197/096734018X15254461646567.

Brulle, Robert J., and Kari M. Norgaard. "Avoiding Cultural Trauma: Climate Change and Social Inertia." *Environmental Politics* 28, no. 5 (January 2019): 886–908. https://doi.org/10.1080/09644016.2018.1562138.

Crowe, Jessica A., and Ruopu Li. "Is the Just Transition Socially Accepted? Energy History, Place, and Support for Coal and Solar in Illinois, Texas, and Vermont." *Energy Research & Social Science* 59 (January 2020). https://doi.org/10.1016/j.erss.2019.101309.

Della Bosca, Hannah, and Josephine Gillespie. "The Coal Story: Generational Coal Mining Communities and Strategies of Energy Transition in Australia." *Energy Policy* 120 (September 2018): 734–740. https://doi.org/10.1016/j.enpol.2018.04.032.

Flood, Kate, Marie Mahon, and John McDonagh. "Assigning Value to Cultural Ecosystem Services: The Significance of Memory and Imagination in the Conservation of Irish Peatlands." *Ecosystem Services* 50 (August 2021). https://doi.org/10.1016/j.ecoser.2021.101326.

Finnish Energy. *Finnish Energy Attitudes 2022*. Helsinki: Energiateollisuus, 2023. Accessed June 13, 2024. https://energia.fi/wp-content/uploads/2022/12/Energia-asenteet_2022_FINAL.pdf.

Government of Finland. *Government Programme of Prime Minister Matti Vanhanen's Second Cabinet*, 2007. https://julkaisut.valtioneuvosto.fi/bitstream/handle/10024/164721/hallitusohjelma-Vanhanen_II.pdf?sequence=1&isAllowed=y.

Government of Finland. *Programme of Prime Minister Sanna Marin's Government 2019*, 2019. Accessed June 13, 2024. https://julkaisut.valtioneuvosto.fi/bitstream/handle/10024/161931/VN_2019_31.pdf?sequence=1&isAllowed=y.

Heikkilä, Raimo, and Tapio Lindholm. "Soidensuojelun vaikeat vuosikymmenet." ["The Difficult Decades of Peatland Conservation"]. *Terra* 120, no. 2 (2008): 95–106.

Heikkilä, Raimo, and Tapio Lindholm. "Saarijärven Paavon henki Suomen soilla – Katsaus soiden käyttöön Suomessa." ["A Review of Peat Use in Finland"]. *Terra* 127, no. 4 (December 2015): 181–190.

Intergovernmental Panel on Climate Change (IPCC). *2006 IPCC Guidelines for National Greenhouse Gas Inventories: Volume 2.* www.ipcc-nggip.iges.or.jp/public/2006gl/vol2.html.

Karkinen, Katri. "Sitkeä sissi: Turkistarhaaja yhteiskunnan muutoksessa." ["Fur Farmer amidst Societal Change"]. PhD diss., University of Joensuu, 2009.

Konttinen, Esa, Miikka Salo, and Möttönen Sakari. "Turvetuotannon sosiaalisen toimiluvan menetys Saarijärven reitillä." ["The Loss of SLO for Peat Extraction in the Saarijärvi Route"]. *Alue ja ympäristö* 47, no. 1 (2018): 62–78. https://doi.org/10.30663/ay.60607.

Korhonen, Timo, Petri Hirvonen, Jussi Rämet, and Sirpa Karjalainen. *Turvetyöryhmän loppuraportti* [*Final Report of the Working Group on Peat*], 2021. https://julkaisut.valtioneuvosto.fi/bitstream/handle/10024/163045/TEM_2021_24.pdf.

Korkiasaari, Jouni. *Suomalaisten Ruotsiin suuntautuneen siirtolaisuuden yhteiskunnalliset syyt 1900-luvulla.* [*Societal Reasons for Finns' Emigration to Sweden in the Twentieth Century*]. Turku: Siirtolaisinstituutti, 2006.

Kortelainen, Jarmo, and Teijo Rytter. "EU Policy on the Move: Mobility and Domestic Translation of the European Union's Renewable Energy Policy." *Journal of Environmental Policy & Planning* 19, no. 4 (2016): 360–373. https://doi.org/10.1080/1523908X.2016.1223539.

Laasasenaho, Kari, Anu Palomäki, and Risto Lauhanen. "A Just Transition from the Perspective of Finnish Peat Entrepreneurs." *Mires and Peat* 28 (2022). https://www.mires-and-peat.net/article/128812-a-just-transition-from-the-perspective-of-finnish-peat-entrepreneurs.

Lähde, Ville. "Politics in a World of Scarcity." In *The Politics of Ecosocialism: Transforming Welfare,* edited by Kajsa Bergnäs, Teppo Eskelinen, Johanna Perkiö, and Rikard Warlenius, 55–67. London: Routledge, 2015.

Laihanen, Mika, Antii Karhunen, and Tapio Ranta. "Future Role of Domestic Biomass and Peat in National Energy Security." *International Journal of Energy and Environment* 12, no. 3 (2021): 149–158. www.ijee.ieefoundation.org/vol12/issue3/IJEE_03_v12n3.pdf.

Latvala-Harvilahti, Pauliina. "Kohtaamispaikkana suo: suotaide kulttuuriperintö–ja luontoarvojen yhdistäjänä." ["Mire as a Meeting Point: Peatland Art as the Bridge-Builder for Cultural Heritage and Nature Values"]. *Terra* 134, no. 2 (2022): 83–97. https://doi.org/10.30677/terra.112015.

Laurén, Kirsi. "Suo karnevalistisena tilana." ["Mire as a Carnivalist Space"]. *Terra* 134, no. 2 (2022): 99–112. https://doi.org/10.30677/terra.112016.

Laurén, Kirsi, Virpi Kaukio, and Pauliina Latvala–Harvilahti. "Soiden muuttuvan kulttuuriperinnön jäljillä." ["Tracing the Changing Cultural Heritage of Peatlands"]. *Terra* 134, no. 2 (2022): 83–97. https://doi.org/10.30677/terra.116743.

Lempinen, Hanna. "'Jos se ei riitä, siitä riidellään': Energiaturvallisuuden kieli ja kuvakieli turve–energian markkina–argumentteina." ["'What Is Scarce Will Be Fought for': Language and Images of Energy Security as Marketing Arguments for Energy Peat"]. *Alue ja ympäristö* 42, no. 2 (2013): 53–63. https://aluejaymparisto.journal.fi/article/view/64786.

Lempinen, Hanna. "'Suo siellä, vetelät housussa täällä': Turve 3.0 suomalaisen hyvinvointivaltiona pelastusrenkaana?" ["Peat 3.0 as the Salvation of the Finnish Welfare State?"]. *Alue ja ympäristö* 46, no. 1 (2017): 40–45. https://aluejaymparisto.journal.fi/article/view/64906.

Lempinen, Hanna. "'Barely Surviving on a Pile of Gold?' Arguing for the Case of Peat Energy in 2010s Finland." *Energy Policy* 128 (2019): 1–7. https://doi.org/10.1016/j.enpol.2018.12.041.

Lempinen, Hanna. "Suomalainen turvesuo keskellä maailman muutosta." ["Finnish Peatlands at the Centre of a Changing World"]. *Alue ja ympäristö* 51, no. 1 (2022): 200–202. https://doi.org/10.30663/ay.115336.

Lempinen, Hanna. "Pohjoissuomalaisen turvesiirtymän koettu energiamaisema." ["Experienced Energyscape of the Northern Finnish Peat Transition"]. *Alue ja ympäristö* 52, no. 2 (2023): 29–46. https://doi.org/10.30663/ay.129036.

Lempinen, Hanna, and Annukka Vainio. "Lost in Transition: Peat Workers' Experiences of Finland's Low Carbon Transition Policies." *The Extractive Industries & Society* 15 (2023). https://doi.org/10.1016/j.exis.2023.101312.

Livesey, Sharon M. "Global Warming Wars: Rhetorical and Discourse Analytic Approaches to Exxonmobil's Corporate Public Discourse." *Journal of Business Communication* 39, no. 1 (2002): 117–146. https://doi.org/10.1177/002194360203900106.

McCauley, Darren, and Raphael Heffron. "Just Transition: Integrating Climate, Energy and Environmental Justice". *Energy Policy* 119 (2008): 1–7. https://doi.org/10.1016/j.enpol.2018.04.014

McGrath, Margaret, and Helen McGonagle. "Exploring 'Wicked Problems' from an Occupational Perspective: The Case of Turf Cutting in Rural Ireland." *Journal of Occupational Science* 23, no. 3 (2016): 308–320. https://doi.org/10.1080/14427591.2016.1169437.

Ministry of Agriculture and Forestry. "Suot ja turvemaat." ["Mires and Peatlands"]. Accessed on October 20, 2023. https://mmm.fi/luonto-ja-ilmasto/suot-ja-turvemaat.

Ministry of Economic Affairs and Employment. "Working Group on Peat Proposes Ways to Help the Sector in Transition and to Increase Role of Peat in Security of Supply." Press release on 31 March 2021. https://tem.fi/en/-/working-group-on-peat-proposes-ways-to-help-the-sector-in-transition-and-to-increase-role-of-peat-in-security-of-supply.

Ministry of Economic Affairs and Employment. "National Energy and Climate Strategy." https://tem.fi/documents/1410877/2769658/National+Energy+and+Climate+Strategy+2013.pdf/630dc2c6-4a23-4f2e-b3043cd69daf8265/National+Energy+and+Climate+Strategy+2013.pdf?t=1464241221000.

Möttönen, Sakari, Mikka Salo, Tapio Litmanen, and Esa Konttinen. "Ympäristönsuojelulaki vuotaa: turvekiista tunnustuskamppailuna." ["Environmental Protection Law Is Leaking: Peat Dispute as a Conflict for Recognition"]. *Alue Ja ympäristö*, 52, no. 1 (2023): 77–96. https://doi.org/10.30663/ay.115243.

Myllyntaus, Timo. "Farewell to Self-sufficiency: Finland and the Globalization of Fossil Fuels." In *Energy, Policy, and the Environment. Modelling Sustainable Development for the North*, edited by Marja Järvelä and Sirkku Juhola, 31–44. New York: Springer, 2011.

Orasmaa, Tuuli. *Maaseudun tulevaisuus*. [*The Future of the Countryside*]. Tampere: Vastapaino, 2023.

Pellervo Economic Research (PTT). *Turvetoimialan vaikutukset talouteen ovat suurimmat paikallisella tasolla* [*Impacts of the Peat Industry Are at Their Greatest at the Local Level*]. *PTT Policy Brief 1/2021*. www.ptt.fi/wp-content/uploads/media/turvetoimiala_policybrief_012021.pdf.

Pellervo Economic Research (PTT). *Bioenergia työllistää maakunnissa*. [*Bioeconomy Brings Employment to Regions*]. www.motiva.fi/files/7384/PTT_2013_04_24_Bioenergia_tyollistaa_maakunnissa.pdf.

Pettersson, Sanna K., Ville Hallikainen, Arto, Naskali, Sinikka Rovanperä, and Seija Tuulentie. "Ympäristökonfliktit Suomessa: mistä on kiistelty ja miksi?" ["Environmental Conflicts in Finland: What Has Been Argued about and Why?"]. *Terra* 129, no. 2 (2017): 87–107.

Prontera, Andreas. "Energy Policy: Concepts, Actors, Instruments and Recent Developments." *World Political Science Review* 5, no. 1 (2009): 1–30. https://doi.org/10.2202/1935-6226.1063.

Raivo, Petri. "The Finnish Landscape and Its Meanings." *Fennia* 180, no. 1–2 (2002): 89–98.

Ratamäki, Outi, Pekka Jokinen, Eerika Albrecht, and Antti Belinskij. "Framing the Peat: The Political Ecology of Finnish Mire Policies and Law." *Mires and Peat* 24 (2019): 1–12. www.mires-and-peat.net/article/128910-framing-the-peat-the-political-ecology-of-finnish-mire-policies-and-law.

Ruostetsaari, Ilkka. "Stealth Democracy, Elitism, and Citizenship in Finnish Energy Policy." *Energy Research & Social Science* 34 (2017): 93–103. https://doi.org/10.1016/j.erss.2017.06.022.

Ruuskanen, Esa. *Suosta voimaa ja lämpöä: Turve Suomen energiapolitiikassa 1940–2010.* [*Heat and Power from Peat: Peat in Finland's Energy Politics 1940–2020*]. Jyväskylä: Vapo, 2010.

Scrase, J. Ivan, and David G. Ockwell. "The Role of Discourse and Linguistic Framing in Effects in Sustaining High Carbon Energy Policy: An Accessible Introduction." *Energy Policy* 38, no. 5 (2010): 2225–2233. https://doi.org/10.1016/j.enpol.2009.12.010.

Sippola, Matti. "Kiinnekohtia aluepolitiikan 50–vuotiselta taipaleelta." ["Points of Reference in the 50 Years of Regional Politics"]. *Terra* 128, no. 2 (2016): 89–95.

Soimakallio, Sampo, Paula Sankelo, Maria Kopsakangas-Savolainen, Camilla Sederholm, Karoliina Auvinen, Tero Heinonen, Annika Johansson et al. *Turpeen rooli ja sen käytöstä luopumisen vaikutukset Suomessa.* [*The Role of Peat and Impacts of Transition in Finland*]. Helsinki: Sitra, 2020. www.sitra.fi/app/uploads/2020/06/turpeen-rooli-ja-sen-kaytosta-luopumisen-vaikutukset-suomessa-tekninen-raportti.pdf.

Statistics Finland. *Väestötieteen perusteet. Suomen väestö vuosina 1749–2060* [*Population of Finland 1749–2060*], 2016. https://tilastokoulu.stat.fi/verkkokoulu_v2.xql?course_id=tkoulu_vaesto&example_id=1&lesson_id=12&page_type=esim&subject_id=1.

Statistics Finland. *Energy in Finland*, 2022. www.doria.fi/bitstream/handle/10024/185778/yene_efp_202200_2022_25869_net.pdf.

Sverigefinnarnas Arkiv. *Joukkomuutto Suomesta Ruotsiin.* [*Mass Migration from Finland to Sweden*]. 2016. https://arkisto.org/wp-content/uploads/2016/12/vikvagg.pdf.

Sztompka, Piotr. "The Trauma of Social Change." In *Cultural Trauma and Collective Identity*, edited by Jeffrey Alexander, 155–195. Berkeley: University of California Press, 2004.

Tanskanen, Minna. *Näkyvän takana. tutkimus metsäojitetun suomaiseman kulttuurisuudesta.* [*Behind the Visible: A Research on the Cultural Dimensions of Drained Forest Peatland Landscapes*]. Joensuu: University of Joensuu, 2000.

Tilastokeskus. "Energian kokonaiskulutus energialähteittäin." ["Overall Energy Consumption by Energy Source"]. 2023. https://stat.fi/tilasto/ehk.

Vainio, Annukka, Vilja Varho, Petri Tapio, Anna Pulkka, and Riikka Paloniemi. "Citizens' Images of a Sustainable Energy Transition." *Energy* 183 (2019): 606–616. https://doi.org/10.1016/j.energy.2019.06.134.

United Nations Framework Convention on Climate Change (UNFCCC). "United Nations Framework Convention on Climate Change." 1992. https://unfccc.int/resource/docs/convkp/conveng.pdf.

United Nations Framework Convention on Climate Change (UNFCCC). "Kyoto Protocol to the United Nations Framework Convention on Climate Change." 1997. https://unfccc.int/sites/default/files/kpeng.pdf.

Whalley Hammell, Karen R. "Belonging, Occupation, and Human Well–Being: An Exploration." *Canadian Journal of Occupational Therapy* 81, no. 1 (2014): 39–50. https://doi.org/10.1177/0008417413520489.

Williams, Stephen, and Andréanne Doyon. "Justice in Energy Transitions." *Environmental Innovation and Societal Transitions* 31 (2019): 144–153. https://doi.org/10.1016/j.eist.2018.12.001.

World Energy Council. *World Energy Resources*: *Peat*, 2013. www.worldenergy.org/assets/images/imported/2013/10/WER_2013_6_Peat.pdf.

Part 2

Living with Energy Landscapes

4 Northern Scotland's Late Oil-Fuelled Industrialisation

Labour Mobility and Community Transformation since the 1970s

Ewan Gibbs

In 1972, Rab Wilson tossed a coin and allowed the outcome to determine his future. Rab was a 27-year-old construction worker who was born and bred in Edinburgh. He had grown frustrated with insecure jobs and low wages. Rab thought there were opportunities to build houses in southern England. However, another boom also attracted his attention. He had heard that men were earning good money in the Scottish Highlands working in the new offshore oil industry. As the coin fell, it impelled him to drive north. The choice was made.

Speaking 50 years later at his home in Easter Ross, Rab recalled that he had a 'wee Lambretta' at the time, 'so I decided to jump on the scooter and up I came.' Rab initially worked at the MK Shand pipe coating plant in Saltburn, earning 50 pence an hour heating pipes in hot sand at a new facility that serviced the demand for pipelines that would bring oil ashore. 'There was a squad of us there, and it was lovely weather, and I thought this is a brilliant job, this is.'[1] Rab soon obtained a position across the firth at the Nigg yard established by Highland Fabricators, or 'HiFab,' which better matched his background in the scaffolding trade. Soon, he was earning 85 pence an hour 'slinging pipes' and operating 30-tonne hydraulic cranes to build huge oil platforms to be installed in the North Sea.[2]

Rab was just one of tens of thousands of workers who were drawn to work in Northern Scotland's new oil industry. There was a huge demand for pipe manufacturing and laying, oil platform construction, and workers to build the terminals that would process the oil once it came ashore. This chapter examines two instances of the booming onshore industrial activity associated with offshore oil and gas, which have had enduring impacts on the physical and economic landscapes of Northern Scotland. It begins with a discussion of rig manufacturing at Nigg. Next, it examines the oil terminal located at the Sullom Voe inlet on Shetland, an archipelago of islands that forms the northernmost part of the United Kingdom, with the Atlantic Ocean to the west and the North Sea to the east. Nigg and Sullom Voe were positioned at opposite ends of the oil extraction setup. The rigs, which were built at Nigg, were towed to the offshore oilfields to drill for oil, which was eventually pipelined to terminals, such as Sullom Voe, before being loaded onto tankers. Rig construction at Nigg commenced in 1972 and plans for establishing Sullom Voe

DOI: 10.4324/9781003473534-7

began around the same time, with the first flow of oil in 1978, ahead of the formal opening and substantial completion of the site in 1981.

The drive for North Sea oil production had profound and lasting impacts. For two centuries before the 1970s, Northern Scotland had been characterised by relatively small settlements, outwards emigration, and depopulation. Even as Scotland's Central Belt between Ayrshire in the southwest and Fife in the north-east exploited its coalfields to power steelworks, factories, and shipyards, further north, the economy remained largely dependent on agriculture, fishing, and artisanal production.[3] The analysis presented here draws on oral history interviews recorded in Easter Ross and Sutherland in the Highlands and in Shetland[4] in 2022, with some recorded remotely. Interviewees were largely former fabrication yard and terminal workers and members of their families, but they also included individuals who had lived through the changes brought to the locality. The archival research for this study draws on records maintained by the local government, the British government's Scottish Office, and the Department of Energy, as well as from unions and the oil industry.

The oil developments at Nigg and Sullom Voe marked a significant rupture in the history of both places, as is outlined below, but also fit within a larger pattern of changes to the physical landscape and economic and social life associated with rural modernism in postwar Britain. Energy infrastructures became increasingly concentrated outside of large cities in the form of giant coal-fired power stations, new nuclear facilities and expanding oil refineries.[5] Within the Scottish Highlands, the development of hydroelectricity animated debates about the purpose of the landscape and its appropriateness for industrial development. These pitted conservationist arguments, often from landed interests, against social democratic ends which prized the value of electricity generation in providing essential services and dignified employment.[6] Linda M. Ross' research demonstrates that the Dounreay nuclear facility in Caithness, to the north of Nigg, was characterised by 'social fusion' along with nuclear fission, which anticipated the changes brought by the oil industry.[7] Shetland had also undergone considerable shifts and booms in economic activity, before oil, associated primarily with the archipelago's strategic position, which incentivised the Royal Navy and Royal Air Force to each maintain a presence on the islands, including at Sullom Voe.[8] The coming of oil, though, was associated with a larger shift in population and the economic structure of Northern Scotland.

Oil extraction is a capital-intensive sector, which is often concentrated in locations distant from urban centres. This industry's self-presentation as a leader in the use of advanced technology has also tended to obscure the pivotal role of manual labour in its operations. Labour historians have begun correcting the 'marked decline of interest' in oil workers and their communities, which characterised scholarship in this field in the 2000s and 2010s.[9] This disinterest was encouraged by the decline of organised labour in the energy industries as well as the increased concentration of production offshore, globally. The North Sea is an important example of early pioneering offshore oil production, which demonstrates the importance of worker mobility and labour agency. Nigg and Shetland reveal the connections between offshore and onshore industrial activity. The human stories explored in this chapter unearth the strong historical foundations of North Sea oil and gas. Both individual and collective stories of local and incoming workers illuminate the meanings and memories that undergird North Sea oil production. They foreground the 'dazzlingly futuristic architecture of industrial terraforming in often rural and peripheral sites of eastern Scotland and the northern archipelagos.'[10] These narratives and the fluid social relations that animated the North Sea oil boom challenge the boundaries drawn between threatened indigenous communities and the damaging activities of extractive companies and their itinerant workforces. The analytical approach adopted here embodies Dominic Boyer's call for assessments of energy transformations that 'recalibrate the capacity of "the local."'[11] The meanings attached to place and belonging in Northern Scotland were remade by geographies of labour migration, industrial investments, and production for international oil markets. They were defined from below as well as from above through the moulding of community structures and relationships between incomers and residents from the pre-oil era over half a century. This included the remade physical landscapes brought by the oil industry. For more than half a century, rigs have sat in the Cromarty Firth and the Sullom Voe inlet has been a hive of activity and a dominant feature of the skyline of the north of Shetland's Mainland island.

The construction of oil platforms and terminals became a national priority for the British government in a testing economic context. The discovery of North Sea oil in the late 1960s and early 1970s held out hopes for prosperity and energy security.[12] Incoming workers and locals saw opportunities in the social changes which oil-fuelled industrialisation promised to areas that had been excluded during earlier waves of industrialisation fuelled by coal. These changes were not, however, universally embraced. Objections, which paralleled the earlier opposition to hydroelectricity, ranged from how oil developments would change the natural landscape to the impact of thousands of new residents.

Oil was integrated into a reformed social fabric relatively quickly, but the boom years were brief. After peaking in the mid-1980s, employment in Nigg declined before the yard was eventually closed in the early 2000s, leading to enduring dislocation. Shetland experienced a more gradual slowdown of its still operational oil terminal. The analysis in this chapter is presented in two sections. The first focuses on the influx of capital and labour in the 1970s, and the second on the community's response and changes in the social structure. The concluding section discusses how oil has moulded expectations relating to the transition to renewables, with both locations currently emerging as leading centres of wind power generation.

The Influx of Capital and Labour

Shetland and Easter Ross emerged as strategic locations in the early 1970s following the initial discovery of offshore oil. The Cromarty Firth in Easter Ross had already been identified as a centre for industrial development in the 1960s as part of the project of rural modernisation in the Highlands.[13] The British Aluminium Company (BACo) began building a smelter at Invergordon in 1968, not far from the sites that MK Shand and HiFab developed soon after. Workers began to arrive in Easter Ross from depressed regions in Scotland, spurred by job losses and factory closures.[14] BACo was dwarfed by the new oil industry. Aluminium smelting employed around 1,000 workers, but HiFab employed 4,500 at its peak in the mid-1980s.[15] Including subcontractors and white-collar staff, around 6,000 people worked at the yard in early 1984.[16] Between 1971 and 1981, the population of the counties of Ross and Cromarty grew by around 25% to approximately 65,000 residents. Alexander Mather's 1987 account of economic and social change in Ross and Cromarty emphasised how 'within a few short years, the eastern part of Ross and Cromarty had changed from [being] an area with a shortage of industrial jobs to one in which the local population could not hope to satisfy [the] labour demand.'[17] New arrivals from Clydeside and Tyneside with shipbuilding backgrounds in West Central Scotland and North East England, respectively, held expectations of union recognition as well as craft working norms and workplace autonomy for skilled workers.[18] This situation was distinctive in the context of the eastern Scottish Highlands, which had only just begun hosting large-scale heavy industry through aluminium smelting.

At Nigg, big things were happening in a small space. Brown and Root, a Texan oil services firm, was the majority shareholder in HiFab, formed in early 1972. The company's early brochures boasted of technological prowess and meeting engineering challenges: 'as new supplies of hydrocarbons are found, big answers for development are needed quickly.'[19] North Sea construction powerfully shaped the corporate identity of Brown and Root. A 1997 company history, whose title profiled the firm as *Offshore Pioneers*, quotes a senior marine engineer on North Sea projects, describing the 'vast technological gaps that yawned.' HiFab conquered these challenges, building platforms with parameters scaled five to ten times above those of earlier platforms constructed in the Gulf of Mexico.[20] Craig Mackay began an apprenticeship at HiFab in the late 1970s. He recalled 'the Americans' as bringing a new attitude, which affronted Scottish orthodoxies. Craft demarcations cherished by shipyard workers were challenged, and working Sundays was embraced, as the high wages on offer undermined the traditional Highland observance of the Sabbath.[21]

Craig also emphasised shared learning. American managers were 'good guys' who paid 'good bonuses,' but 'they were out of their league' when it came to building platforms, given that their previous experience was with jack-up rig technology in the Gulf of Mexico: 'They hadn't a clue when they came here. They didn't know what the North Sea was like.'[22] The novel North Sea 'giants' were celebrated in a 1976 film produced for the British multinational oil and gas company, British Petroleum (BP), to mark the beginning of production during the previous year. The narrator boldly announced that unlike ships, 'platforms can only stand and fight' in a struggle against the elements: 'we must use weapons the sea does not possess. Scale of imagination of patience. Scale of initiative. Scale of investment, of risk and the scale of size forced on them by nature.'[23]

Treacherous conditions pivotally shaped the platforms built at Nigg, but the more localised geography of the Cromarty Firth was equally important. Its landscape was transformed by the construction of the massive graving dock, which was required for building platforms on the scale the North Sea demanded. Sue Jane Taylor, an artist who grew up on the Black Isle, not far from Nigg, explores North Sea oil and gas in her work. She has childhood memories of the Cromarty Firth as 'a quiet, pink, sandy shore where local families picnicked and bathed', before HiFab 'engulfed' Nigg Bay.[24] Taylor's recollections concur with those of John Mackenzie, the son of a Nigg farm worker who worked as a joiner with a local construction firm in the early 1970s. Although John found a well-paying job at the yard, he nevertheless recalled its construction as destructive and scarring. There were permanent and lasting consequences from these changes which damaged amenity and differentiated oil rig manufacturing from the construction of hydroelectric dams through its exclusion of locals from pristine areas frequented by visitors. The sand dunes had previously been a beauty spot enjoyed by holidaymakers from Glasgow seeking recreation from a polluted and noisy industrial city. John described the harshness of the working environment created by the vast new dock as follows:

> The site conditions could be horrendous at times. Water was a problem. It was a flat site; there used to be big pools, and in [the] summertime it was dry and dusty. It would be like the Sahara Desert in the summertime. Because you're wide open to the Firth, the wind blowing down the Firth, there's dust and everything blowing. They used to have a big tank at the back with a tractor and it would be spraying water everywhere to keep the dust down. Then in the wintertime, it was the reverse, and trimming drains to get the water in, snow and everything.[25]

These deprivations were worth bearing for others for whom HiFab presented an opportunity to return home. Tommy Patterson moved to Australia with his wife Margaret during the late 1960s so that he could ply his trade as a plumber and heating engineer under more favourable circumstances than those they left behind in Dingwall, a small town on the southern edge of the Cromarty Firth. In Western Australia, Tommy found work offshore and, through a contact there, learnt of opportunities at Nigg. Margaret, who today is a non-party affiliated local councillor for Dingwall, emphasised that Tommy's wages helped them support a family of eleven. Margaret welcomed the permanent transformation of Dingwall and other towns in the vicinity of Nigg created by the yard: 'It's changed the area forever, and that's not a bad thing. That's a good thing. Absolutely wonderful. People, they could [now] afford to live [here].'[26] Margaret underlined that a major change in the landscape of the Cromarty Firth was the building of new homes for workers by the council as the population swelled with new arrivals.

These experiences were mirrored in Shetland. Commencing operations at Sullom Voe (Figure 4.1) was an urgent priority for the British government. In December 1977, a year before oil first flowed through the terminal, its incompletion was estimated to be costing the United Kingdom £200 million per month on its balance of payments at a time when this was a crucial economic indicator.[27] Changes in Shetland's population reflected the intensity of building efforts. The

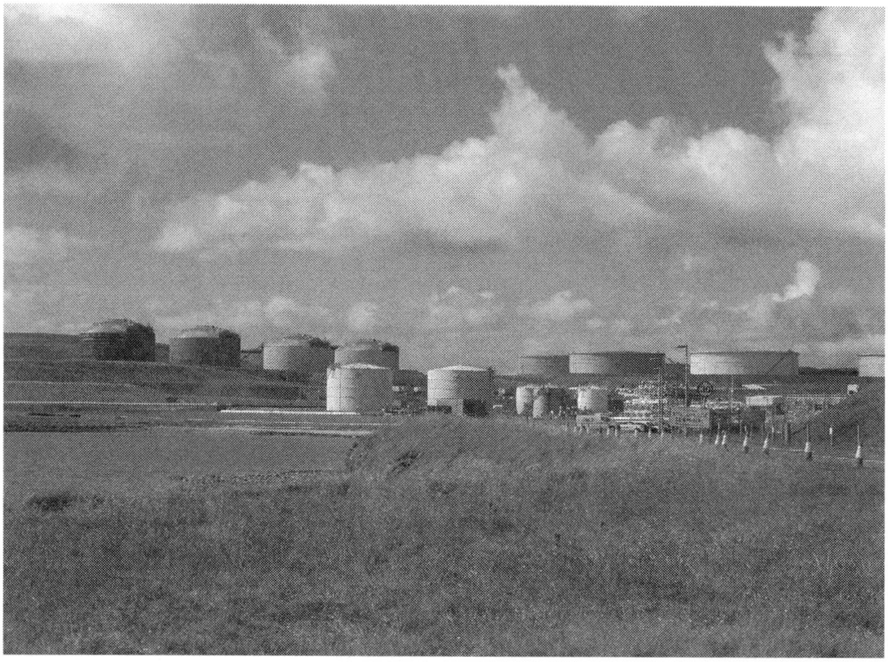

Figure 4.1 Sullom Voe terminal tanks, Shetland. Photograph by Mat Tuck, 2016.

permanent population grew from 17,000 to 23,000 between 1971 and 1981. The island's population swelled with up to 7,000 temporary workers, peaking at 30,000. The increase in the number of permanent residents proved enduring.[28] Graham Smith was among those who benefited from employment at Sullom Voe. As a student at Stirling University, he spent 'a long summer' working for the civil engineering contractors, Christiani & Nielsen, during the early 1980s. After his last stint in Shetland, Graham 'lived well for a year' on the money he earned and was able to buy a coveted hi-fi system and speakers.[29]

Sullom Voe changed the lives of Shetlanders too. Drew Ratter described growing up engrossed in 1960s transatlantic pop and rock music. Shetland experienced modernity and economic growth, encouraged by prosperous markets for its traditional knitwear and fish processing industries. These changes brought alterations to the landscapes, which included the expansion of fish processing plants. They were part of the larger process of modernisation in Northern Scotland, underpinned by support from the Highlands and Island Development Board (HIDB), which was formed in the mid-1960s under Harold Wilson's Labour government.[30] Nonetheless, Drew emphasised that Shetland remained a relatively low-wage economy before 1974, a 'tremendously [important] key date. I often say [that] in 1974, everything changed. Just totally. That was the pivotal year.'[31] Finding himself between jobs, Drew signed on at the Labour Exchange:

> Somebody I knew there sent me to go and see a company called Land and Marine, who needed somebody to go and drive their van. So, I went and drove their van. I'd left a job where I think I was making about seventeen pounds a week and immediately started making a hundred pounds a week. And it got significantly better than that later. The really fundamental thing as I said was anybody that wanted a job could have had two.[32]

The impact of oil remained visible in Shetland and withstood the end of the construction boom following the completion of the terminal in the early 1980s. Positive outcomes included long-term increases in the population and employment along with the benefits that oil-related activity brought to municipal budgets. These are related to the favourable conditions that the Shetland Islands Council (SIC) achieved in negotiations with both the British government and the oil companies grouped in the Sullom Voe Association (SVA), with BP responsible for terminal construction and subsequent operations.[33]

By the early 1980s, a conclusive sense of transformation was attached to the economic changes brought to Easter Ross and Shetland by influxes of workers and capital. The SIC published Elizabeth Marshall's *Shetland's Oil Era* in 1981 to accompany the formal opening of the oil terminal at Sullom Voe by Queen Elizabeth II. A foreword by David Howell, the Secretary of State for Energy, celebrated the opening of both the Brent and Ninian pipelines. Production at the Ninian field relied on platforms built at Nigg. Howell hailed their significance along with the 'twin pipelines [which] now converge from afar, arteries carrying the life blood, oil, on which industrial society today depends.'[34] Marshall herself was a history graduate who found employment doing research and policy work for the SIC before she later launched an energy consultancy. Her trajectory is emblematic of the opportunities that oil created in Shetland. At the time, these opportunities extended beyond the immediate offer of employment in construction and the offshore industry. They also incorporated white-collar and public-sector roles, which oil income and population growth helped to sustain.

Absorption and Change

Oil development stoked concern among local political leaders, who feared that their localities lacked the capacity to absorb development at this scale and intensity. In 1974, James Hunter, a historian of the Highlands, crofting activist, and policy-maker, described the 'massive oil-related industrialisation' of the Cromarty Firth. It was, he argued, characterised by 'all the social problems generally associated with rapid industrial growth: distortion of local labour force, large-scale in-migration, lack of social service provision, land speculation, [and] disruption of the natural environment.'[35] These were marked by a much quicker pace and a larger scale compared to the changes associated with the hydroelectric dam, the schemes that the Board sponsored in Shetland or even the Invergordon smelter. Similar fears shaped responses in Shetland. In 1974, a formative year, Edward Thomason, who had recently been the Council Convener, wrote to Ian Clark, the Clerk of the

Council, who played a central role in the terminal negotiations. Thomason was concerned about whether Shetland 'could emerge from this period of industrial development [of] a native economy so ill-balanced or rundown that there would be no real basis on which to deploy the Reserve Fund.' This fund would be accrued by the Council from oil industry-induced disturbance payments, rents, and through municipal control of the ports.[36] Reflecting on this period half a decade later, anthropologist Anthony P. Cohen characterised Shetland as experiencing an 'exercise in sudden industrialisation,' which obligated a reckoning even before the full extent of terminal construction or offshore activity had become clear.[37]

There is an evident tension between the broadly positive retrospective accounts of former rig builders and terminal workers and these contemporary political and scholarly perspectives. The relationship between them, however, can be understood in the determination of politicians like Thomason to ensure that Shetland was not overwhelmed by the interests of the oil companies and the British government's priorities. An understanding of a traditional culture under threat became a political resource for Shetlanders in negotiating the threats and opportunities created by oil developments. It was marshalled whilst navigating economic change and forging a sense of locality, which spanned the pre-oil period to include incomers who were happy to be regarded as part of the communities being changed by powerful economic forces.

The prospect and realities of oil-fuelled industrialisation were experienced as hopeful and even liberating by incomers and some longstanding residents of Easter Ross and Shetland but as a threatening imposition by others. Both positions centred on the long-term economic, social, and environmental impacts of oil. In 1969, a long-time resident wrote to Willie Ross, the Secretary of State for Scotland and champion of the HIDB, from her home in Balintraid, a settlement that underwent major expansion associated with the oil industry. She complained about plans for oil development on the Cromarty Firth, objecting that 'the contractors with their giant earth-moving equipment are tearing the heart out of Inverbreakie farm near Invergordon.'[38] Opposition was encouraged by hysterical predictions of population growth embraced by the County Council, which released plans catering for population growth up to 250,000 people. This fivefold increase was modelled on earlier experiences of industrialisation in the Scottish coalfields.[39]

By contrast, a 1973 study conducted by the Crofters Commission, a government agency, favoured the opportunity provided by oil investment, condemning the prevailing perspective:

> Whether explicitly or implicitly, [the view is] that the Highlands and Islands should remain as they are for the enjoyment of urban holidaymakers and nature lovers: with the inescapable corollary that the crofting population must make do with a depressed standard of living or migrate in search of better opportunities elsewhere.[40]

The Commission suggested that sites such as Nigg Bay, which had been enjoyed by holidaymakers from industrialised areas in Scotland, ought first and foremost to serve the economic needs of local people. If that meant sacrificing amenity to some extent to encourage economic sustainability in an area that for centuries had suffered underdevelopment, it was a price worth paying.

There were class implications to these statements, with those advocating crofting interests communicating the perspective of small farmers struggling to sustain their livelihoods. These sentiments were aptly demonstrated when a crowd of 700 people gathered at Nigg in January 1972 to watch a film showing hosted by Phillip Southwell, a petroleum engineer and the president of Brown and Root UK. An unemployed man who travelled from Tain, ten miles to the north, told the *Press and Journal* that he had found the film hopeful for the future of Easter Ross given that the area was badly in need of jobs. He emphasised that 'this crowd proves just how hungry the area is for work.'[41] The fulfilment of these hopes was demonstrated in the career of Jerry Strang who had qualified as a hotel manager and subsequently managed hotels in the Highlands, Aberdeenshire, and in England. Yet, Jerry's trajectory demonstrates the logic in the conclusion of the Crofters Commission that tourism did not hold out a viable economic alternative for the Highlands. Upon starting as an entry-level tradesman's helper at Nigg in 1980, Jerry found that he was earning 'far more' than he had in the precarious tourism trade. He rose to become the 'time man' at the yard, responsible for measuring performance and workers' completion of tasks. After this, Jerry worked offshore in the North Sea and the Middle East.[42]

In Shetland, the terminal companies were anxious to demonstrate that they offered employment conducive to the archipelago's traditions and economic structures built around farming, fishing, and knitwear. Pressured by local political concerns over both the durability of oil and the damage it could do to Shetland's longer-term economic prospects, the companies made accommodations. In 1977, union representatives argued that 'a particular problem arose with local operatives who run crofts', requesting that consideration be given to them for necessary absence without penalty by the building firms. It was agreed to 'allow locals to have a weekend off work at the same frequency as periodic leave arrangements for travelling men, without penalty.'[43] Subsequently, *Shetland Oil News*, which was funded by the industry, reported that shift work allowed fishermen and crofters to continue their occupations.[44]

Friction was generated by the change that the terminal brought to the landscape, which was magnified by the highly visible environmental impact of its operations. In May 1984, John Williamson, who lived in Ollaberry, northwest of the terminal, reported 'a huge black cloud' over the strait between the Mainland and the North Isles. Williamson, a former merchant navy seaman, wrote to the Industrial Pollution Inspectorate suggesting that BP ought to be subjected to the same penalties applied to owners of polluting ships. He bemoaned the impact of oil, reflecting that 'before the oil came to Shetland, we had lovely, clean fresh air, but I am afraid that isn't the case any longer.'[45] The smoke was a threat to Williamson's greenhouses, where he had previously been able to grow figs, nectarines, cherries, grapes, and plums.[46] The Inspectorate warned BP that they had breached their operating agreement by emitting black smoke for longer than the agreed four-hour window, which could provide grounds for prosecution.[47] BP officials were apprehensive. The terminal manager worried that the Inspectorate's involvement could encourage hostile 'press campaigns' and public opposition.[48]

These sensitivities were reinforced by the company's dependence on the archipelago. Shetland's position in relation to North Sea oil discoveries, which were concentrated in the East Shetland basin, created a powerful logic, reinforced by the technologies available in the 1970s. Economic necessity incentivised the building of a terminal to serve the East Shetland oil fields as an alternative to piping oil to the Scottish mainland or offshore loading onto tankers.[49] Alastair Hamilton arrived in Shetland in the late 1970s to work in the field of economic development after completing his studies at Oxford University. He recalled that Shetland was able to deal with the oil companies on the terms that it did during the 1970s because:

> The technological situation was different too. A lot of people would say if all [of the] East Shetland [basin] was discovered now, it would have gone straight to Peterhead [on the Scottish mainland] because the technologies for pipelines etcetera are different. And offshore loading as well. I mean all of that from Shetland's economic standpoint. It all came at exactly the right time.[50]

Alastair joked that, having earlier completed a master's dissertation on Shetland and oil, he had contributed to an influx of social scientists who followed the building workers:

> There was a joke going around [on] the south[ern] Mainland [island] in the late seventies that every household consisted of two adults, two children, and a social anthropologist. And we had a social anthropologist living with us. A guy from San Francisco. And at that time there were at least half a dozen others living in the south[ern] Mainland. From Belfast. All over.[51]

Historian of Shetland, Ann Black, has critically assessed the impact of this interest. She argues that a view of Shetland as an isolated 'peripheral' location not only misrepresents the history of seafaring and cosmopolitan trade, but it also overlooks the negotiating position of the SIC and the bargain it was able to drive with the central government and the oil companies. Fluctuations, such as the nineteenth-century herring trade and both World Wars, led to periodic upswings in economic activity in Shetland, which, as an Atlantic archipelago, has a long history of mercantile activity. Nevertheless, the portrayal of Shetland as an isolated set of islands newly in the glare of big business was a powerful rhetorical device, which helped the SIC to secure uniquely favourable terms from the oil industry and the British government.[52]

Shetland's political and cultural autonomy strengthened the hand of the SIC in its negotiations with the oil companies.[53] Directly comparing Shetland with Nigg, Alastair Hamilton argued that Shetland had

> a different psychology, I guess. I grew up in Inverness, and I know Easter Ross fairly well. And I can't imagine folk in Easter Ross acting like folk here [in Shetland] did. We're an island, and I think that's kind of fundamental to the psychology of the whole thing. We're an identifiable [group]. We're not part of a larger whole.[54]

One outcome of the SIC's bargaining was the housing provided to workers in camps and aboard accommodation ships to reduce the impact of the construction site.[55] The camps and the ships were largely a temporary addition to Shetland's landscape, just as the resident workforce was to its population. Jacqueline Birnie, who grew up in Mossbank, recalled how one of these camps in this vicinity was dismantled during the early 1980s. Jacqueline's grandfather worked to maintain roads and also kept a croft, and her extended family included other close male relatives who worked in construction and skilled trades jobs; none of them worked in the oil industry. Previously a small settlement comprising just six crofts, Mossbank became a major population centre as the site of new council housing built for longer-term residents who had found operations-focused jobs at the terminal. Some buildings from the camp era have survived, including the Fraser Peterson Centre, which was originally built for camp workers but repurposed as a leisure centre. Jacqueline reflected on the difficulties of spending her childhood in Mossbank. She 'felt very much an outsider in my own home,' surrounded by the children of oil workers who 'came and went.' Jacqueline was even excluded from activities at the Fraser Peterson Centre, which primarily catered to terminal workers and their families. Nonetheless, she was sanguine about these changes, saying, in relation to the oil influx, that 'we would be a much smaller population without that.'[56]

Sullom Voe changed Shetland's gender balance, too. Before the discovery of oil, it was a society marked by male seafaring and therefore a female majority onshore. Since the 1970s, however, men have outnumbered women in the archipelago.[57] These transformations were fictionalised in Robert Alan Jamieson's novel *Thin Wealth*, which features the story of an old sailor who reluctantly gives up his job navigating the world with the merchant navy to take up a new job in local oil shipping.[58] James Paton's family story resonates with Jamieson's fiction. His father had been a whaler, sailing with Norwegian crews in the Southern Ocean 'feeding starving Europe' during and after the Second World War. As a young man, James found work at the terminal site. He recalled hauling huge sacks of salt, each weighing 50 kg, 14 hours a day and earning what he regarded as very high wages in return. The money supported James' studies in chemistry at Robert Gordon's Institute of Technology in Aberdeen. After he dropped out of the programme, he found work at the now operational terminal as a lab technician in 1983, again commanding a good salary of around £13,500, which was more than one and a half times the average salary of a British man at the time.[59] Similarly, Mathew Nicholson described his father's work at Sullom Voe as a means to an end, which allowed him to live in Shetland, where his parents were from. As a 'shift chemist,' Mathew's father worked long, demanding, and irregular hours: 'He absolutely hated the place. He stuck at it because it's a good job and it's well paid.'[60]

Oil also created opportunities for women in and beyond Shetland. Rosa Stepanova moved permanently to Shetland from West Germany not long after she and a friend had hitchhiked there from northern England in the mid-1970s and were dropped off by a lorry driver who was transporting material to the Sullom Voe site. Rosa considered the influx of male workers (known colloquially as 'bears') to have contributed to an atmosphere of 'sexual liberation' which engulfed young women on the islands. However, this also had a darker, more coercive and violent dimension, which she observed at the terminal canteen where she worked:

> There was a strange sort of ritual; it happened after every lunchtime when some of the bears would stay behind, and the kitchen ladies would sit round the table and then they would throw a little something at each other. And it always ended in, well, what nowadays we call a sexual assault. And some of it was consensual ... like the manager in the canteen, she was one of the women who encouraged this. But I was terrified. ... And there were young girls, and it usually ended up that, you know, one of these, or two, or three, might come in and one of the women on the table, and sort of putting hands up their skirts, and it was just, it was so vile.[61]

Other interviewees too recalled the highly gendered and sexualised atmosphere in the camps. As a student, who was also active in socialist and feminist politics at Glasgow University, Isabel Mitchell took up a summer job as a chambermaid during the early 1980s. She remembered that the camps felt 'very rough and male,' but also that she reacted against the worst instances of objectification and sexism. During her final shift, she recalled that she 'just got my big bin bag and I took the lot out, emptied every room of all the pornography, in the bin and took it to the skip. I took the lot.'[62]

Employment in the oil industry furthered some women's independence, providing a release from the drudgery of the home-based knitting industry along with superior earnings. For example, in 1977, the school-aged daughter of BP's terminal site manager was able to earn a weekly wage of £90.90 cleaning the Sullom Voe camps, which was more than double the earnings of some SIC employees.[63] Doreen Thomason pursued a longer career at Sullom Voe. Her mother was a home-based knitter on Fetlar, one of the North Isles, but Doreen found work as a typist at Sullom Voe in the early 1970s, not long after she became a teenage single mother. BP wages paid for her to take her children on holidays and enabled her to live comfortably whilst benefiting from company travel perks. Doreen reflected on the major differences between her lifestyle and that of her parents' generation: 'It's a totally different life. These people hardly ever went off the island.'[64] Oil transportation led to a tenfold increase in traffic at Shetland's main airport at Sumburgh. Additionally, in 1978, a new airport was opened at a former airfield from the Second World War era at Scatsta, close to Sullom Voe.[65] Scatsta subsequently closed in 2020 after oil traffic declined, and Sumburgh became the centre for helicopters supplying offshore bases.

A new physical and social landscape of camps, accommodation ships, and toxic masculinity also reshaped life on the Cromarty Firth. The cruise ships, Highland Queen and Hermes, were both launched at Fairfield's shipyard in Glasgow in 1930 but were repurposed to house HiFab workers. They were moored side by side, next to the graving dock. In February 1974, C. Manson, the Surveyor General at the Department of Trade and Industry, wrote in alarmed terms to the Scottish Office. HiFab saw the floating beds as an expedient but temporary way to deal with an influx of workers required to complete large contracts. Manson, though, was 'extremely worried for the safety of the men accommodated,' adding that the 'elderly ships' 'may well not comply with the current international requirements for sea-going passenger ships of their class.'[66] A report from the previous year pointed to missing fire axes aboard the ship, indicating that they had been taken by workers who had used them to smash fittings and doors on the ship.[67]

HiFab emphasised the long-term work opportunities at Nigg for Highlanders as well as permanent migrants. In 1975, R. A. Henderson, HiFab's company secretary, wrote to Jimmy Milne, the general secretary of the Scottish Trades Union Congress, affirming the company's commitment to Nigg. Henderson highlighted mutual reliance between Easter Ross and his firm. He noted that the fabrication yard was uniquely 'well suited to undertake the work that would provide continuity of employment and prosperity in this remote part of Scotland.' Conversely, 'dependence of the area on the continuity of work at Nigg is without parallel anywhere in the United Kingdom.'[68]

Eileen Wilson, the wife of Rab, who had let the toss of a coin decide their life trajectory, explained the gravity of demographic changes in the locality and her sense of identity, which was invested in her husband's job and their joint relocation to Nigg from Edinburgh:

> You were a Nigg worker's wife. There was a certain camaraderie there. At one stage, the incomers outnumbered the local population. … There was a playgroup started; it was started by incomers, and it was run by incomers, all voluntary, and it was there for preschools. But over the years, it gradually shifted. The local mothers started being involved. It was them integrating with us instead of us integrating [with] them.[69]

Eileen's reflections reveal that the extent of the population increase around Nigg had similar impacts to those experienced by Jacqueline Birnie in Mossbank. These were associated with changing geographies that transformed the landscape. New housing accommodated permanently relocated incomers who built new communities which mushroomed in much expanded settlements. A large segment of the settled population had arrived in the area attracted by the oil industry and depended on it for work. In 1979, after seven years of rig building at Nigg, the Scottish Trades Union Congress made representations to the Offshore Supplies Office, a government agency formed to secure North Sea work for British-based firms. The union confederation underlined the obligation to workers who had relocated on the promise of economic prosperity:

> A high proportion of the Nigg workers are emigres from Clydeside and elsewhere, who have permanently settled in the Invergordon area, it would be a particularly bitter social tragedy, as well as an economic problem, if this yard were to close.[70]

Nonetheless, mass redundancy struck Nigg in 1984. Employment at the yard reached its peak during the construction of the Tension Leg Platform for Conoco Inc. (Figure 4.2). By July 1984, 2,800 workers at the yard had been made redundant, with just 1,066 remaining. The yard continued to operate with fewer employees until 2000.[71] Scottish Office officials concluded that 'for skilled and unskilled workers alike, prospects are extremely poor.'[72] Liz Sangster remembered 'the big payoff' having a catastrophic impact in the surrounding area. The town of Tain, ten miles north of Nigg, where Liz grew up during the 1970s, went from being a 'boom town' to a 'dole town.' Her parents had separated years earlier after her father Willie's alcohol dependency became worse, facilitated by his earnings as a HiFab fabricator. As a teenager, Liz nursed Willie, who responded to redundancy by drinking even more heavily and ceasing to eat. Having lost his job in the spring, Willie passed away before the end of the year.[73]

The rhythms of activity in Shetland have been distinctive, characterised by flows of oil and gas rather than the cycles of platform contracts. Sullom Voe was shaped

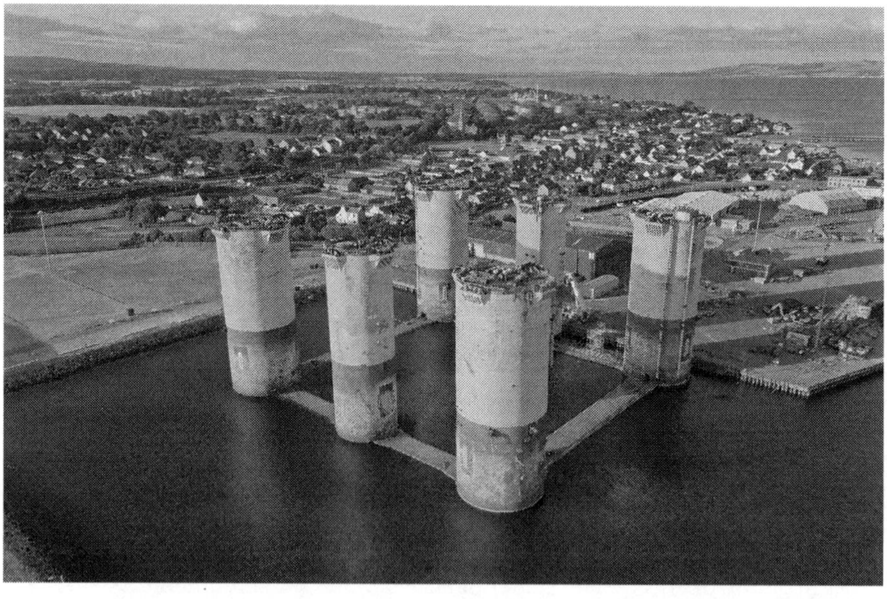

Figure 4.2 The Hutton Tension Leg Platform rig in the Cromarty Firth at Invergordon. The rig returned to the Cromarty Firth from 2009 before being dismantled in the early 2020s. Photograph by Glen Wallace, 2021.

by oil and gas prices and expectations relating to the production of the North Sea fields, which supply the terminal. Andrew Blackadder arrived in Shetland in 1980. He served as a trustee of the Shetland Amenity Trust, an organisation funded with an endowment originating in the revenues that SIC accrued from the terminal. In 2022, Andrew explained that the continuation of the oil industry into the twenty-first century defied expectations that the fields would be exhausted by 2000, but 'the oil industry basically threatened to leave Shetland, and they were going to get a much-reduced deal with the council to stay.'[74]

During the mid-1990s, the terminal's future operation was negotiated on much less favourable terms for the SIC, considering the lower oil prices and the ageing fields and infrastructure, which lowered throughput and increased unit costs.[75] Yet rising oil prices and the exploitation of the West Shetland Basin in the twenty-first century have given a 'new lease of life' to the terminal. The construction of a gas plant by TotalEnergies in the 2010s led to a surge in the workforce, with an influx of over 2,000 workers into Shetland, replicating dimensions of the earlier construction phase. This influx included construction workers from former industrial regions in Scotland, like Clydeside and Fife, and from North East England.[76]

Arthur Spence was the terminal manager during the gas plant's construction. He first visited Sullom Voe in the mid-1970s when he was a young worker installing portacabins. As the terminal manager during the 2010s, Arthur was charged with overseeing the handing over of the facility by BP to its current owners, EnQuest PLC. This process marked a shift in ownership from a major multinational oil and gas company to a smaller independent one. Arthur explained changes in how Sullom Voe was managed, indicating its shrinking profile and downgrading in terms of managerial authority and territorial significance: 'I was the last on-island terminal manager. After that, they made [*sic*] two operations managers on rotation, so they didn't stay on the island. One was in Norway and one was down in Aberdeenshire.'[77]

These changes mirrored a shift in oil labour patterns. In Andrew Blackadder's view, BP was characterised by a 'resident' relationship to Shetland, but they were replaced by companies who 'treated Shetland like an offshore oil rig. They'd just fly in for two weeks [and then] fly off; and they'd just accommodate people in hotels and then they'd fly off to where they came from.'[78] This shift can be distinguished from the earlier phase, which led to significant permanent migration to Shetland. These patterns of temporary residence were different, too. They were not marked by work camps and the enduring reshaping of geography of the Mainland during and after the terminal construction phase. However, it also reinforces the perspective of some of the earlier temporary workers. David Watt was an unemployed Glaswegian graduate with a teaching degree when he found work servicing the Sullom Voe camps. Before he set foot on Shetland, 'I think I just regarded it as offshore. I didn't fully twig what that was about.'[79] It was only after living on the islands that David fully understood that Shetland was in fact a long-settled place with a distinctive society and culture.

Conclusion

Hydrocarbon development led to lasting changes to Shetland and Easter Ross. The graving dock at Nigg and the huge oil and gas terminal at Sullom Voe can be seen as part of the North Sea oil and gas industry's contribution to rural modernisation in postwar Britain. These were major permanent changes to the landscapes of Northern Scotland, with lasting social and economic implications. Within a Scottish context, North Sea oil development extended prior debates over industrialisation in the Highlands, including the expansion of hydroelectricity and the development of nuclear activities in Dounreay. The North Sea industry made a larger impact over a more expansive geography but posed similar dilemmas between scenic beauty and conservation versus the economic benefits of energy production.

Rig manufacturing propelled a new Highland industrial labour market. Sullom Voe offered advantages for indigenous as well as incoming workers, including women who found temporary as well as long-term opportunities in the oil industry. The archipelago did not undergo an industrial depression like the one that swept Easter Ross in the 1980s. Cyclical oil-related economic activity was more manageable in Shetland because it was broadly tied up with the comings and goings of temporary workforces. Yet, as *Shetland Times* journalist Jonathan Wills pointed out, there were limitations relating to the settlement achieved by the SIC. By 1985, the Council had earned just three days' worth of the value of the 1.2 million barrels of oil then flowing daily through Sullom Voe.[80] The benefits of oil were distributed unevenly, with the profit from the North Sea overwhelmingly flowing to British Treasury coffers and the shareholders of companies headquartered far from Sullom Voe and Nigg.

Memories of Northern Scotland's oil-fuelled industrialisation during the late twentieth century constitute a living history that reveals how individual lives and community structures were remade in the context of an expanding new energy infrastructure. Labour mobility and population influx lie at the centre of individual and collective memories of the oil industry take-off during the 1970s. These movements helped to reverse long-term depopulation trends. Oil-related work also presented important economic opportunities for locals who had often been concentrated in traditional industries that paid lower wages. In the most intensely affected areas, there was a strong sense of upheaval and change, which was particularly marked in northern settlements of the Shetland mainland, such as Mossbank. Here, the influx of workers felt disorientating to locals, whilst new community structures primarily configured by incomers were also visible in Easter Ross. In each case temporary work camps and accommodation vessels became part of the landscape in the 1970s and 1980s as labour demand outstripped local capacity and housing. These social impacts compounded the losses experienced through environmental destruction around the Nigg graving dock and the anxieties of Shetland farmers induced by Sullom Voe's flaring. Yet, in the longer term, these changes proved integrative and comparatively inclusive. The more enduring presence of new housing for permanent residents who moved for employment in the oil industry signifies these changes. Oil employment helped to deliver on the hopes of local supporters such as the Crofters Commission, who saw in it an opportunity to obtain prosperity and a more robust economy.

These earlier histories are foundational for perspectives on a new energy transition in the Highlands and Islands that is once again transforming the local and regional economic and social fabric. Previous oil and gas experiences have proven formative to expectations, hopes, and disappointments associated with the shift towards a renewable energy economy, with both the Cromarty Firth and Shetland emerging as leading centres of Scottish wind power. Those outlooks demonstrate how transition, when given a quantified localised form, can be experienced as a form of continuity rather than as an existential change in energy-rich regions. Residents and workers see a strong but unfulfilled potential in the repurposing of the skills, infrastructure, and natural geography of the Cromarty Firth towards manufacturing turbines. They have been left disappointed that the former Nigg yard is used to inventory rather than produce the large offshore structures, which are playing an increasingly important role in Scotland's energy mix. During an interview recorded following the withdrawal of the yard's then owner, Global Energy Group, from the plan for a 'Nigg Offshore Wind' plant in the summer of 2022, a worker in his twenties said he was not optimistic about a future in turbine building 'if we are just going to be storing them.'[81] Confirmation in 2024 that Sumitomo Corporation intends to open a heavy cable factory employing more than 150 manufacturing workers is some consolation.[82] It will provide employment and activity closer to that associated with the MK Shand pipe works than the platform manufacturing pioneered at the Nigg yard. In 2025, Mitsui purchased the Port of Nigg from Global Energy, creating renewed hopes for an upturn in local wind turbine manufacturing.[83]

In Shetland, the controversy surrounding the Viking Wind Farm, a large onshore development, has centred on amenities and social disruption, indicating parallels with earlier concerns over Sullom Voe. These concerns also extend to control, ownership, and benefits. Local objections increased after a stake originally owned by the Council and then taken over by the Shetland Charitable Trust was divested, leaving the project entirely in the hands of private owners, who did not promise the same sort of sustained benefits included in the earlier Sullom Voe agreement.[84] Once again, the landscapes of the small archipelago of Shetland are evidently central to the objectives of big business and government. Prior experience conditions suspicion but also the hope and expectation that there ought to be rewards for the population of these islands from renewables, even though the energy will be exported, and most profits will go to multinational owners.

Notes

1 Rab and Eileen Wilson, interview with author, Alness, June 21, 2022.

2 Rab and Eileen Wilson, interview with author, Alness, June 21, 2022.

3 Perchard and Mackenzie, "Too Much," 4.

4 The term 'Shetland' as opposed to 'the Shetlands' or 'Shetland Islands,' used here to refer to the archipelago, reflects the terminological preferences of my interviewees.

5 Ross et al., "Introduction," 1–2.
6 Munro and Ross, "Contested Energy," 42–43.
7 Ross, "Dounreay," 97.
8 Black, "The Impact of External Shocks," 215.
9 Atabaki, Bini, and Ehsani, "Introduction," 1.
10 MacDonald, "Dynamic Positioning," 66.
11 Boyer, *Energopolitics*, 6.
12 Kemp, *Official History*, 665.
13 McCullough, "Resolving," 427–429.
14 Perchard, "A Little Local Difficulty?" 296.
15 Background Briefing for Meeting between Lord Gray, Scottish Office, and Mr Dick Morris, Chairman of Brown and Root (UK) Ltd, April 13, 1984, National Records of Scotland, Edinburgh (hereafter NRS), Scottish Economic Policy (hereafter SEP) 4/4667.
16 "Redundancy Warning to 3000 at Oil Yard," *Glasgow Herald*, February 7, 1984, NRS, SEP 4/4667.
17 Mather (eds) *Third Statistical Account*, xviii.
18 Taylor, *Oilwork*, 27.
19 Brown and Root-Wimpey, Highland Fabricators Ltd: Nigg Fabrication Yard Progress Report, July 1973, Glasgow Caledonian University Archive Centre (hereafter GCUAC), Records of the Scottish Trades Union Congress (hereafter STUC), Box 396, Visit to Nigg Bay folder.
20 Pratt et al., *Offshore Pioneers*, 224.
21 Craig MacKay, interview with author, Brora, June 23, 2022.
22 Craig MacKay, interview with author, Brora, June 23, 2022.
23 John Armstrong, *Sea Area Forties* (1976), National Library of Scotland Moving Image Archive, 4225.
24 Taylor, *Oilwork*, 24.
25 John Mackenzie, interview with author, Uddingston, June 14, 2022.
26 Margaret Patterson, remote interview with author, July 22, 2022.
27 P.H. Grattan to Dickson Mabon, Subject: Sullom Voe, 7 December 7, 1977, The National Archives, Kew (hereafter TNA), POWE 63/1758.
28 Hill et al., "Oil Development," 21–23.
29 Graham Smith, "Sullom Voe (Shetland) in the 1980s"; Graham Smith, remote interview with author, September 1, 2022.
30 Wills, *A Place*, 11.
31 Drew Ratter, interview with author, Northmavine, June 29, 2022.
32 Drew Ratter, interview with author, Northmavine, June 29, 2022.
33 *Shetland's Oil Monies and Their Governance* (SIC, 2008), Shetland Archives, Lerwick (hereafter SA), 4.5315.
34 Marshall, *Shetland's Oil Era*, 1.
35 James Hunter, "The Social Impact of Oil-Related Industrialisation in the North of Scotland, Easter Ross: Initial Survey," March 10, 1975, SA, D53/2/1/4/6/3.
36 Edward Thomason to Ian R. Clark, August 19, 1974, SA, D53/2/1/4.1; Thomason, *Island Challenge*.
37 Cohen, "Oil and Cultural Account," 164.
38 Jean Forsyth, to Willie Ross, January 9, 1969, NRS, DD12/3241.
39 Community and Resource Planning Unit, *An Investigation*, 59.
40 Crofters Commission, North Sea Oil Developments and Their Possible Effect on Crofting, October 23, 1973, Inverness, GCUAC, STUC, Box 396.

41 "Fantastic Scenes at Ross Hall as Rig Firm 'Shows the Flag,'' *Press and Journal*, January 25, 1972, SEP 4/2436.
42 Jerry Strang, interview with author, Dornoch, June 24, 2022.
43 Meeting at Lerwick Hotel, May 22, 1977, TNA, POWE 63/1756.
44 "The Right People for the Job," *Shetland Oil News* no.6 January 19, 1979, 3, SA, D21/22/3.
45 Letter to G. Stott, Her Majesty's Industrial Pollution Inspectorate, Edinburgh, from John Williamson, Lubba, Ollaberry, May 1984, NRS, Scottish Development Department (hereafter DD) 9/534.
46 Drew Ratter, "Black Reek," *Shetland Times*, NRS, DD9/534.
47 BBC Radio Shetland – Good Evening Shetland, May 24, 1984, NRS, DD9/534.
48 C. J. Dilloway, Shetland, to G. Stott, Edinburgh, May 28, 1984, NRS, DD9/534.
49 Fenwick, "Shetland Islands," 226–227.
50 Andrew Blackadder and Alastair Hamilton, interview with author, Lerwick, July 1, 2022.
51 Andrew Blackadder and Alastair Hamilton, interview with author, Lerwick, July 1, 2022.
52 Black, "The Impact of External Shocks," 5, 232.
53 Nicholson, "These Groups of Islands."
54 Andrew Blackadder and Alastair Hamilton, interview.
55 P. H. Grantham, "Visit to Sullom Voe by Mr Walker," August 2, 1977, NRS DD9/533; Wills and Arthur, "Sullom Voe 1981," 6.
56 Jacqueline Birnie, remote interview with author, July 29, 2022.
57 Abrams, *Myth*, 1–3.
58 Jamieson, *Thin Wealth*.
59 James Paton, interview with author, Lerwick, June 29, 2022; Office of National Statistics, "Average salary between 1980 and 1997."
60 Mathew Nicholson, interview with author, Lerwick, July 1, 2022.
61 Rosa Stepanova, interview with author, Lerwick, 29 June 2022.
62 Isabel Mitchell, interview with author, Lerwick, July 1, 2022.
63 P. H. Grantham, "Visit to Sullom Voe by Mr Walker," August 2, 1977, NRS DD9/533.
64 Doreen Thomason, interview with author, Lerwick, June, 29 2022.
65 Black, "The Impact of External Shocks," 228–229.
66 C. Manson, Surveyor-General, DTI, to W.K. Fraser, Scottish Home and Health Department, Subject: Dormitory Ships at Nigg Bay, February 25, 1974, NRS, SEP 1/79.
67 E.W. Macintyre, Northern Area Fire Brigade to J. J. Haughney, Scottish Home and Health Department, October 30, 1973, NRS, SEP 1/79.
68 R. A. Henderson to Jimmy Milne, November 20, 1975, GCUAC, STUC, Box 396, North Sea folder.
69 Rab and Eileen Wilson, interview.
70 Papers for presentation to Offshore Supplies Office for Meeting to be held on Thursday July 26, 1979, GCUAC, STUC, 303 Energy Sub Committee.
71 Energy Voice, "The History of the Nigg Yard."
72 HiFab Redundancy Notice, May 3, 1984 and HiFab Redundancy Notice, July 18, 1984, both in NRS, SEP 4/4667.
73 Liz Sangster, interview with author, Glasgow University, May 18, 2022.
74 Andrew Blackadder and Alastair Hamilton, interview.
75 *Shetland's Oil Monies and Their Governance* (SIC, 2008), SA, SA 4.5315.
76 Highlands and Islands Enterprise, Impact Analysis.
77 Arthur Spence, interview with author, Glasgow University, July 20, 2022.
78 Andrew Blackadder and Alastair Hamilton, interview.

79 David Watt, interview with author, Glasgow University, May 31, 2022.
80 Jonathan Wills, "Ten Years of Oil = Three Days' Worth for Shetland," *Shetland Times*, January 3, 1985, NRS, SEP 1/740.
81 Global Energy Group worker, remote interview with author, June 22, 2022.
82 Thomas, "Sumitomo commits to £350m."
83 BBC News "Japanese investor completes takeover of Highland port"
84 Marter, "Change of Viking Energy Ownership."

Bibliography

Abrams, Lynn. *Myth and Materiality in a Woman's World: Shetland, 1800–2000.* Manchester: Manchester University Press, 2010.

Atabaki, Touraj. Elisabetta Bini, and Kaveh Ehsani. "Introduction." In *Working for Oil: Comparative Social History of Labour in the Global Oil Industry*, edited by Touraj Atabaki, Elisabetta. Bini, and Kaveh Ehsani, 1–10. London: Palgrave Macmillan, 2018.

BBC News, "Japanese Investor Completes Takeover of Highland Port", *BBC News*, July 30, 2025. www.bbc.co.uk/news/articles/c3ez8d5jnygo.

Black, Barbara A. "The Impact of External Shocks upon a Peripheral Economy: War and Oil in Twentieth Century Shetland." PhD diss., University of Glasgow, 1995.

Boyer, Dominic. *Energopolitics: Wind and Power in the Anthropocene.* Durham: Duke University Press, 2019.

Cohen, Anthoy P. "Òil and Cultural Account: Reflections on a Shetland Community." In *The Social Impact of Oil in Scotland: A Contribution to the Sociology of Oil*, edited by Ron Parsler and Dan Shapiro, 164–176. Westmead: Gower Publishing, 1980.

Community and Resource Planning Unit, *An Investigation into Industrial Development in Easter Ross and the Oil Refinery at Nigg Point*. London, 1976.

Energy Voice. "The History of the Nigg Yard." *Energy Voice*, October 19, 2011. www.energyvoice.com/renewables-energy-transition/19530/the-history-of-the-nigg-yard/.

Fenwick, J. M. "The Shetland Islands and the Impact of Oil." In *Onshore Impacts of Offshore Oil* edited by William J. Cairns and Patrick M. Rodgers, 226–234. London: Applied Science Publishers, 1981.

Highlands and Islands Enterprise. Impact Analysis of Sullom Voe Terminal and Job-Related Reductions, 2015. www.hie.co.uk/media/6430/impactplusanalysisplusofplussullomplusvoeplusterminalplus-a2565095.pdf.

Hill, Archie E., Carole L. Seyfrit, and Mona J. E. Danner. "Oil Development and Social Change in the Shetland Islands, 1971–1991." *Impact Assessment and Project Appraisal* 16 (1998): 16–25.

Jamieson, Robert A. *Thin Wealth: A Novel from an Oil Decade.* Edinburgh: Polygon, 1986.

Kemp, Alex. *The Official History of North Sea Oil and Gas: Vol.1: The Growing Dominance of the State.* London: Routledge, 2013.

MacDonald, Graeme. "Dynamic Positioning: North Sea Petroculture's Backwash." In *Cold Water Oil: Offshore Petroleum Cultures*, edited by Fiona Polack and Danine Farquharson, 61–96. London: Routledge, 2021.

Marshall, Elizabeth. *Shetland's Oil Era.* Lerwick: Shetland Islands Council, 1981.

Marter, Hans J. "Change of Viking Energy Ownership Had no Impact on Land Leases." *Shetland News*, September 11, 2020. www.shetnews.co.uk/2020/09/11/change-of-viking-energy-ownership-had-no-impact-land-leases/.

Mather, Alexander S. (eds) *The Third Statistical Account of Scotland: The County of Ross and Cromarty*. Edinburgh: Scottish Academic Press, 1987.

McCullough, Katie L. "Resolving the 'Highland Problem': The Highlands and Islands of Scotland and the European Union." *Local Economy* 33 (2018): 421–437.

Munro, Alan, and Duncan M. Ross. "Contested Energy: A Long-Term Perspective on Opposition to Renewable Power Developments in Scotland." *Scottish Affairs* 75 (2011): 37–61.

Nicholson, Mathew. "'These Groups of Islands are Different': Identity and Constitutional Change in Orkney and Shetland, 1966–1990", *Revue LISA* 21 (2023). https://doi.org/10.4000/lisa.15558.

Office of National Statistics. "Average Salary between 1980 and 1997." www.ons.gov.uk/aboutus/transparencyandgovernance/freedomofinformationfoi/averagesalarybetween1980and1997.

Perchard, Andrew. "A Little Local Difficulty? Deindustrialization and Globalization in a Scottish Town." In *The Deindustrialized World: Confronting Ruination in Postindustrial Places*, edited by Steven High, Lachlan MacKinnnon and Andrew Perchard, 284–312. Vancouver: University of British Columbia Press, 2017.

Perchard, Andrew, and Niall Mackenzie. "'Too Much on The Highlands?' Recasting the Economic History of the Highlands and Islands." *Northern Scotland* 4 (2013): 3–22.

Pratt, Joseph A., Tyler Priest, and Christopher J. Castaneda. *Offshore Pioneers: Brown & Root and the History of Offshore Oil and Gas*. Amsterdam: Elsevier, 1997.

Ross, Linda M. "Dounreay: Creating the Nuclear North." *The Scottish Historical Review* 100 (2021): 82–108.

Ross, Linda M., Katrina Navickas, Matthew Kelly, and Ben Anderson. "Introduction." In *New Lives New Landscapes Revisited: Rural Modernity in Britain*, edited by Linda M. Ross, Katrina Navickas, Matthew Kelly, and Ben Anderson, 1–14. Oxford: Oxford University Press, 2023.

Smith, Graham. "Sullom Voe (Shetland) in the 1980s," *Graham Smith*, https://grahamsmithprofile.wordpress.com/other-life/sullom-voe-shetland-in-the-1980s/.

Taylor, Sue J. *Oilwork: North Sea Diaries*. Edinburgh: Birlinn Ltd., 2005.

Thomas, Allister. "Sumitomo Commits to £350m Highlands Cable Factory as Funding Package Agreed." *Energy Voice*, February 23, 2024. www.energyvoice.com/renewables-energy-transition/wind/uk-wind/scotwind/548492/sumitomo-highlands-cable-factory/.

Thomason, Edward. *Island Challenge*. Lerwick: *The Shetland Times*, 1997.

Wills, Jonathan. *A Place in the Sun: Shetland and Oil, Myths and Realities*. St John's: Memorial University Press, 1991.

Wills, Jonathan, and Erik Arthur. "Sullom Voe 1981: A *Shetland Times* Special Colour Report," *Shetland Times*, May 15, 1981.

5 The Cultural Landscapes of Dam Building in Switzerland

Secondary Infrastructure and its Territorial Archive at the Grande Dixence Dam, 1950–1965

Rune Frandsen

> Yesterday, at 17:15, our canton was struck to the core by a terrible catastrophe. The bottom [part] of the Allalin Glacier, which is very steeply-sloping and hangs over a cliff overlooking Mattmark, collapsed onto the excavation site of the dam under construction. It is known that more than 800 workers are currently working for Elektrowatt and [operate] under the orders of the Mattmark Dam Construction Consortium. Most of these workers live near the village of Saas Almagell. The construction site, which was pulverised by several hundred cubic metres of ice, consisted of several barracks with a canteen, a workshop, stores, two large repair sheds (for heavy equipment such as Euclid trucks, Trax, giant excavators, etc.), offices, crushers, and conveyor belts.[1]

The Mattmark Dam bars the remote Saas Valley in the alpine canton of Valais in Switzerland. A cross is located here at a stone's throw from the restaurant and visitor centre, engraved with the inscription: 'TOTE UNTER DEM GLETSCHER, LEBENDIGE IM GEDÄCHTNIS' (Figure 5.1) [2] It memorialises the 88 victims who died when the Allalin Glacier, overlooking the Mattmark construction site, collapsed onto the workers' barracks on August 30, 1965. A majority of the workers were *saisonniers* (seasonal workers) from Italy. According to reports on the incident, had the glacier collapsed half an hour later, up to 700 lives could have been lost.[3]

Roger Bonvin, the Federal councillor overseeing the inquiry that followed the disaster, made the following statement: '[N]ever have those of my generation, or the previous one, experienced such a vast catastrophe.'[4] While this statement may be true, it does not take cognizance of the fact that the conditions for an accident of this magnitude had never before converged in such a way. The Mattmark Dam was only one among many such dams built in Switzerland in the decades following World War II. This ongoing industrial process of turning the Alps into spaces of extraction went hand in hand with the development of a swarming landscape of labour inhabited by those building the dams. Sending so many workers up into that environment only increased exposure to such risks.

The press coverage of the catastrophe was remarkable in its illumination of the time of construction as a specific moment of infrastructure failure. It has been argued that disruption is what suddenly renders infrastructure visible.[5] It draws

DOI: 10.4324/9781003473534-8

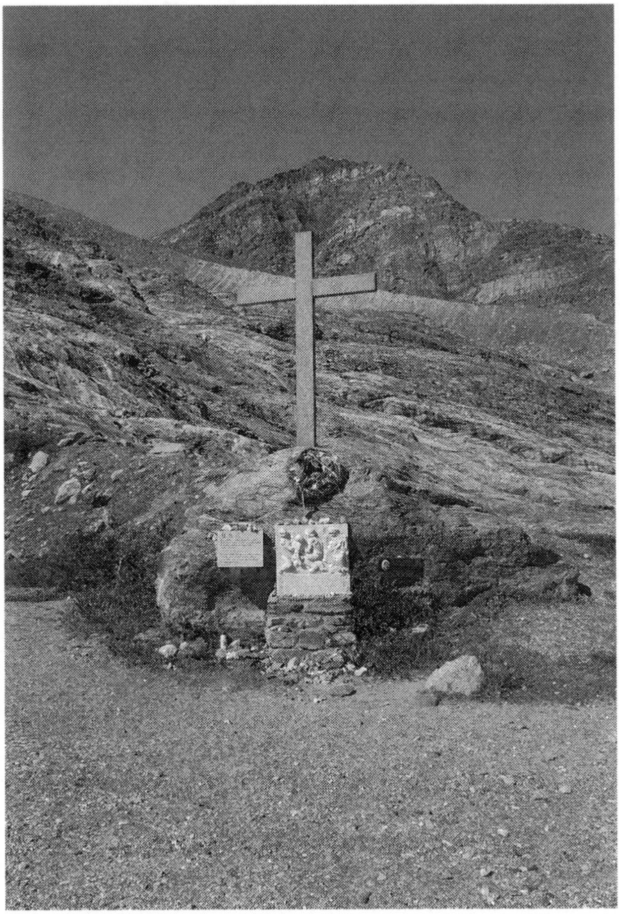

Figure 5.1 Memorial for the victims of the tragedy that occurred on August 30, 1965 at Mattmark in Switzerland. Photograph by the author, 2021.

attention to a materiality otherwise forgotten. The diagnosis of the disruption necessarily entails investigating the minute assemblages enabling the functioning of the infrastructure. The infrastructure rendered visible at Mattmark was not, however, that of the dam nor that of the accessory parts required for its correct functioning. Rather, the incident put under scrutiny what I describe in this chapter as *secondary infrastructure:* the workers who were building the dyke, the barracks they lived in, the machines they used, and the roads on which they travelled. This term draws attention to the spaces of temporary labour hidden in the shadow of the dam. The focus on the material conditions of the transformation of the Alps was, however, short-lived. During the trial that followed the tragedy, the workers, accused of biting the hand that fed them, saw the dismissal of their appeal against the acquittal of the construction companies.[6] The disaster, they were told, was unforeseeable, even though ice blocks had already come loose in the days preceding the event.[7]

The Mattmark memorial stands out as an oddity in the architectural heritage associated with the construction of large dams in Switzerland following World War II. There is little to commemorate the contributions of the workers employed to build the alpine dams. While large infrastructure projects like Mattmark—or the Grande Dixence Dam, further west in the Rhône Valley—are increasingly being listed and protected as monuments of national or cantonal importance in Switzerland, the material traces associated with their construction are either left to disintegrate within the landscape, or they are actively removed as part of 'renaturation' campaigns.[8] For instance, a study that analysed appraisals of four dams that were inventoried by the Office for Historical Preservation in the canton of Valais has shown that these listings connect the dams to canonical histories of construction and engineering at the expense of the workers.[9] Only on the listing for Mattmark are the workers mentioned. It would seem that 88 workers had to die for their contribution to be valued. The question that arises is were it not for the dams, what might the heritage of these workers have been had a tragic accident not occurred? Could there not have also been more positive monuments?

To address this question, this chapter seeks to uncover the landscapes of labour required for the construction of those dams as cultural landscapes. Perceived as temporary, most of the components of this landscape have now disappeared, and the workers have therefore been deprived of representation in the heritagisation process. To make this point, two conceptual framings are deployed in this chapter. The first entails contextualising the construction of large dams in Switzerland within a national strategy for achieving energy autarky and showing the recurring means, namely the secondary infrastructure, mobilised to build this energy infrastructure. An examination of the extent of these networks required for the seamless supply of material, labour, and capital reveals how the intense wave of dam building in Switzerland in the wake of World War II increased the human presence in these high alpine environments, thereby statistically increasing the risk of such accidents. Here, Lisa Björkman's notion of 'infrastructure as method' is useful for tracing the various networks enlisted for the construction of these great dams.[10] An examination of some of the key elements of this secondary infrastructure reveals that the building of dams was processual. Their construction at various sites did not occur as singular phenomena; rather, they were all interconnected through their use of common infrastructure.

The Grande Dixence Dam and its water-harvesting network, located less than 50 km from Mattmark, will serve as a case study. The tragedy of 1965 draws attention to a much larger phenomenon, best represented by this dam, given its sheer size. This 'bigger sister' dam to Mattmark is still the tallest concrete gravity dam worldwide and was built by Énergie Ouest Suisse (EOS), the largest supplier of electricity to western Switzerland at the time. The gigantic scale of the dam and of its adduction network, a network of tunnels augmenting the volume of water brought to the reservoir from neighbouring watersheds, necessitated a density of infrastructure that would accommodate virtually any possible configuration of interactions between the terrain and the newly built infrastructure. Furthermore, a recently published account of the tragedy by Toni Ricciardi, Sandro Cattacin, and Rémi Baudouï has shed light on the profit-oriented organisation of the construction site of Mattmark.[11] By contrast, the Grande Dixence Dam retains the aura of the heroic conquest of nature through Swiss savoir-faire, which merits unpacking and revision.

Next, building on the notion that infrastructures are relational,[12] I present a spatial reading of the assemblage of diverse networks of this secondary infrastructure, conceived as a *taskscape*[13] and as the lived space of the workers hired at these gigantic construction sites. The concept of a taskscape was first introduced in 1993 by Tim Ingold as a way of emphasising the interconnectedness between two spheres, nature and society, traditionally considered to be separate. Ingold conceptualised a taskscape as a mesh of interconnected actors, human and non-human, reacting and attuning to each other's actions and to the surrounding environment. The concept of a taskscape enables the bridging of two main strands of literature, which have tackled the social aspects of dam construction sites in Switzerland.

The first strand of literature is based on the established argument that the dam-building industry relied on tried and tested infrastructural networks to the extent that it could be considered an infrastructure in itself. Proponents of this logic in the recent literature include architectural historians. For instance, Marc Angélil and Cary Siress have critiqued *Opération Béton*, a short documentary directed by Jean-Luc Godard in 1958 and dedicated to the Grande Dixence Dam, which was built between 1953 and 1961. They note that Godard 'basically failed to see the dam as one specific concretization of more diffuse "discourse, institutions, regulatory decisions, laws, administrative measures, scientific statements." '[14] Sarah Nichols makes a similar argument in her investigation of the organisational arrangements required for the seamless delivery of cement for that same dam.[15] These authors echo an older scholarship, which highlighted the transnational and structural dimensions of hydropower transactions between Switzerland and Europe.[16]

The second strand of literature, which stems mainly from the fields of social studies and local history, provides accounts of high alpine infrastructure construction sites and of the often-temporary settlements formed for their construction.[17] A few of these studies focus specifically on post-World War II dam construction sites.[18] All of these studies, however, are limited by their localisation agenda. Consequently, they do not consider the wider logic or mobility behind the formation of these settlements.

If, as Angélil and Siress and Nichols argue, the dam construction sites were in fact repeatable occurrences supported by the same infrastructure, then so were the temporary settlements. The concept of a *taskscape* is used in this chapter to bridge the two strands of literature described above.

A focus on secondary infrastructure leads to the redefining of the temporal and geographical scales of the dam-building industry. This specificity of industrial modes of production becomes apparent from the movements traced in this chapter. The dam-building taskscape was not limited to Switzerland; its reach was transnational. As discussed in the second part of this chapter, which focuses on the people—largely 'guest workers' from Italy—inhabiting this taskscape, it has a long industrial prehistory in Switzerland prior to its expansion in the decades following World War II.

The chapter ends with reflections on what remains of this dam-building taskscape; reflections that draw on what Peter van Wyck has termed the *territorial archive.*[19] The traces that are left also importantly testify to the contributions of thousands of workers who were employed at the dam construction sites. They attest to the existence of this ephemeral landscape, which completely belies the highly romanticised images of the Alps, thereby rendering a more multifaceted image that shows what they can *also* be.[20]

Construction of the Large Dams of Valais

In the three decades following World War II, over 100 large dams above 15 m in height and impounding more than three million cubic metres of water were built in Switzerland.[21] This propulsion of infrastructure construction can be traced to a study by the Federal Bureau for Waters. Striving for energy autonomy, the Bureau published the results of the study in six issues between 1932 and 1945 titled *Switzerland's available hydropower, considered from the point of view of storage possibilities for winter power generation* (author's translation into English).[22] These writings provided guidelines for the country's hydroelectric future and steered the as-yet uncoordinated efforts of electricity producers towards a common goal of harnessing as much hydropower as possible. The study systematically investigated the potential of the country's remaining streams and rivers to generate hydropower, starting with the waterbeds closest to the industrial centres of northern Switzerland. The last three issues specifically targeted the 'white coal' contained in the glaciers of the Alps, focusing on the watersheds of Graubünden (1942), Ticino (1943), and Valais (1945), the three mountainous and southernmost cantons bordering Italy and France, which had the greatest untapped potential.

Three record-breaking dams were erected quasi-simultaneously in the canton of Valais. Towering at an impressive height of 285 m, the Grande Dixence was the tallest arch-dam at the time in Switzerland. The wall itself was completed in 1961 in a remote valley, the Val des Dix, and its water-harvesting network in 1965. Further east, the previously mentioned Mattmark Dyke, built between 1960 and 1967, was the largest, although not the tallest, rockfill-dam in the country.[23] These dams all operate in conjunction with extensive networks of water collection tunnels, known as adductions, which bring water from valleys, where dam construction is not possible, to neighbouring reservoirs. The water is then stored in the reservoirs until the onset of winter when they are turbined. The practical consequences of creating this new artificial hydrography catapulted thousands of workers into the Alps to pour the concrete of the dams; dig the adductions, which now link together most of the side-valleys of Valais; or assemble the machines transmuting water and gravity into electricity. For the construction of the Grande Dixence network alone, estimates place up to 25,000 workers at the construction site, with at least 30 temporary settlements erected in this territory according to the reach of this infrastructure.

This can be seen, for instance, in Figure 5.2, which shows at least three sites. Le Chargeur, located at the base of the dam, was the headquarters for this construction site. On the right side of the dam, at the level of the crowning, Blava was the main concrete production site. On the far right, the Prafleuri Valley was the main aggregate extraction site. It was here that the workers were housed, mostly in prefabricated wooden barrack-huts, which were dismantled upon completion of the construction.[24]

Networks were established to ensure the seamless provision of labour, energy, communication, and tools and materials required by the workers, team leaders, engineers, and multiple other tradespersons involved in the implementation of this monumental project. Roads, tunnels, bridges, cable cars, and power lines

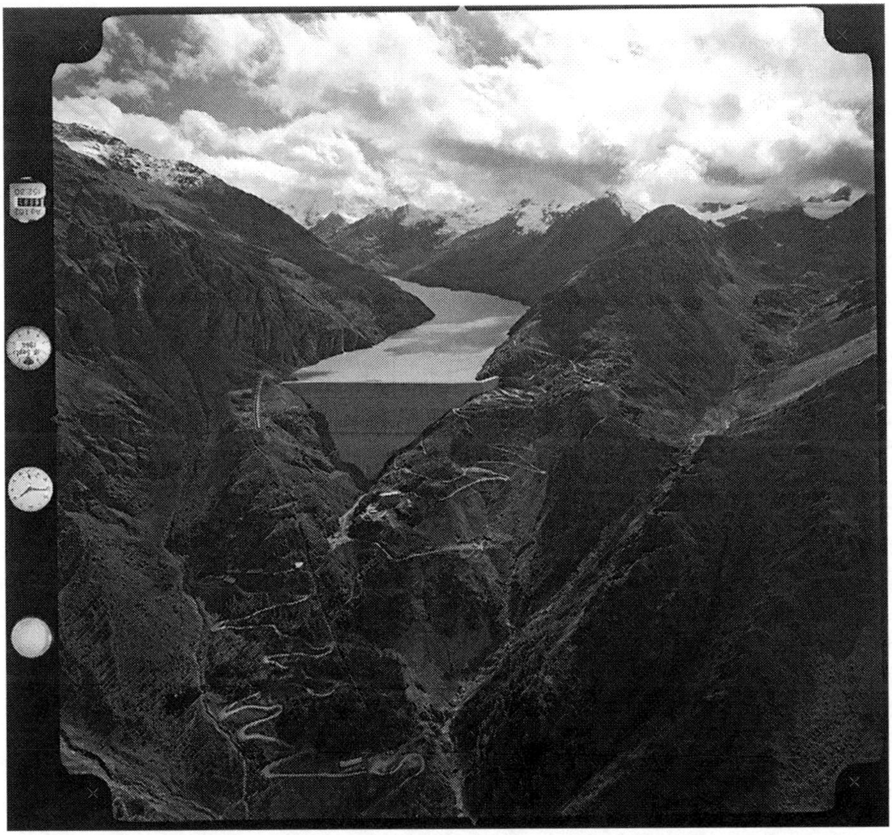

Figure 5.2 An aerial photograph of the Grande Dixence Dam built between 1953 and 1961 in Valais, Switzerland.

Source: ETH-Bibliothek Zürich, Bildarchiv/Stiftung Luftbild Schweiz/Fotograf: Swissair Photo AG/ LBS_P1-648985/CC BY-SA 4.0.

were planned to enable access to these remote locations. Thus, a swarming landscape developed in these precarious locations; one which had little to do with the traditional lifestyles associated with the imaginary of alpine valleys. In other words, two sets of infrastructure came about with the construction of the Grande Dixence Dam. The primary infrastructure included the dam itself, the tunnels, and the hydroelectric plants. The 'secondary infrastructure' comprised the various networks that made the construction of the hydroelectric network in its finished state possible.

Debra Johnson and Colin Turner define secondary infrastructure as 'the technologies that directly facilitate access to networked infrastructure.'[25] For the sake of precision, we could add that they are the technologies that directly facilitate access to *the functions* of networked infrastructure. Thus, they include vehicles for using the road infrastructure and computers for accessing the Internet. In the case of the hydroelectric network, they could include the water intakes of the various streams tapped by the adduction network. While the notion of primary infrastructure remains unchanged in this chapter, secondary infrastructure is conceived as an *infra*-infrastructure, that is, a substrate-providing material, transport, labour, and capital, all of which were required for the construction of the entire primary infrastructure of Grande Dixence.

Jean-Luc Godard's documentary, *Opération Béton* portrays the richness of the infrastructural networks required to construct the dam.[26] In what appears to be the first documentary that Godard directed, his camera follows the production of concrete from the excavation of moraine—material comprising a mass of rocks and sediment left behind by a moving glacier—below the Prafleuri Glacier all the way to its re-solidified form as it is poured in layers on what eventually becomes the Grande Dixence. The construction process followed a 'just in time' logic. Therefore, the constant pouring of concrete on the dam during the summer, referred to as concrete-pouring campaigns because of their military-like organisation, implied continuous staffing and operations at Prafleuri as well. About 315 people (including the workers and the people necessary for the settlement's operation) were stationed there.[27] Each day, a minimum of 15,000 tonnes of material had to be extracted from the moraines and processed within the cleaning and crushing facilities to obtain the aggregate for concrete production. A tunnel equipped with a conveyor belt, a funicular as well as water, gas, and electricity conduits served as an umbilical cord connecting Prafleuri and Blava, the site where concrete, which was poured on the dam, was produced.[28] In Blava, the aggregate would again pass through sorting and crushing facilities before being mixed with locally sourced water and cement shipped from Sion using two parallel 21 km-long cable cars. The concrete paste was transported on one of the four *blondins* (cable cranes), spanning the entire width of the valley above the dam to reach the pouring site. A total of six million cubic metres of concrete were poured on the dam under construction between 1953 and 1961.

While the Grande Dixence Dam is illustrative of the effects of such projects on the alpine landscape, it is important to bear in mind that more than 100 large dams were built during the three decades following World War II. Notwithstanding variations between the individual dam construction sites, the modus operandi of dam construction was predicated on the provision of heavy materials that were foreign to the valleys where they were built. Whether these materials were locally sourced or shipped from the valley floor, their procurement was only made possible through the superimposition of an industrial infrastructure on these rugged terrains.

The Taskscape of Dam Building

The secondary infrastructure at the Grande Dixence reveals that the individual construction sites were not supported by independent networks for the provision of labour or material; instead, many of these networks were shared among apparently distinct projects. This is a characteristic of infrastructure networks: they form relationships with their users, who rely on them and use them repetitively across time and space.[29] The interconnectedness of multiple secondary infrastructure networks can be conceived as an assemblage of all of the dam construction sites of that time linked together. Ingold's concept of a *taskscape* can help us to conceptualise this assemblage.

This concept stems from Ingold's discussion on the notion of landscape. He argues that rather than understanding a landscape as a finished result and a fixed image, it should be seen as a 'process of becoming of the world as a whole,' which is constantly shifting and evolving.[30] Thus, a taskscape aptly describes the assemblage of dam building given the selective reuse of parts of the existing infrastructure, the reconfiguration of existing infrastructure networks required for the construction sites and their adaptation to the needs of specific geographical locations, or the building of entirely new networks where necessary. Not all dam projects were built simultaneously, but construction sites gradually proliferated in the Alps. Over a 30-year period, a series of dam construction sites followed, one after the other, so that with the completion of the Gigerwald Dam in the canton of St. Gallen in 1976, most of Switzerland's glaciers had been incorporated into its hydroelectric infrastructure.

Énergie Ouest Suisse (EOS) a ubiquitous actor in dam building, which repeatedly used the same barracks across multiple construction sites, illustrates the argument presented here. EOS was the largest electricity provider in French-speaking Switzerland since its creation in 1919. The company built the Cleuson Dam between 1947 and 1951 in a valley parallel to that of the Grande Dixence. However, the completion of this hollow-gravity dam was somewhat delayed because of an expert debate concerning structural aspects of its design. In 1951, EOS transferred the barrack-huts needed for the construction of the Salanfe Dam, lower in the Rhône Valley. The construction at this site had started in 1949, but its pace slowed down because of cumulative delays in Cleuson, and lasted until 1952. At the beginning of 1953, the barracks were transferred to the construction site of the Grande Dixence at Le Chargeur, the headquarters at the construction site. The pouring of concrete, which was the most labour-intensive part of the building process, had already started at the site, and the housing deficit necessitated the construction of a housing block to alleviate the shortage. This example reveals that three apparently separate events were actually ordered chronologically within the biography of the construction company tasked with building those dams.

The construction of the Grande Dixence network replicated at a smaller scale the processual nature of dam building of that period in Switzerland. Besides the Grande Dixence Dam itself, two smaller dams make up the complex along with the extensive adduction network totalling approximately 100 km in length. In terms of the phasing of the construction, it began at the dam's construction site in 1950 and progressively extended out towards the Rhône Valley and the Mattertal. The network's capillaries were not completed until 1965.

An examination of the progress of a company named Losinger illustrates the processual nature of dam construction. This company had been hired to dig the 'Grand Collecteur,' that is, the main tunnels linking the Ferpècle and Arolla Valleys to the Grande Dixence Lake in the Dix Valley. To complete this task, the company established two temporary settlements in 1951. The first, located at the southern extremity of the lake, was named Cheilon. The second, directly overlooking the Arolla Valley, was named P4.[31] Losinger dismantled the P4 settlement in 1955, following the completion of the 'Grand Collecteur.' Components of the P4 settlement were then relocated to Bricola, in the Ferpècle Valley, which was one valley to the west of Arolla. There, the second part of the Grand Collecteur, linking the Arolla and Zermatt Valleys, was being dug. Following the completion of this section of the construction in 1959, Losinger re-established the P4 settlement, as more work was required on the first section of the Grand Collecteur. These works had not yet been planned at the time that the first iteration of the P4 settlement was built.[32]

In the above examples of Losinger and EOS, although the physical locations of the settlements differed, the same barrack-huts, machines, and probably even the same workers were moved around. Although the settlement may have seemed temporary at a particular location, the logic behind their establishment was one of serial ephemerality. This re-evaluation of the notion of temporariness is however only possible by shifting the point of observation from that of a given place to one that follows the logic of movement of a material or human flow. With this in mind, our understanding of this industrial heritage also shifts: a bigger structure is at play here. Evidently, the process of transforming the Alps into spaces of extraction is still at work today. However, precisely because of this perceived ephemerality, it remains mostly an invisible process. Dam building should be viewed as part of a continually evolving organism, rather than as a single event in time.

Secondary Infrastructure as the Lived Space of Workers

Infrastructures are relational.[33] They are given different meanings by the people who use them, build them, or live beside them. Accessing them and their functions along with their representations and uses depends, among other things, on their socio-economic or geographical contexts. This can be directly illustrated by the construction of the Grande Dixence Dam. For the company, the construction of the dam was a capital investment, which generated revenue through the sale of electricity that it eventually produced. For the workers who fabricated its form, it was a livelihood.[34] The elasticity of the concept of infrastructure is essential to recentre attention on the workers. The space inhabited by the workers at the Grande Dixence and within its network was in large part contingent on this secondary infrastructure. It also extended far beyond the temporary settlements. The secondary infrastructure was both a function and a (lived) space.

Paradoxically, in striving to achieve energy autarky, Switzerland relied heavily on an imported workforce. A large proportion of this labour force was supplied under the A-Permit regime, a short-term residence permit that was contingent on the guest-worker quota. However, this was merely an institutionalised response to a much longer history of labour exchange between Switzerland and neighbouring countries, primarily Italy, which was formalised through the 1931 Federal Law on the Residence and Establishment of Foreigners. The industrialisation of the Alps cannot be conceived outside of the contribution of these foreign workers. The major alpine crossings that were used prior to 1931, such as the Gotthard, the Lötschberg and the Simplon tunnels, prompted the development of what have been described as ghetto communities inhabited by labour hired to work on their construction sites.[35] Likewise, the (old) Dixence and Barberine Dams, which were the first large dams to be built in Switzerland, relied heavily on the support of foreign workers—again primarily Italians—for their construction. By the time that the construction of the large dams in Valais commenced, the labour procurement channels and infrastructure necessary for their construction had evolved into an efficient ecosystem.

The following section attempts to reconstruct the path that Italian workers may have taken, availing of the infrastructure necessary to reach Switzerland and travel up to the dam construction site. It aims to show that these infrastructures were also the lived spaces of these workers. As political scientist William Walters reminds us: 'all migrations involve journeys, and those journeys are more often than not mediated by complex infrastructures, authorities and norms of transportation.'[36] To this, we could add that those infrastructures (vehicles, ports, and roads) are spaces in themselves or, in the case of norms of transportation and conventions of use, they drastically influence these spaces.

Some of the most common methods of procuring seasonal work were through word of mouth, informal networks, and letters sent to construction companies.[37] Construction companies would also send their representatives across the border to towns or villages in Italy. Their arrival would most often be announced by the municipality officials, and they would conduct interviews to fill the quotas of workers required for the next season. For this reason, workers on a construction site were sometimes all from the same area or village, comprising siblings working with uncles and cousins. The province of Belluno in northern Italy seems to have been an important labour reservoir for the Mauvoisin construction site located around 20 km south-west of the Grande Dixence Dam in the parallel valley of Val de Bagnes.

Most seasonal workers employed at the Grande Dixence complex travelled there by train, as cars only become sufficiently affordable for the workers to purchase at a later date. The main entry point into Switzerland would have been through the

Simplon Tunnel to the train station in the city of Brig.[38] As discussed above, the digging of this tunnel, which took place in two phases—the first from 1898 to 1905 and the second between 1913 and 1921—had already been carried out largely by Italian migrant workers. This is indicative of the extensive role of migration in the industrial prehistory of dam building, evidencing a continuity in the labour procurement channels whose legal framework evolved in line with the nature of the construction projects.

Once the workers reached Brig, a major obstacle impeded their onward journey to the high alpine valleys: the *Grenzsanität*, or border sanitary control. This service, instituted in 1920, was officially aimed at protecting the Swiss population against the great epidemics of the time: smallpox, cholera, exanthematic typhus, plague, and syphilis.[39] Radioscopic exams were introduced to scan for tuberculosis and were later replaced by radiography. Sanitary control posts were established along the Swiss border in the entry cities, notably Chiasso, Brig, Geneva, Basel, Schaffhausen, and Buchs. Up to 2,000 workers could arrive on a single day in Brig at the beginning of the new season.[40] The control measures have been described as humiliating in the literature and news reports and were perceived as infringements of the fundamental right to maintain the privacy of one's own body.[41] From Brig, depending on which construction site the workers were travelling to, a train took them to Visp, Sion, or Martigny. Once there, a passenger bus organised by the construction companies would take them up to the various building sites. There, they inhabited the barrack-huts, walked through the tunnels to the mining front, and staffed the working stations. The working site thus became their home.

Here, the notion of secondary infrastructure can be deployed to conceptualise the lived space of the workers. This aids us in understanding how the geographical scale of this taskscape is expanded through the functions of these flows, creating a mesh—using Tim Ingold's word—of spatial locations that are not necessarily contiguous.[42] The seasonal working permit exemplifies this. For sociologist Delia Castelnuovo-Frigessi, one of the first Italian intellectuals to raise awareness on the issue in the 1970s, the rights and duties of workers under the A-Permit were entirely negative.[43] They were allowed to stay for a maximum of nine months in Switzerland, after which time they had to head back to their home country for at least three months. This 'rotation principle' hindered permanent settlement in Switzerland.[44] During those nine months, the workers were not allowed to change employers. Their partners and children were not allowed to migrate to the country with them. They were only eligible to obtain a B-permit (annual permit) after they had completed ten shifts.[45] The seasonal workers paid taxes for services they could not use in practice, such as schools for their children. Yet, despite paying social security taxes, they could not request unemployment benefits because their stay in the country was tied to their working contract.

Swiss historians, Gérald and Silvia Arlettaz have discussed the difficulty of establishing the exact numbers of foreign workers employed in the canton of Valais because the census methods applied did not generate conclusive records.[46] A similar shortcoming can be observed for the seasonal workers employed at the Grande Dixence Dam construction site. The presumed lists of workers employed on the various construction sites either never made it to the archives or were discarded.[47] Indications of how many workers were involved in each case are therefore based on circumstantial evidence.[48] In Valais, historian, Myriam Evequoz-Dayen notes a direct correlation between the construction of the hydroelectric infrastructure, the engineering work it required, and the entry of foreign workers in the canton.[49] There, these seasonal workers worked alongside local inhabitants of the alpine valleys, who found in these industrial enterprises new earning possibilities which progressively superseded their traditional livelihoods. However, while it is no secret that this blooming Swiss industry would not have been possible without the seasonal workers, accounts of their contribution, especially in the case of the Grande Dixence Dam's construction, are rare.[50]

Ultimately, however, the question of precisely how many guest workers contributed to the Grande Dixence project is irrelevant here. Rather, the intention is to show how systemic reliance on guest workers, whose involvement extended beyond the actual construction of the dams in Valais, redefined the temporal and geographic scales of the landscape. This extensive reliance on seasonal workers also aptly illustrates Ingold's notion of a taskscape, leading to a conception of the dam-building industry as a project with transnational consequences.

Villages in distant locations such as southern Italy became integrated into the economic and demographic domain of the Swiss dam-building industry through their responses to events and projects in Switzerland. Although many Italian women were employed as housewives or maids, or on the construction sites as cooks and cleaning staff, men accounted for the majority of the seasonal workers in the country. This meant that villages in Italy were periodically emptied of large sections of their male populations.[51] This situation required adaptations of the domestic and professional organisation of 'those who stayed,' namely women, children, but also individuals with disabilities, whose entry into Switzerland was barred by the border sanitary control regime.[52] In the workers' home countries, during their absence, a lot of their duties and responsibilities had to be assumed by those who remained behind.[53] Thus, the migration of their mostly male counterparts affected them equally. The effects on local communities in the canton of Valais, such as residents of the village of Hérémence, which was also an important reservoir of labour for the construction of the Grande Dixence, have been described by Marie-France Vouilloz-Burnier.[54] However, echoing the lack of attention given to foreign workers working on the large dams, and despite the attention given to the phenomenon of 'those who stayed' elsewhere in Europe in migration studies since the beginning of the 1970s, such consequences of the 'guest-worker' system in Switzerland remain to be investigated.

A second aspect of the taskscape concerned the remittances that the seasonal workers sent back home and the income they brought when they returned, which contributed to the local economies of their native regions. As Alex Mayenfish has shown in his documentary film, *Saisonniers en Suisse: une vie à la dure (Seasonal workers in Switzerland: a tough life)*, the foreign workers modernised their homes or built new ones, while remaining in a precarious situation of dependence on the renewal of their contract for the next season to complete these improvement projects in their villages.[55]

The Territorial Archive of the Secondary Infrastructure

Because of the perceived ephemerality of most components of the secondary infrastructure, a lot of its materiality has vanished. The barracks in which the workers were housed were dismantled. Following their use in Blava, the Blondin ropeways were shipped to the construction site of the Manicouagan Dam in Canada. Some traces, however, still remain. The concrete buttresses seen in Figure 5.3 provide one such example. They were built to prevent the cliff overhanging the Bertol settlement from collapsing onto the barrack-huts clustered against it. Together with the roads and bridges built to access the construction site, which have since assumed other functions, they now constitute what could be termed a *territorial archive* of the construction sites.[56]

The documentation of the territorial archive circumvents difficulties entailed in attempts to grasp a putative 'temporary' phase of the landscape because ephemeral or not, dam building marked the terrain in the same way as did age-old practices. Despite the supposed ephemerality of the dam's construction period, this phase imprinted an additional layer on the land. In this layer, all previous phases or states

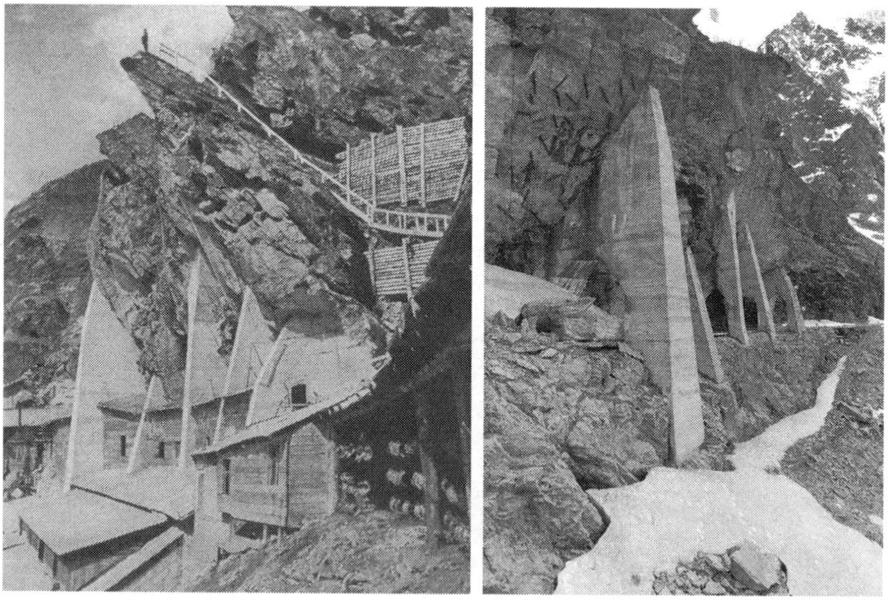

Figure 5.3 The temporary settlement at Bertol, above Arolla, in 1956 and today.

Source: left: Frank Gygli, Grande Dixence (Lausanne: Marguerat, 1956), 36 and right: the author, 2021.

are expressed with the same simultaneity today, regardless of the duration of the activities that made those marks in the first place. Examples of this layer include the age-old agricultural practices that are expressed in the present through what remains of terrace-supporting walls or hay barns or the medieval roads over mountain passes that stand today in immediate proximity to the ruins of so-called temporary settlements. The ephemerality of the dam-building phase ultimately bears no relevance to the marks it leaves on the landscape.

Accessing those traces in the present, however, poses a difficulty. This is not so much because they cannot be accessed, although one must be prepared for hour-long walks. They are present in the terrain for anyone to see. However, the failure of historical reconstructions means that the traces do not (yet) have meaning. Without a narrative associated with these remnants, they are at risk of being forgotten along with the stories of the workers who contributed to these construction sites. While the dams, characterised by their monumental scale and technological prowess, have been registered as national or regional monuments, the perceived temporariness of the secondary infrastructure requires the requisite knowledge for there to be awareness of the traces of their former existence. Ephemerality has meant that the contributions of the thousands of workers digging the tunnels have gone unnoticed.

A process-based reading of the construction site, as has been presented here, can shed light on what may be left of these construction sites. However, the nature of this territorial archive, for instance, kitchen tiles cemented on a concrete slab far above the last inhabited hamlet (Figure 5.4) also reveals that these infrastructural networks are not simply remnants of a mechanical apparatus. They are also the remains of a space where people once lived, slept, or cooked. The work of reconstruction and narration of secondary infrastructure is not only an act

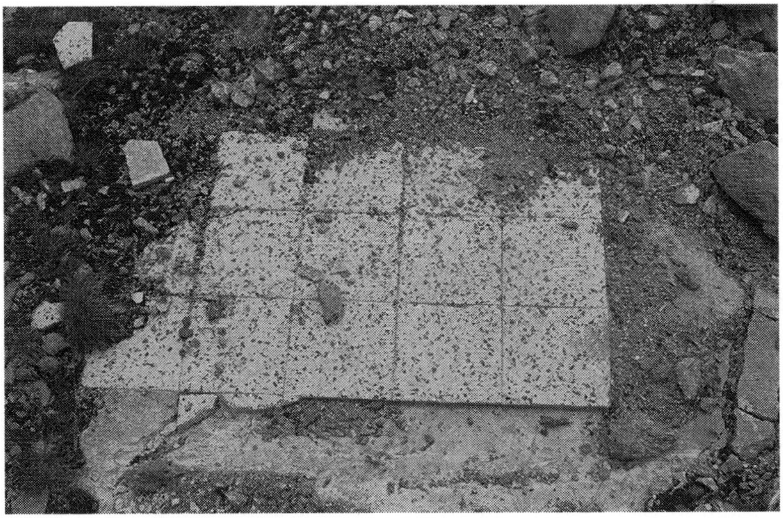

Figure 5.4 The remains of the Bertol settlement. Photograph by the author, 2021.

of remembrance; it also has the potential to redefine alpine landscapes as spaces inhabited for industrial purposes by thousands of workers, local and foreign.

Conclusion

The question that arises is what should be done with those traces in the present. In December 2022, the Alpiq Group, which is the mother company to EOS and one of Switzerland's two largest electricity producers, announced its intention to build a solar farm in the Prafleuri Valley, located in the Alps of Valais.[57] An argument put forward in the press release in support of the project was that the facility would be located at the site of a former aggregate quarry, which was the source of the concrete used to construct the Grande Dixence Dam (Figure 5.2). Alpiq defended the choice of the location, stating that 'a relatively flat area of 350,000 m^2 has already been largely remodelled by human activity.'[58] In other words, this was no longer a natural landscape. The press release went on to say that 'the pre-existing infrastructure on the site is one of the project's key assets: road access already exists, as does a medium-voltage power line that reaches the dam area.'[59]

Alpiq's argumentation is illuminating in two ways. Firstly, the reasoning presented can be described as a logic of continuity. Following the completion of the Grande Dixence Dam in 1961, the aggregate extraction site and access roads were abandoned. Yet with this new project, their meaning was suddenly restored. The secondary infrastructure and its remnants are mobilised both materially—in the case of the access roads which regained their original function—and discursively. The solar farm is rhetorically inserted into a historical lineage of industrial use of the site. This shows the endurance of the assemblage of secondary infrastructure well beyond the construction of the large dams during the decades following World War II. The components of this assemblage evidently remain as part of the energy landscape, undergoing further changes.

Secondly, while Alpiq's argumentation is clearly not disinterested and can be criticised for its uncritical reading of the site's history, it at least attempts to reveal an alternative image. In a place mostly marketed for its untouched nature, documentation of the secondary infrastructure provides important leads for an updated reading of what Switzerland's alpine landscapes have also been: an industrial territory dedicated to the building of resource-extracting infrastructures.

These secondary infrastructures, though planned as temporary, were of crucial importance, shaping the current appearance of the Alps. This was a landscape inhabited in large part by foreign seasonal workers, sometimes denied basic human rights and now deprived of even a right to history. Rendering this cultural landscape visible in the present through the documentation of the territorial archive reasserts the importance of the thousands of bodies needed for the construction of the national hydroelectric networks. The dam is not the only built heritage of these workers. The traces left behind are also important witnesses to this contribution to the image and existence of this ephemeral landscape, which stands in complete contradiction to the prevailing highly romanticised images of the Alps. Whether we like it or not, the Alps are changing, and it seems important to find a way to reconcile the images of this 'sublime' landscape inherited from the Enlightenment with its rich industrial past.

Notes

1 "La terrible question," 1. English translation by the author.
2 "Dead under the glacier, living in memory."
3 Ricciardi, Cattacin, and Baudouï, *Mattmark*, 89.
4 "M. Roger Bonvin fut l'un des pionniers du Mattmark," 9.
5 Star, "The Ethnography of Infrastructure," 381.
6 Ricciardi, Cattacin, and Baudouï, *Mattmark*, 13.
7 Ricciardi, Cattacin, and Baudouï, *Mattmark*, 126 and 79.
8 Frandsen, "Shadow Territory," 217.
9 The movements observed in Valais, described by Frandsen in "Notes on the Inventory Practice," 40 could probably be extrapolated to the rest of the country and the Alps.
10 Björkman, "Infrastructure as Method" in *Pipe Politics*, 1–30.
11 Ricciardi, Cattacin, and Baudouï, *Mattmark*.
12 Star, "The Ethnography of Infrastructure," 380. See also Graham and Marvin, *Splintering Urbanism*.
13 Ingold, "The Temporality of the Landscape,"157. This approach is also inspired by Lefebvre, *La Production de l'Espace*. See also Easterling, *Extrastatecraft*.
14 Angélil and Siress, "Epilogue—Environment-Making," 883.
15 Nichols, "Opération Béton," 377.
16 See, for instance, Schnitter, "La Contribution de La Suisse" ("The Contribution of Switzerland"). See also E.S., "Alfred Stucky (1892–1969)"; Paquier, "Les Effets"; and Steigmeier, *Power on*.
17 Ricciardi, Cattacin, and Baudouï, *Mattmark*; Torricelli and Cambin, "I Villaggi Dei Tunnel"; Pfander, "Helvetia Barbieri."
18 Logean, *Du berger au mineur*; David Jollien, "La Vie de Chantier."
19 Van Wyck, *The Highway of the Atom,* 35.
20 I rely here on Doreen Massey's notion of *positive heterogeneity*. See Massey, *For Space*, 12.
21 Swiss Committee on Dams, "Liste des barrages suisses" ("List of Swiss Dams"). This is in addition to the 33 large dams built before that year. See also International Commission on Large Dams, "Constitution Statutes," 3.
22 Here, the two most relevant issues for the current chapter are cited: Département fédéral des postes et de chemins de fer, "Les Forces Hydrauliques … Première Partie" and Département fédéral des postes et de chemins de fer, "Les Forces Hydrauliques… Sixième Partie."
23 Swiss Committee on Dams, "Liste des barrages suisses." The Göscheneralp Dam is taller than the Mattmark Dyke.
24 For a disambiguation of the term barrack-hut, see van Pelt, "Labor Service," 507.
25 Turner and Johnson, "Infrastructure and Territoriality," 5.
26 Godard, *Opération Béton.*
27 Desmeules, "Les Installations de Chantier," 12.
28 The construction of this tunnel warrants attention, as it unexpectedly documents the living conditions on such a construction site. Frank Gygli refers to the workers sometimes

sleeping in this gallery in case of bad weather, as to not to go back to the main camp on foot. See Gygli and Bolomey, *Grande Dixence,* 114.

29 In the words of Star, infrastructures have 'reach or scope, which may be either spatial or temporal. Infrastructure has reach beyond a single event or onsite practice. Star,' "The Ethnography of Infrastructure," 381.

30 Ingold, "The Temporality of the Landscape," 164.

31 It was also sometimes referred to as the Vignette settlement.

32 Frandsen, "Shadow Territory," 99–102.

33 Star, "The Ethnography of Infrastructure," 380. See also Graham and Marvin, *Splintering Urbanism*, 177, 184.

34 Among the seasonal workers recruited for the construction of the large dams, the Italians were the most numerous, but workers from countries such as Germany, France, or Spain also contributed.

35 Burkhalter and Sumi, *Der Gotthard*; Pfander, "Helvetia Barbieri," 68.

36 William Walters, "Migration, Vehicles, and Politics," 470.

37 For more discussion on this topic, see Simone, "Ritornello: 'People as Infrastructure.'" Simone's description of people gathering information via social networks is reminiscent of Ingold's 'taskscape,' especially given their shared use of the word 'attunement.'

38 Though not situated on the Swiss–Italian border, the Simplon tunnel provided direct access from Italy to Brig in the Rhône Valley. The border control post was therefore installed in this city.

39 Santiago, "Le Service Sanitaire," 157.

40 Perrig, "Der Grenzsanitätdienst," 1589.

41 Santiago, "Le Service Sanitaire," 189.

42 Ingold, "The Meshwork."

43 Castelnuovo-Frigessi, *La condition immigrée*, 30.

44 Widmer, *Le Rôle de La Main-d'œuvre Étrangère*, 38.

45 In the case of Valais, an exception could be made, and workers could stay up to three years if there was the assurance that they would then return to their home country. I have not, however, encountered any such instances in my research.

46 Papilloud et al., *Le Valais et Les Étrangers XIXe – Xxe*, 14.

47 Scholars in the field of migration studies have circumvented this methodological challenge by collecting oral histories and histories from below. See, for instance, Tayob "Transnational Practices of Care and Refusal." While these methods also do not provide exact numbers, the focus on individuals serving as representatives delivers richer and more complex accounts of the processes of migration, which also destabilise the notion of migrant identity. See also Mouffe, "For a politics of nomadic identity."

48 The Medical Borders Project led by Jelena Martinovic of the EDHEA school of Art, Sierre (Valais) was able to trace some this history by accessing the x-rays of the *saisonniers* submitted to a medical examination upon entering Switzerland.

49 Evequoz-Dayen, "Le Valais et Les Étrangers Depuis 1945," 132.

50 In the available sources and literature, the same figure—60%—is cited as the proportion of local residents (*Valaisans*) at the Grande Dixence construction site. This number was first mentioned by Vincent Favre in the construction site newspaper, *Le Choucas*: Favre, "Les hommes du barrage," 39. The 60% claim has also found its way into more recent scholarship, for example, Logean's work on the construction site of the Grande Dixence wall: Logean, *Du Berger au Mineur*, 27. This number was in turn reproduced in Jollien's work published in 2015: Jollien, "La Vie de Chantier" 95. Sarah Nichols, too cites this percentage in her 2021 dissertation: Nichols, "Opération Béton," 375.

51 Vuilleumier, "Le mouvement," 104. Vuilleumier provides a ratio of 139 men for 100 women.
52 Santiago, "Le Service Sanitaire," 169–170.
53 See, for instance, O'Loughlin, "Missing Men?" 2–3.
54 Vouilloz-Burnier, *A l'ombre de la Dixence*. The entire book is dedicated to this issue.
55 Mayenfisch, *Saisonniers En Suisse*.
56 Van Wyck, *The Highway of the Atom*, 35.
57 Alpiq, "Lancement d'un projet solaire."
58 Alpiq, "Lancement d'un projet solaire." See also Commune Hérémence, "Parc Solaire d'Hérémence."
59 Alpiq, "Lancement d'un projet solaire."

Bibliography

Alpiq Group. "Lancement d'un projet solaire d'envergure à Prafleuri" ["Launch of a large-scale alpine solar project in Prafleuri"]. Accessed October 8, 2025. www.grande-dixence.ch/files/mm_221219_projet-prafleuri_fr.pdf.

Angélil, Marc, and Cary Siress. *Mirroring Effects: Tales of Territory*. Berlin: Ruby Press, 2019. https://doi.org/10.1080/02513625.2020.1851914

Björkman, Lisa. "Infrastructure as Method" (Supplementary Chapter for *Pipe Politics, Contested Waters*). Durham: Duke University Press, 2015. https://assets-us-01.kc-usercontent.com/f7ca9afb-82c2-002a-a423-84e111d5b498/886227dd-c7bc-400b-b0d5-0ab275c94041/978-0-8223-5969-2_602.pdf?fm=webp&auto=format&lossless=true.

Burkhalter, Marianne, and Christian Sumi (eds.) *Der Gotthard*. [*The Gotthard*]. Zürich: Scheidegger & Spiess, 2015.

Castelnuovo-Frigessi, Delia. *La condition immigrée: les ouvriers italiens en Suisse*. [*The immigrant condition: Italian workers in Switzerland*]. Lausanne: Editions d'en bas, 1978.

Commune d'Hérémence. "Parc Solaire d'Hérémence," ["Hérémence Solar Park"]. Accessed January 24, 2024. www.heremence.ch/data/documents/pilierpublic/PrafleuriSolaire.pdf.

Département fédéral des postes et de chemins de fer. "Les Forces Hydrauliques Disponibles de La Suisse Considérées Au Point de Vue des Possibilités d'accumulation Pour La Production d'énergied'hiver. Première Partie: Considérations Générales et Possibilités d'accumulation Dans Le Bassin de l'Aar" ["Available Hydraulic Forces in Switzerland Considered from the Point of View of Storage Possibilities for Winter Energy Production. Part I: General Considerations and Storage Possibilities in the Aare Basin"]. Communication Du Service Des Eaux 25, Bern, 1932.

Département fédéral des postes et de chemins de fer. "Les Forces Hydrauliques Disponibles de La Suisse Considérées Au Point de Vue des Possibilités d'accumulation Pour La Production d'énergie d'hiver. Sixième Partie: Possibilités d'accumulation dans Le Bassin du Rhône" ["Available Hydropower in Switzerland Considered from the Point of View of Storage Possibilities for Winter Power Generation. Part Six: Storage Possibilities in the Rhône Basin"]. *Communication Du Service Des Eaux* 30, Bern, 1945.

Desmeules, Jacques. "Les Installations de Chantier Du Barrage de La Grande Dixence" ["Construction Site Facilities at the Grande Dixence Dam"]. *Le Choucas* 13 (April 1955): 12.

E. S. "Alfred Stucky (1892–1969)." *Bulletin Technique de La Suisse Romande* 95, no. 20 (1969): 268–270.

Easterling, Keller. *Extrastatecraft: The Power of Infrastructure Space*. London: Verso, 2016.

Evequoz-Dayen, Myriam. "Le Valais et Les Étrangers depuis 1945" ["The Valais and Foreigners since 1945"]. In *Le Valais et Les Étrangers XIXe – Xxe*, edited by Jean-Henri Papilloud, Silvia Arlettaz, Gérald Arlettaz, Alain Clavien, Maria-Pia Tschopp-Bessero, Marie Claude Morand, and Myriam Evequoz-Dayen, 123–189. Sion: Groupe Valaisan de sciences Humaines, 1992.

Favre, Vincent, "Les hommes du barrage" ["The Men of the Dam"]. *Le Choucas* 49, Septembre (1961): 39.

Frandsen, Rune. "Shadow Territory and Secondary Infrastructures: The Hidden Landscapes of Temporary Labor at the Grande Dixence (1950–1965)." DSc Diss., ETH Zurich, 2023.

Frandsen, Rune. "Notes on the Inventory Practice in the Canton of Valais: The Swiss Dams and the Heritage of the Italian Seasonal Workers." *Kritische Berichte* 52, no. 1 (2024): 34–43.

Gygli, Frank, and Georges Bolomey. *Grande Dixence*. Lausanne: Marguerat, 1961.

Godard, Jean-Luc, dir. *Opération Béton*. Geneva: Actua-Films, 1958. 16mm.

Graham, Stephen, and Simon Marvin. *Splintering Urbanism*. London: Routledge, 2001. https://doi.org/10.4324/9780203452202

Ingold, Tim. "The Temporality of the Landscape." *World Archaeology* 25, no. 2 (1993): 152–174. https://doi.org/10.1080/00438243.1993.9980235

Ingold, Tim. *Being Alive: Essays on Movement, Knowledge and Description*. London: Routledge, 2011, 63–94. https://doi.org/10.4324/9780203818336

International Commission on Large Dams, "Constitution Statutes, 2011." Accessed January 24, 2024, www.icoldcigb.org/userfiles/files/CIGB/INSTITUTIONAL_FILES/Constitution2011.pdf.

Jollien, David. "La Vie de Chantier à a Grande Dixence (1950–1965): Une Fabrique de Héros?" ["Life on the Grande Dixence construction site (1950–1965): A Factory of Heroes"]. MA thesis, University of Fribourg, 2015.

"La terrible question: combien y a-t-il de victimes?" ["The terrible question: how many victims are there?"]. *Le Nouvelliste*, August 31, 1965.

Lefebvre, Henri. *La Production de l'Espace*. [*The Production of Space*]. Paris: Editions Anthropos, 1974.

Logean, Elisabeth. *Du berger au mineur: la construction du barrage de la Grande Dixence (1951–1962) entre paix sociale et crise d'identité*. [*From Shepherd to Miner: The Construction of the Grande Dixence Dam (1951–1962) between Social Peace and Identity Crisis*]. Sierre: Monographic, 2000.

"M. Roger Bonvin fut l'un des pionniers du Mattmark." *Journal de Sierre*, September 7, 1965.

Massey, Doreen. *For Space*. London: Sage, 2005.

Mayenfisch, Alex, dir. *Saisonniers En Suisse: Une Vie à La Dure*. [*Seasonal Workers in Switzerland: A Tough Life*]. Switzerland: Les Documents de la RTS, 2003. www.rts.ch/archives/tv/divers/documentaires/5128347-une-vie-a-la-dure.html.

Mouffe, Chantal. "For a Politics of Nomadic Identity." In *Travellers' Tales, Narratives of Home and Displacement,* edited by Jon Bird, Barry Curtis, Melinda Mash, Tim Putnam, George Robertson, and Lisa Tickner 103–110. London: Routledge, 1994. https://doi.org/10.4324/9780203978986

Nichols, Sarah. "Opération Béton: Constructing Concrete in Switzerland." DSc diss., ETH Zurich, 2021.

O'Loughlin, Bridget. "Missing Men? The Debate over Rural Poverty and Women-headed Households in Southern Africa." *The Journal of Peasant Studies* 25, no. 2 (2008): 1–48. https://doi.org/10.1080/03066159808438665

Papilloud, Jean-Henri, Silvia Arlettaz, Gérald Arlettaz, Alain Clavien, Maria-Pia Tschopp-Bessero, Marie Claude Morand, and Myriam Evéquoz-Dayen, eds., *Le Valais et Les Étrangers XIXe – XXe.* [*The Valais and Foreigners in the XIXth and XXth Centuries*]. Sion: Groupe Valaisan de Sciences Humaines, 1992.

Paquier, Serge. "Les Effets d'une Dépendance Énergétique: La Crise Du Charbon Comme Stimuli de l'utilisation des Ressources Hydrauliques." ["The Effects of Energy Dependence: The Coal Crisis as a Stimulus for the Use of Water Resources"]. In *14/18 La Suisse et La Grande Guerre*, [*14/18 Switzerland and the First World War*] edited by Roman Rossfeld, Thomas Buomberger, and Patrick Kury, 126–142. Baden: Hier und Jetzt Verlag für Kultur und Geschichte, 2014.

Perrig, Hans. "Der Grenzsanitätdienst," ["The Border Sanitary Service"]. *Praxis: Schweizerische Rundschau für Medizin/Revue Suisse de La Médecine* (*Praxis: Swiss Journal of Medicine*) 52 (1965): 1583–1589.

Pfander, Ursula. "Helvetia Barbieri—Alltag Im Provisorium." ["Helvetia Barbieri—A Make-shift Everyday Life"). In *Tiefenbohrungen: Frauen und Männer auf den grossen Tunnelbaustellen der Schweiz 1870–2005.* [*Deep Drilling: Women and Men on the Large Construction Sites in Switzerland 1870–2005*], edited by Elisabeth Joris, Katrin Rieder, and Béatrice Ziegler, 66–87. Zurich: Hier Und Jetzt Verlag für Kultur und Geschichte, 2006.

Ricciardi, Toni, Sandro Cattacin, and Rémi Baudouï. *Mattmark, 30 août 1965: La Catastrophe* [*Mattmark, 30 August 1965: The Catastrophe*]. Zurich: Editions Seismo, 2015. https://doi.org/10.13140/RG.2.1.1296.8164

Santiago, Cynthia. "Le Service Sanitaire de Frontière." ["The Border Health Service"]. *Annales Valaisannes: Bulletin Trimestriel de La Société d'Histoire Du Valais Romand* (2018): 155–203.

Schnitter, Niklaus J. "La Contribution de La Suisse. La Construction de Barrages Dans Le Monde." ["Switzerland's Contribution. Dams Building Around the World"]. *Ingénieurs et Architectes Suisses* 111, no. 13 (1985): 239–241.

Simone, AbdouMaliq. "Ritornello: 'People as Infrastructure.'" *Urban Geography* 42, no. 9 (2021): 1341–1348. https://doi.org/10.1080/02723638.2021.1894397

Star, Susan Leigh. "The Ethnography of Infrastructure." *American Behavioral Scientist* 43, no. 3 (1999): 377–391. https://doi.org/10.1177/00027649921955326

Steigmeier, Andreas. *Power on: Elektrowatt, 1895–1995.* Zürich: Elektrowatt AG, 1995.

Swiss Committee on Dams, "Liste des Barrages Suisses (liste non exhaustive)," ["List of Swiss Dams (non-exhaustive)"]. Accessed May 1, 2023. www.swissdams.ch/fr/les-barrages/liste-des-barrages-suisses.

Tayob, Huda. "Transnational Practices of Care and Refusal." *e-flux Architecture*, 2021. www.e-flux.com/architecture/coloniality-infrastructure/411251/transnational-practices-of-care-and-refusal/.

Torricelli, Gian P., and Lorenza Boschetti Cambin. "I Villaggi Dei Tunnel Nella Regione Del San Gottardo." ["Tunnel Villages in the Gotthard Region"]. In *Der Gotthard*, edited by Marianne Burkhalter and Christian Sumi, 519–561. Zürich: Scheidegger & Spiess, 2016.

Turner, Colin, and Debra Johnson. "Infrastructure and Territoriality." In *Global Infrastructure Networks: The Transnational Strategy and Policy Interface*. Northampton, MA: Edward Elgar Publishing, 2017. https://doi.org/10.4337/9780857934413

van Pelt, Robert-Jan. "Labour Service Barrack-Huts in Germany and the United States, 1933–45." *Zeitgeschichte* 4, no. 45 (2018): 507–539. https://doi.org/10.14220/9783737008518.507

van Wyck, Peter. *The Highway of the Atom.* Montreal: McGill-Queen's University Press, 2010.

Vouilloz-Burnier, Marie-France. *A l'Ombre de la Dixence, Vie Quotidienne des Femmes dans l'Arc Alpin.* [*In the Shadow of the Dixence, Women's Daily Life in the Alpine Arc*]. Sierre: Monographic, 2009. http://dx.doi.org/10.1515/9780773580879

Vuilleumier, Marc. "Le mouvement ouvrier et les travailleurs italiens en Suisse jusqu'à la Première Guerre mondial" ["The Labour Movement and Italian workers in Switzerland up to the First World War"]. *Revue Syndicale Suisse*: *Organe de l'Union Syndicale Suisse* 82 (1990): 104.

Walters, William. "Migration, Vehicles, and Politics: Three Theses on Viapolitics." *European Journal of Social Theory* 18, no. 4 (2015): 469–488. https://doi.org/10.1177/1368431014554859

Widmer, Jean-Philippe. *Le Rôle de La Main-d'oeuvre Étrangère Dans l'évolution Du Marché Suisse Du Travail* [*The Role of Foreign Labour in the Evolution of the Swiss Labour Market*]. Neuchâtel: Groupe d'Études Économiques, 1978.

Deutschland celebrated Honecker's celebratory letter, which 'recognised the good achievements of the community's workers and the responsible activities of local representatives.'[14]

In fact, the 1985 commemorative booklet recounted, the millennial celebration coincided with resettlement notices, and, by the time the booklet itself had been produced, virtually nothing remained of either village. The disorientation many resettlers experienced in Leipzig-Grünau's prefab blocks was powerfully captured in Kurt Tetzlaff's 1983 documentary film *Erinnerung an eine Landschaft – für Manuela* (*Memory of a Landscape – for Manuela*). 'It's over,' an old woman reminisced from the muddy terrain between Grünau's rising slab towers; sharing album photos of her family at the only home she had known for 50 years in Bösdorf, she shook her head at the alienating surroundings at Grünau. 'We have to get used to it,' she sighed, 'here are the keys to the new apartment.'[15] In powerful closing words, the engaged local chroniclers testified in their 1985 booklet:

> Eythra had to vanish for the coal in order to excavate and secure this valuable raw material. The residents of Eythra accepted this necessity. Nevertheless, it was not easy for many of them to give up their *Heimat*, their ancestral home and all the interpersonal relationships. Each former Eythraer happily remembers the beautifully situated [village of] Eythra. With their industrious hands and engagement, residents shaped their *Heimatort* over the decades. Eythraer who live in various residential complexes in Grünau, elsewhere in the Leipzig District, and in the various districts of the Republic work hard to sustain contact with each other. Most of their conversations relate to 'their Eythra.'

Memories are exchanged about the years they experienced together, about the political and cultural highlights, and about family events, because so many Eythraers were related to each other. For all of them, Eythra will remain unforgotten.[16] Whereas Tetzlaff's film offered a not-so-cloaked commentary about the material and immaterial costs of pit mining for an East German audience, the *Heimat* book was a written form of heritage preservation explicitly designed to anchor living memories of a dear home and its 'necessary' destruction. The absence of any stated long-term relevance once the *Erlebnisgeneration* had passed away is telling. The book was supposed to help them cope with their loss. Could it speak to future generations, however, who had no memory of the people and spaces it so intimately described?

The quandary embedded in the *Heimatbuch*'s closing pledge of commemoration spoke to how dogged formal and informal efforts evolved over the course of the next two decades: from short-term obsessions with memorialisation to a long-term waning of heritage output. For instance, the practice of taking images of the village before and during its destruction testified to an ongoing personal and communal commemorative process of dealing with loss. Throughout the late 1980s, the young amateur photographer and activist Thomas Thiel and Eythra pastor Horst König took pictures of the empty, plundered and decaying structures of the dying village (Figure 6.1).

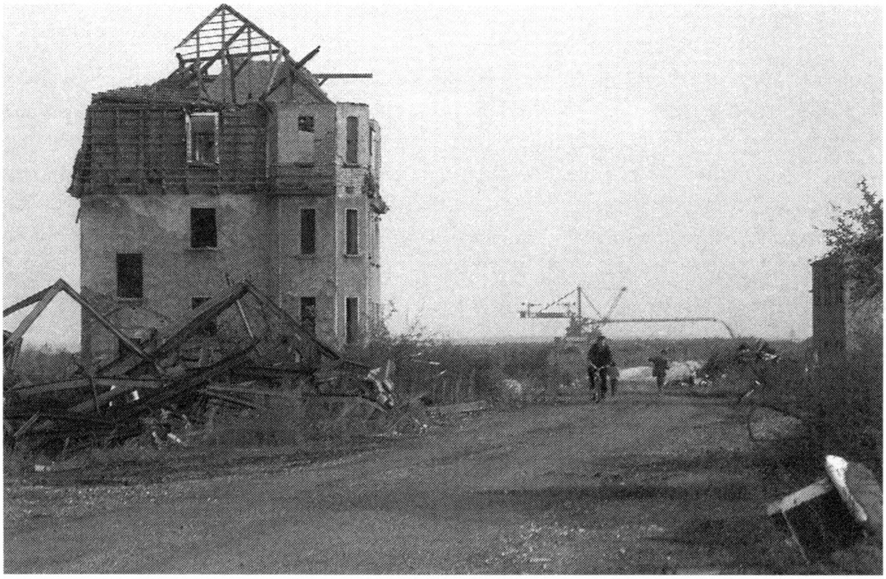

Figure 6.1 Eythra undergoing devastation in 1988. A stacker is in the background dumping overburden. Photograph by Thomas Thiel.

In July 1989, Thiel and König produced an exhibition for a Protestant church conference in Leipzig. Living memories of the violent suppression of the East German uprising of 1953 warned them that any opposition against ongoing environmental damage was risky, not least as it shortly followed the crackdown on Tiananmen Square in China. As the next five months saw Leipzig take a leading role in the 'peaceful revolution' against East Germany's SED dictatorship, the exhibition was accessible in a chapel in the Leipzig neighbourhood of Lößnig, offering testimony to the ecological and cultural devastation that was sweeping Leipzig's South.[17]

Relics salvaged from the devastated communities provided another commemorative opportunity. In the Leipzig-Grünau high-rise district that became home to two-thirds of the two villages' resettlers, the Protestant Pauluskirche was founded on April 1, 1978, and built from 1981 to 1983—practically simultaneous to the arrival of the Eythraer. After the Eythra church celebrated its last service in 1983, relics from the church were diverted to Grünau to console and root those who had lost their *Heimat*. In addition to church bells from Magdeborn and baptismal implements from Bösdorf, Eucharist utensils from Eythra enjoyed regular use in services heavily attended by displaced residents. Attention to these traces has lessened, however, as members who remember the vanished villages diminish in number. Testifying to long-term challenges, there appears to be minimal interest in this heritage by most of the congregation today.[18]

When civic initiatives halted the Zwenkau pit's further expansion into the adjoining Eichholz forest after the end of SED rule, an opportunity arose to commemorate Eythra on a fragment of land at the brink of what later became the Zwenkauer lake. In the early eighteenth century, the petty nobility in Eythra had established a grand promenade lined with linden trees south of the medieval palace towards an artificial 'Trianon' temple ruin with three Doric pillars (completed in 1790). Locals had denigrated the pillars as 'gallows,' until they were disassembled and presumably lost when the noble family was expropriated and fled to the West in 1945.[19] The palace itself, having served as a kindergarten since World War II, was destroyed along with the village to make way for the coal pit in the 1980s. When the Eythra *Heimatverein* (local association) discovered the pillars in a depot in 1996, its engaged members suddenly celebrated them as an integral piece of the lost *Heimat* and proudly restored them at their former location (Figure 6.2). This last relic from the vanished village was valorised at its inauguration on August 11, 2002, by Rudolf Binsack, a surviving member of the noble family that had owned the palace from 1935 to 1945, as 'the only building from Eythra that remains intact. It is standing again, and even on the same place as before, halfway to Zwenkau [from Eythra]. As a result, it is a sign of unity for the people from both places, and it testifies to the Eythraers' indefatigable will.'[20]

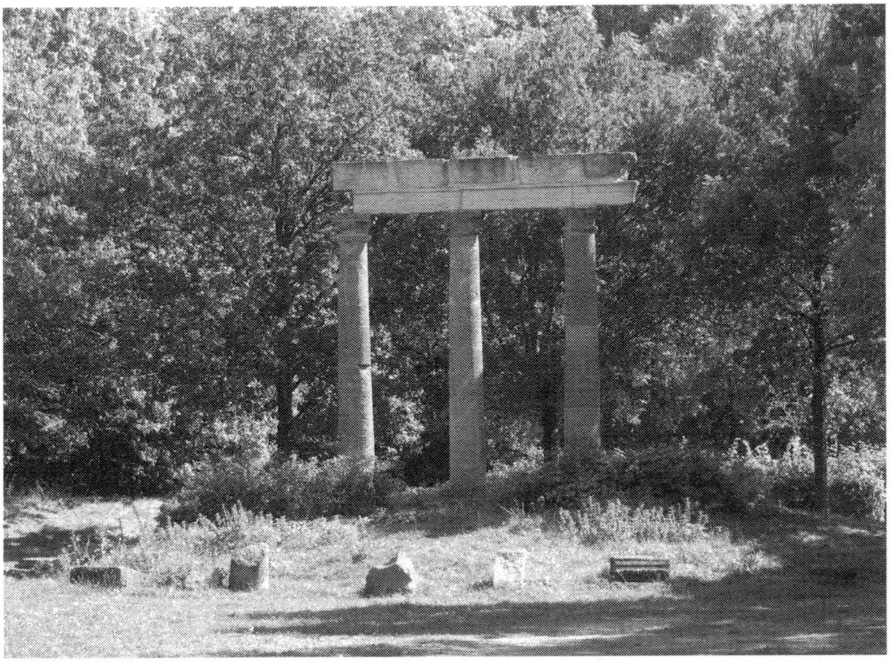

Figure 6.2 Eythra Trianon Monument. Photograph by the author, July 2022.

Such construction of memory filled a small-run commemorative picture book about Eythra in 2012 that was financed by the city of Zwenkau, Landkreis Leipzig (the regional county adjacent to Leipzig), the Zwenkau Heimat and Museum League and the Bösdorf steelworks, which had been moved on account of mining. Here diverse personages with ties to the vanished place took the chance to highlight their strongest memories. Regional church official Ekkehard Vollbach, the son of an Eythra pastor, celebrated the rescue of a fifteenth-century altar from the village for exhibition in Leipzig's Stadtgeschichtliches Museum (Leipzig Urban History Museum).[21] Roswitha Krieg, born in Eythra and living in West German Augsburg since 1960, had sustained close written contact with former neighbours and regretted the village's 'downfall' due to coal that had 'long since been sent into production and billowed from smokestacks.' She concluded: 'the place is lost for all times. Only those born later could fail to miss it!'[22] This twin memorial message was as poignant as it was fateful: even Eythraers like Krieg or Binsack who had spent their lives far away sustained ties with their lost *Heimat*. But it would no longer matter for future generations. Loss inevitably transformed into permanent erasure, because the physical essence of the lost *Heimat* had no history in lived memory.

The obscure 2012 picture book is a rare case in which the neighbouring city of Zwenkau (itself threatened with destruction in the most radical mining plans of the late 1980s) expended funds and attention on its vanished neighbouring village. By contrast, neither Eythra nor Bösdorf (nor even the mining heritage) won more than a passing mention in an earlier, 1994 *Heimatbuch* devoted to Zwenkau itself. In its obsession with idiosyncratic everyday activities and settings from a halcyon yesteryear (with whole pages devoted to a cheese shop or individual grocery stores), the mining pit and neighbouring villages had fallen outside of official memory, apart from the closing mention that 'in 1987 the pit mine demanded tribute from this place as well and erased Eythra from the map.'[23] Although Zwenkau's mayor observed that commemoration of vanished places was 'overdue,'[24] the 2012 Eythra book had no successors and is itself very hard to find. In like manner, though the resettlers have gathered at an annual reunion since 2004 to exchange old postcards and anecdotes, drink beer and catch up, they seldom visit the Trianon or Zwenkau lake, where token memorials offer scarce compensation for the surrounding, uniform post-mining landscape that has nothing to say about the cultural landscapes of their memories.[25]

Because the resettlers who lost their *Heimat* tend to be dispersed and are passing away, the lead role in any remaining commemoration has increasingly fallen to a small group of typically older Zwenkau residents in the city *Heimatverein*. Led by Zwenkau hobby chronicler and schoolteacher Dietrich Wünschmann, they collected relics such as nineteenth-century border markers that local people had squirrelled away before 1989. In 2005, they volunteered to reinstall them next to the Trianon reconstruction to remember the former Prussian-Saxon border that had run west of Zwenkau beside Eythra.[26] It was also the Zwenkau Heimatverein that in 2012 collected donations for the further restoration of linden trees that had lined the 650-metre promenade from the Trianon towards the palace. Where today the promenade suddenly ends at the lake—the closest land to former Eythra—the Heimatverein established a monument to the vanished village in 2014 on one of the many boulders that miners had taken from the pit as it was excavated. The Lindenallee itself had once stretched another 200 metres into the lake towards the palace.[27] Similar to German expellee communities from areas east of the postwar border, commemoration among resettlers failed to gain substantial traction with subsequent generations. The last resettlers in the region today tend to have only childhood memories that bear no relation to post-mining landscapes.

As the last witnesses pass away, they leave behind books, photographs, and some memorials, which will only survive if those without memory of the lost places adopt them as part of their own heritage. The Trianon site itself hosts occasional weddings; otherwise, the bicyclists racing to circle the lake's pathway may take a few photographs (though usually not), but seldom realise that the ensemble is supposed to speak to a vanished village. In casual conversation with Leipziger cyclists passing by the Eythra and Bösdorf monument stones overlooking the lake on September 4, 2022, I encountered neither memory nor interest in the meaning of the memorial stones—only the repeat of a prevalent false rumour that the East German regime (as early as Walter Ulbricht) had intended to destroy all of Leipzig to harvest coal. Such a fantastic myth allows Leipzigers to use the fate of vanished villages as a means of reinforcing their own victim narrative. The vanished villages themselves were apparently irrelevant to them. Meanwhile, the lookout area facing the lake where Eythra and Bösdorf had once existed (decorated with pieces of excavator buckets) is badly vandalised and bereft of informational plaques. Its commemorative value was apparently of little priority.

With the passing of a generation, the urgency of commemoration under communism and immediately after has been receding, and only traces remain of the former world before mining—with little meaning for those who hike or bike around the lake today from Zwenkau or Leipzig. Memorial projects devoted to a surviving town like Zwenkau embrace nostalgic, quaint and 'unchanging' *Heimat* imaginaries populated by microscopic family histories, tales about individual buildings, and archaeological records said to prove the city's ancient rootedness. Yet the latter seldom mention that the very depth of archaeological knowledge would have been impossible without the decimation of a neighbouring village like Eythra, where so much excavation had been undertaken by experts amid the ruins. Indeed, it was only because the village was destroyed that local archaeologists had access to retrieve pottery fragments and even the rare remains of a wooden well. Such a retrieval process was far more thorough in nearby Breunsdorf, whose destruction took place immediately after 1989, though here again, the village itself is lost forever.[28] Such collections of Neolithic remnants inform scholarly understanding of prehistoric times. But the 'recultivated' monocultural landscapes of fields, canalised rivers and human-made lakes impoverish multifaceted memory. With the passing of communicative memory, future histories will probably convey the full history of Eythra and Bösdorf as an aside to clarify the origins of Lake Zwenkau, which consumes the space once occupied by thousand-year-old villages whose histories abruptly ended.

Mölbis

While Eythra vanished from the map, the village of Mölbis survived but was affected by the extreme air, water, and noise pollution downwind from the Braunkohlenveredlung Espenhain (VEB BKE), Espenhain's carbo-chemical plant, which produced electricity, plastic, tar, benzine, paraffin, and numerous other products from the coal harvested out of the surrounding pits.[29] Once environmental data was publicly accessible in 1990, the district hygiene institute estimated that over the previous year, Mölbis had surpassed the maximum limit for sulphur dioxide exposure on 85 days, leading to increased danger of tumours, skin disease and cancer among adults and especially children. From 1974 to 1989, upper respiratory illness in Borna County increased by 211%, with particular problems for children, many of whom suffered from bronchitis and flaking skin. While numerous residents could not work due to respiratory and circulatory problems, sulphuric gases ate into stucco walls, foundations, windows, doors, roofing, drainpipes, chimneys, wooden frames, and other structural elements, such that buildings in the region aged twice as quickly as in less damaged areas.[30] The two bakeries, three inns, clothing stores, hair salon, clubs, and pretty much every other business had long since closed, as had the school in 1973, because an increasing number of families had fled from the village. It was small wonder, then, that by 1990, only 353 villagers remained—roughly one-third of the peak population in 1946 (Figure 6.3).[31]

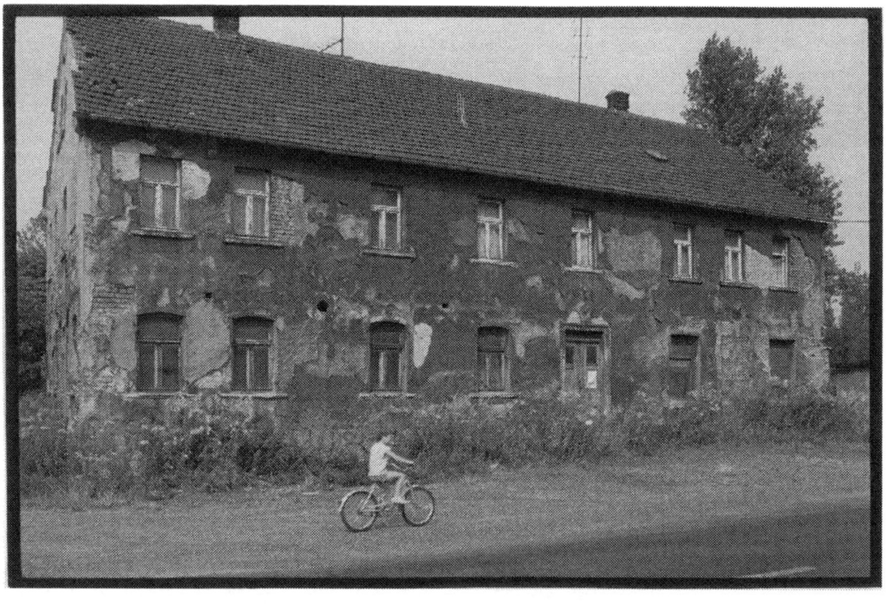

Figure 6.3 Mölbis in 1990. Photograph by Heidi Vogel-Hennig, Archiv Bürgerbewegung (ABL) 039-001-007.

By the time Karl-Heinz Dallmann became Mölbis's pastor in 1986, residents were freely debating whether they should abandon the village, even though the pit was slated to end at the settlement's edge, rather than undermine it entirely. After a resident mailed a particularly incendiary complaint to the Politburo in Berlin in 1987, central plans locked into place that decreed the village's imminent end (on the excuse of coal excavation, but in reality to end the scandal of extreme pollution).[32] In an unpublished vignette in 1988, Dallmann lamented the 'demoralising' quality of the regime's sudden decision. Whereas villages slated for erasure to make way for coal had years to prepare, for Mölbis 'it's suddenly the case of adjusting to a completely new situation within two years.' The outcome, he concluded, would be displacement, the decimation of the village and an unchanged environmental disaster, as 'the natural world will still be covered with filth, soot, toxins and gas.' He signed his testimony as 'the last pastor of Mölbis to the end!'[33] So advanced were SED plans by January 1989 that Borna County's planning commission chairman promised Mölbiser at a village meeting that 'a chronicle for Mölbis is to be created,' presaging a state-sponsored memorial approach akin to *Heimat* books for other villages devastated from mining. Residents themselves were fed up with the pollution and chiefly outraged that the regime only meant to move them to nearby towns and villages that also suffered smog problems. 'Whoever has held out for all of these forty years has a right to be resettled further away,' one villager declared, to which another enjoined: 'We'll fight for that!' Whatever the villagers' ultimate fate, Mölbis itself was set for demolition no later than June 30, 1992.[34]

Free choice suddenly became possible amid the revolutionary atmosphere of 1989. When resettlement was put to a vote in January 1990, 77% of the dwindling village population elected to leave. As villagers reflected in 1994, the situation had been unbearable, even though they loved their *Heimat*. 'When I ride my bike to the plant in the morning, I know how many Mölbiser already made their way ahead of me,' a resident recalled. 'The traces from bicycle tires were exposed in the coal dust that had fallen in the night.' Residents had to walk outside covering their faces before they went indoors. '*Heimat* is *Heimat*,' a Mölbiser recalled, 'but the gas kept getting worse.'[35]

In February 1990, Pastor Dallmann gave an interview in the newspaper relating his dream that one day the Espenhain plant would be restored as an environmentally friendly facility, the ruins would vanish, new homes would appear, and everything would be green and healthy. A resolution by the district council of ministers on February 8, 1990, called for either reconstructing or moving Mölbis 'to be discussed and decided in consultation with the residents' at state expense, as well as for closing the Espenhain plant by December 1991.[36] Already by August 1990, however, the carbo-chemical installations at Espenhain had been shut down due to pressures from the market economy.[37] Meanwhile, professional investigation proved that the village soil had not been seriously contaminated, while air pollution radically diminished each week. Now the Mölbiser had sleepless nights because of their 'torn feelings' about whether to stay or go, Dallmann recalled, 'rather than due to the smell of the gas.'[38] The verdict was a 93% vote in favour of staying in December 1990. The planting of the *Hoffnungslinde* (Hope Linden tree) was ultimately followed by complex federal and state funding for new infrastructure and diverse investments and loans that enabled villagers to restore their homes (Figure 6.4).[39]

Amid the feverish pace of reconstruction, West German finance official on loan, Horst Frese, nonetheless found the time and space to begin thinking about commemoration. His overriding goal was to help local people decipher the labyrinth of Western bureaucratic tangles so they could get access to much-needed funds and supplies. By January 1994, he had secured 200,000 DM from the Bundesstiftung Umwelt (Federal Environmental Foundation) for the creation of the Umwelt- und Industriegeschichtliches Informationszentrum (Environmental and Industrial History Information Centre) in the village's decrepit Orangery—the only structure that remained of the baroque palace so needlessly demolished by Soviet and SED authorities in 1948. The outcome was an effectual museum to

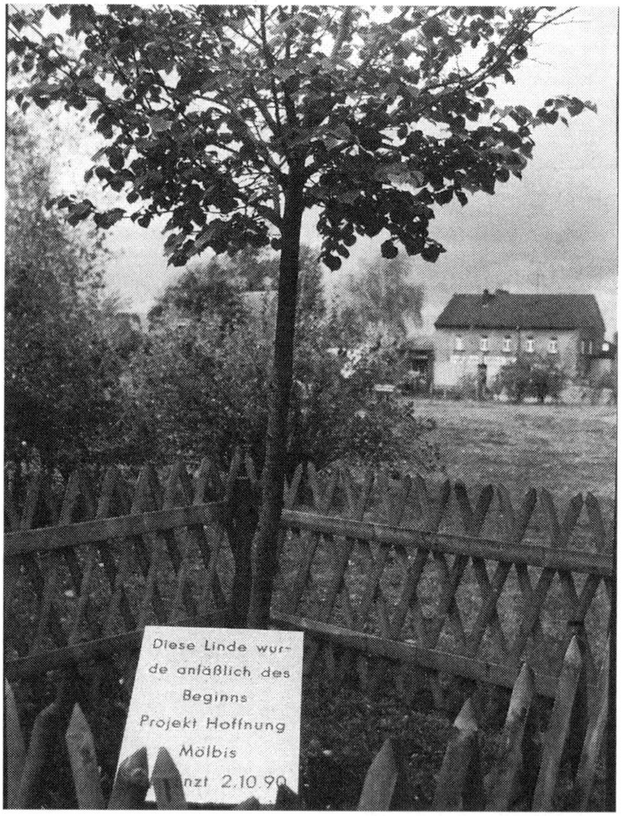

Figure 6.4 Hope Linden, 1992.

Source: Archive Horst Frese.

the grassroots ecological and political movements that had transformed the village from a symbol of filth to an emblem of hope.[40] Reproductions of protest letters to the regime, grassroots assessments about ecological damage exhibited at environmental church services in Mölbis and nearby villages, photographs of the village's decrepitude right after 1989, and even a surviving banner from protest marches in Mölbis testified to the rich history of ecological protest that had accumulated around the symbolic village. Leipzig publisher Thomas Nabert applauded the outcome in 1996, asserting, 'a permanent exhibition there could sustain memory of the pollution and exploitation of the environment, as well as the opposition that formed against it since the early 1980s, while simultaneously offering space for multifaceted events.'[41]

Documentation and graphic historical photographs on permanent display in the Orangery testify not just to the 1980s ecological protest scene that had centred in Mölbis, but also the many dreams and architectural plans that furthered the village's restoration after 1989. An October 1990 spatial plan was drafted by Leipzig's urban planning office and spearheaded by architect Peter Schmelzer, who was born near Mölbis but active since the 1950s in West German Stuttgart. Schmelzer's plan had even intended a reconstruction of the palace on its former site at the centre of the village in place of an industrial site soon slated for closure.[42] The goal was to attract a landscaping school or general conference centre.[43] Although the palace never returned, plans circulated for the construction of new housing, shops and a *Heimat* museum. Unemployment was supposed to end thanks to a new industrial park.[44]

Such initial dreams for restoring Mölbis on the basis of lost historical landmarks rapidly dwindled, however, amid an economic transformation that introduced an influx of new residents and preoccupied old Mölbiser with everyday concerns. 'Most people forget that each house they pass in Mölbis is either a new construction or was a ruin not long ago,' a resident expressed to a visiting social worker in 1995. Among complaints about unemployment and traffic noise, one resident even groused about all the freshly planted linden trees—by default including the symbolically significant Hope Linden: 'Cars under the lindens get so filthy, downright sticky, I tell you.' She expressed the desire for evergreens instead.[45] Short-term rejection of the Hope Linden as an emblem for Mölbis's remarkable history could not have been more clearly expressed.

Once the village survived, the threat of its devastation and near destruction due to coal mining dissipated. And the exhibition became a mere heritage display tucked away in a typically locked Orangery that few visited. As regional landscape planning official Andreas Berkner reflected in 1997: 'when visiting Mölbis at the northern side of the Halde, it is hardly imaginable that [this was] once "the filthiest village in Europe",' or that resettlement had ever been 'seriously up for debate.'[46] Some have even debated clearing out the exhibition to remake the Orangery into just another meeting hall. But it remains intact to this day.

Despite some new jobs from a recycling centre as well as a few additional small industries, most Mölbiser found employment due to their village's proximity to Leipzig; the population doubled because Mölbis turned into a commuter settlement. As a sleepy suburb, Mölbis was hardly a remarkable place once the scaffolding had vanished, the outside officials had departed, and press attention had dwindled. Already in 1993, *Neues Deutschland* went so far as to call the restored village Saxony's 'prestige object.'[47] 'The village profited from its location in Leipzig's *Speckgürtel* [commuter belt],' *Deutschlandfunk* noted in 2014, which is why 'the number of residents has nearly doubled since the *Wende* [the turning point of 1989] from 324 to about 600.'[48] Fearing a loss of identity and a plunge into anonymity, Wende-era mayor Dietmar Haym took a less rosy view of the many new settlers, concluding: 'Mölbis is a large *Siedlung* [settlement] with the appearance of a village. We exist.'[49] Once Mölbis was integrated into Espenhain in 1998–1999 and then (with Espenhain) into Rötha in 2015, this anonymity only increased.[50]

As the Hope Linden grew at the village centre, however, short-term preoccupation with everyday life nonetheless evolved to evince moments of long-term interest in decrypting the meaning of a village landscape that has largely survived and thrived. For the 775th anniversary of Mölbis in 2005, a parade of fire trucks and school ensembles gave way at a certain point to commemoration of the village heritage as the 'filthiest village in Europe' and then its restoration as an emblem of a hopeful future.[51] New residents have also adopted this heritage as their own. Tobias Thieme, whose family was the first to move into the village in 1992, recently recalled his proud memory of the optimism back then and his desire to sustain it as a part of village identity.[52] Speaking at the Orangery, resident Pascal Németh (later mayor of Rötha) recalled moving to Mölbis as a schoolboy in 1993 and then choosing to return with his family in 2009 to obtain more affordable housing than he could find in Leipzig. The restoration of Mölbis became an essential part of village heritage for him, to the point that he joined forces with a regional official in 2021 to inaugurate the project 'Neugestaltung Schlosspark Mölbis – ein attraktives Dorfzentrum' (Redesign of Mölbis Castle Park – an Attractive Village Centre) funded by the European Union's LEADER programme to strengthen the economy and identity of rural regions. Twelve historical plaques, designed by graphic artist Christoph Mäder, were erected across the former palace grounds near the Orangery to depict eight centuries of village development, featuring its remarkable story of near-erasure and revival. The narrative of the village's bygone past was also given visual interpretation in an off-kilter bronze sculpture of village structures (the church, vanished palace, parsonage, etc.), villagers, and various farm animals created by Jürgen Raiber—himself a resident in Mölbis since 1995.

On one of my visits to Mölbis, I encountered the young new resident Thomas Gödicke, who with his family lives in one of the restored historic houses that featured in so much photography when the village was coated in soot at the end of SED rule. Like human beings in most social contexts, many of his neighbours have shown little interest in their historical surroundings, but Gödicke and his fellow engaged residents have been able to dig deeper into the village heritage precisely because in Mölbis so many traces are yet extant. With pride, he showed me around the village, took me into the Orangery, and drove me up the Halde Trages, an adjoining former slag heap that is now a forested mound with interpretive paths and a lookout tower. History was all around us. One just had to know where to look.

Because, unlike Eythra, Mölbis was not erased, and it sustains long-term commemorative potential for residents both new and old.

Conclusion

On October 23, 2004—the anniversary of his family's expropriation in the midst of the communist land reforms—Eythra nobleman Rudolf Binsack planted a linden tree amid the linden rows near the Trianon. Comparing the loss of *Heimat* from mining with the expulsion of Germans from the lost eastern territories after World War II, Binsack tried to conclude on the upbeat note: 'Better to lose your property and keep your life than the other way around.'[53] Unlike those forced to resettle from Eythra, many expellees had died in the midst of their expulsion from the *Heimat*.[54] By failing to mention victims of Nazism, such as Jews, his evocation of history was deeply problematic. But a comparison of the memorial experiences by both groups of *Umsiedler* is not without merit. By contrast with Germans expelled in the aftermath of World War II from the former German eastern territory of Silesia—who I have found regularly shared stories about travel back to visit the villages, cities, and cultural landscapes they cherished in nostalgic memory—villagers whose homes were demolished and cleared to make way for mining pits have rarely bothered visiting 'recultivated' landscapes. Like expellees, many have resided in memories of what they imagined from a halcyon past eternally lost. But unlike expellees' interchange with lost spaces, the very *geology* they had known was permanently transformed. Not only buildings and monuments, but also the entire cultural landscape was wiped away, today replaced by a new one covered with lakes, forests, extensive industrial farms, solar arrays, slag heaps and one of Europe's largest garbage dumps. Only through painstaking recultivation has it been possible to get plants to grow from the acidic soils that humans drew up from the depths in their greed for coal. It will prove even harder to cultivate a new cultural landscape on the unstable, homogenised soils where centuries-old villages have been erased forever.

Of course, it is hard to measure the 'dissipation' of a memory landscape, because the process necessarily involves interaction between witnesses and spaces at the biographical level; it differs from person to person. Nonetheless, past research experience has shown me that erasure can heighten the urgency of short-term memory, because those who still harbour memories know that long-term anchoring of commemorative value will be challenging, if not impossible. For sixteen years, I have joined the many cyclists alongside joggers and other recreation seekers in the *Neuseenlandschaft* lake district south of Leipzig. Like everyone else, I've cycled past the memorial stone for Magdeborn or the rest area sign for Cröbern without ever stopping to ponder what any of it might signify. In casual conversations with current residents of Mölbis, Dreiskau-Muckern, Kahnsdorf, or other towns that might well have been destroyed but later evolved into prosperous commuter towns, I have also traced a general trend of amnesia about the surrounding post-industrial landscape, even though this transformation has deeply determined the economic and social health of their seldom cohesive, largely atomised settlements. And in each of these surviving communities, a few engaged local people—most of them transplants—have started the process of commemorating the region's remarkable past via architectural and monumental artefacts, even as memory fades with the shift in generations.

Notes

1 I wish to thank Corinne Geering, Rebecca Mitchell, Torsten Meyer, and participants at the conference 'Cultural Landscapes of Memory' at the Bergbau-Technik-Park in Leipzig on September 15–19, 2022, for helpful comments. Research for this chapter was funded in part by a grant from the German Academic Exchange Service (DAAD) and by American University in Washington, DC. I wish to extend particular thanks to those who made their personal materials available to me, including Horst Frese, Dietmar Haym, Thomas Thiel, and Heidi Vogel-Hennig.

2 For examples from extensive scholarship on ideology in physical space, see Bartetzky, *Nation – Staat – Stadt*; Bevan, *The Destruction of Memory*; Czaplicka and Ruble, *Composing Urban History*; Diefendorf, *In the Wake of War*; Koshar, *From Monuments to Traces*; and Rampley, *Heritage, Ideology, and Identity.* My own research has compared architectural reconstruction under three competing Cold War regime ideologies on territories of the former Third Reich, just as it has explored the politics of nostalgic memory cultures among German expellees forced from their homes after World War II, when Germany's former eastern territories were incorporated into Poland and subjected to sweeping physical erasures and reinscriptions: Demshuk, *Three Cities after Hitler.*; Demshuk, *The Lost German East.*

3 For an analysis of autarchic approaches to brown coal production south of Leipzig under Nazi and then communist rule, see Baumert, *Autarkiepolitik in der Braunkohlenindustrie.*

4 Demshuk, "Architecture beyond Ideology," 420–449.

5 My ongoing project *Cold War Coal Pits: The Politics of Devastation in East and West Germany* compares the long-term development and consequences of open-cast pit mining and refining of brown coal west of Cologne in former West Germany and south of Leipzig, former East Germany's second-largest city.

6 Research on local history associations in the GDR is expansive and growing. For a seminal work, see Palmowski, *Inventing a Socialist Nation.*

7 The term *Erlebnisgeneration* appears extensively in work on Flucht und Vertreibung, as well as works relating to architecture and heritage. See for instance Bartetzky, Karpf and Paulsen, *Architektur und Städtebau*, 7.

8 For numerous examples, see elderly residents depicted in Tetzlaff, *Erinnerung an eine Landschaft.*

9 This and all further translations by the author. Cited in Berkner and Kulturstiftung Hohenmölsen, *Bergbau und Umsiedlungen*, 24–25, 74.

10 This occurred in numerous instances. See reports in: Dr. Poppe, head of the Office for Mining Affairs at the Leipzig District Planning Commission, "Stand der Ortsverlegung Bösdorf," ["Status Bösdorf's Relocation"] May 4, 1981, Saxon State Archive-Leipzig (SächStAL), 20304 Büro für Bergbauangelegenheiten bei der Bezirksplankommission Leipzig, Nr. 49.

11 Pehlivanian, "Ein ganzer Betrieb zieht um," 3.

12 *Aus der Geschichte Eythras*, foreword. Though hard to find in libraries, the liquidation of copies by private persons (probably having inherited the booklet from recently deceased relatives) allowed me to acquire the rare work online.

13 Quoted in *Aus der Geschichte Eythras*, 79.
14 "Adresse an Gemeinde Eythra," 2.
15 *Erinnerung an eine Landschaft*.
16 *Aus der Geschichte Eythras*, 89.
17 Thomas Thiel, interview with author, Leipzig, August 12, 2022.
18 This was especially clear in my attempts to garner any interest from the pastor and members of the parish in my research about commemorative culture relating to destroyed villages, as well as during a presentation by West-Saxon regional landscape planning expert Andreas Berkner in the Pauluskirche on August 31, 2022. Although Berkner's theme of lignite mining, recultivation, and village displacements was (in my view) especially poignant in the setting of this 1983 structure so tied to the story of resettlement, the topic of the Pauluskirche's role as a repository for artifacts from various vanished villages did not feature in the presentation. Moreover, my queries during the Q&A session elicited no interest from members of the congregation who predominated in the audience.
19 Berkner and Kulturstiftung Hohenmölsen, *Bergbau und Umsiedlungen*, 176.
20 Rudolf Binsack, August 11, 2002, quoted in Berkner and Kulturstiftung Hohenmölsen, *Bergbau und Umsiedlungen*, 178.
21 Vollbach, "Die Kirche," 242–246.
22 Krieg, "Spuk– und andere Geschichten," 251.
23 Winderlich and Kolzenburg, *Zwenkau*, 70.
24 Schulz, "Vorwort," 9–10.
25 Difficult to statistically measure or prove, this is borne out by the fact that virtually none of my interview partners took great pains to go back to see the changed landscapes of their vanished villages. A memorial stone or other relic apparently held little appeal, given the lack of the old house, church, market, or topographies that bore any resemblance to what they had known. When I pushed this point at a gathering of about two dozen elderly former residents from diverse places in Neukieritzsch on August 22, 2022, I was repeatedly told: 'Why go back?'
26 Gunter Hänseroth, interview with author, Leipzig, February 3, 2023.
27 Late 1990s plans to retain a 1939 overburden conveyor bridge as an extension of the Lindenallee over the lake, as an industrial monument failed to attract sufficient investment or political will. Berkner and Kulturstiftung Hohenmölsen, *Bergbau und Umsiedlungen*, 179.
28 Friedrich et al., "Aspects of Change," 447–456.
29 For a detailed account of how pollution from Espenhain spurred environmental movements before and after 1989, see Demshuk, *The Filthiest Village in Europe*: a monograph-length treatment featuring Mölbis, whose symbolic resonance–first as an emblem of unsustainable 'filth' before 1989 and then as an emblem of 'hope' through the 1990s–narrates the collapse of East Germany and the arguable successes of German Reunification on a human level.
30 Dorfsanierungs-und Entwicklungsgesellschaft Mölbis mbH, "Projekt Hoffnung," 8–10.
31 Cited in Sperling, "Zur Geschichte des Dorfes Mölbis," 19.
32 Kretzschmar, "Information über einen Eingabevorgang," 33–34.
33 Karl-Heinz Dallmann, "Mölbis!" unpublished transcript, circa 1988, Archiv Bürgerbewegung (ABL) 022.119.003, 99–101.
34 Emphasis in the original. "Gedächtnisprotokoll des Einwohnerforums am 19.1.89 in Mölbis über die Devastierung des Ortes bis zum 30.6.92," [Recollected minutes from

the residents's forum on January 19, 1989, in Mölbis about the devastation of the municipality by June 30, 1992"], ABL 022.119.002.
35 Kabisch, "Mölbiser Erinnern sich," 25.
36 Ministerrat der Deutschen Demokratischen Republik, "Beschluss 13/6/90 über Entscheidungsvorschläge für die Senkung der Umweltbelastung durch die Betriebe Espenhain, Böhlen, Deuben, Rositz und Webau," February 8, 1990, Bundesarchiv, Berlin (BAB) DK 5/6203, 5.
37 Dorfsanierungs- und Entwicklungsgesellschaft Mölbis mbH, "Projekt Hoffnung," 3–4.
38 Dallmann, "Der Pfarrer von Mölbis," 45.
39 For a sketch of this process, see Frese, "Das Projekt Hoffnung," 47–53. A more detailed analysis appears in chapters 11 and 12 of Demshuk, *The Filthiest Village in Europe*.
40 Frese, "Erinnerungen."
41 Nabert, "Freizeit und Erholung," 4.
42 Dorfsanierungs- und Entwicklungsgesellschaft Mölbis mbH, "Projekt Hoffnung," 15–16.
43 Büro für Stadtplanung Peter Schmelzer, *Rahmenplan Vorentwurf, Gemeinde Mölbis*.
44 Dorfsanierungs-und Entwicklungsgesellschaft Mölbis mbH, "Projekt Hoffnung," 14.
45 "Das Projekt Hoffnung," 69.
46 Berkner, "Der Südraum Leipzig," 360.
47 Bahlburg, "Vom dreckigsten Dorf," 5.
48 Riedel, "Mölbis."
49 Dietmar Haym, interview with the author, Mölbis, January 19, 2023.
50 Haym, Foreword, 6–7.
51 VHS video recording in private possession of Dietmar Haym, viewed by the author.
52 These, "Mölbis—Wandlungen," 231.
53 Cited in Binsack, "Das Rittergut Eythra," 94.
54 For statistical estimates surrounding the flight and expulsion of Germans in the aftermath of World War II, see Demshuk, *The Lost German East*, 52–53.

Bibliography

"Adresse an Gemeinde Eythra zum 1000jährigen Jubiläum." ["Address to the Municipality of Eythra on the 1,000th Anniversary"]. *Neues Deutschland,* September 15, 1979, 2.

Aus der Geschichte Eythras: ein Ort, der über 1000 Jahre bestand. [*From the History of Eythra: A Place that Existed for over 1,000 Years*]. n.p.: n.p., 1985.

Bahlburg, Ronald. "Vom dreckigsten Dorf zum Vorzeigeobjekt. Mölbis erlebt seine Wiedergeburt." ["From the Filthiest Village to a Prestige Object. Mölbis experiences its Renaissance"]. *Neues Deutschland* August 11, 1993, 5.

Bartetzky, Arnold. *Nation – Staat – Stadt. Architektur, Denkmalpflege und visuelle Geschichtskultur vom 19. bis zum 21. Jahrhundert.* [*Nation – State – City. Architecture, Monument Preservation and Visual Historical Culture from the 19th to the 21st Centuries*]. Cologne: Böhlau, 2012.

Bartetzky, Arnold, Nicolaus Karpf, and Greta Paulsen, eds. *Architektur und Städtebau in der DDR: Stimmen und Erinnerungen aus vier Jahrzehnten.* [*Architecture and Urban Planning in the GDR: Voices and Memories over the course of Four Decades*]. Berlin: DOM, 2023.

Baumert, Martin. *Autarkiepolitik in der Braunkohlenindustrie. Ein diachroner Systemvergleich anhand des Braunkohlenindustriekomplexes Böhlen-Espenhain, 1933 bis 1965.* [The Politics of *Autarchy in the Lignite Mining Industry. A Diachronic Comparison of Political Regimes Based on the Böhlen-Espenhain Lignite Industrial Complex, 1933 to 1965*]. Berlin: De Gruyter, 2022.

Berkner, Andreas. "Der Südraum Leipzig – Rückblick und Entwicklungslinien seit 1960 und Ausblick auf den Beginn des neuen Jahrtausends." ["Leipzig's South – A Review and Trends since 1960 and a look ahead to the start of the New Millennium"]. *Sächsische Heimatblätter* 43, no. 5 (1997): 353–361.

Berkner, Andreas, and Kulturstiftung Hohenmölsen, eds. *Bergbau und Umsiedlungen im Mitteldeutschen Braunkohlenrevier*. [*Mining and Resettlements in the Central German Lignite Mining District*]. Beucha: Sax Verlag, 2022.

Bevan, Robert. *The Destruction of Memory: Architecture at War*. London: Reaktion, 2006.

Büro für Stadtplanung, and Peter Schmelzer. *Rahmenplan Vorentwurf, Gemeinde Mölbis*. [*Preliminary Draft Framework Plan, Municipality of Mölbis*]. Oktober 1990. Exhibition in Mölbis Orangery. Visited on July 12, 2022.

Czaplicka, John, and Blair Ruble, eds. *Composing Urban History and the Constitution of Civic Identities*. Washington and Baltimore: Wilson Center and Johns Hopkins University Press, 2003.

Dallmann, Karl-Heinz. "Der Pfarrer von Mölbis träumt einen wunderschönen Traum." ["The Pastor of Mölbis Dreams a Wonderful Dream"]. In *Mölbis: "Unsere Zukunft hat schon begonnen!"* [*Mölbis: "Our Future Has Already Begun!"*] edited by Gemeinde Mölbis et al., 44–45. Leipzig: Passage-Verlag, 1995.

"Das Projekt Hoffnung – Meinungen fünf Jahre danach." ["Project Hope – Opinions Five Years Later"]. In *Mölbis: "Unsere Zukunft hat schon begonnen!"* edited by Gemeinde Mölbis et al. Leipzig: Passage-Verlag, 1995.

Demshuk, Andrew. *Three Cities after Hitler: Redemptive Reconstruction across Cold War Borders*. Pittsburgh: University of Pittsburgh Press, 2021.

Demshuk, Andrew. *The Lost German East: Forced Migration and the Politics of Memory, 1945-1970*. Cambridge: Cambridge University Press, 2012.

Demshuk, Andrew. "Architecture beyond Ideology: The Politics of Forgotten Landmarks in Communist East Germany." *Journal of Urban History* 47, no. 2 (March 2021): 420–449.

Demshuk, Andrew. *The Filthiest Village in Europe: Grassroots Ecology and the Collapse of East Germany*. Ithaca: Cornell University Press, 2026.

Diefendorf, Jeffry. *In the Wake of War: The Reconstruction of German Cities after World War II*. New York: Oxford University Press, 1993.

Dorfsanierungs- und Entwicklungsgesellschaft Mölbis mbH. "Projekt Hoffnung." ["Project Hope"]. Unpublished Manuscript, December 1990.

Frese, Horst. "Das Projekt Hoffnung und die Dorfsanierungsgesellschaft." ["Project Hope and the Village Restoration Society"]. In *Mölbis: "Unsere Zukunft hat schon begonnen!"* edited by Gemeinde Mölbis et al., 47–53, Leipzig: Passage-Verlag, 1995.

Frese, Horst. "Erinnerungen: Als Leihbeamter beim Rat des Bezirks Leipzig (9/1990-12/1993)." ["Memoire: As an Official on Loan in Leipzig's District Council (9/1990–12/1993)"]. Unpublished Manuscript, December 28, 2022. Private archive of Horst Frese.

Friedrich, Christiane, Maria Cladders, Harald Stäuble, Denise Girardelli, and Thomas Tischendorf. "Aspects of Change in the Bandkeramik Settlement Area of Eythra, District Leipzig, Saxony." *Anthropologie* 53, no. 3 (2015): 447–456.

Haym, Dietmar. Foreword. In *Mölbis. 25 Jahre Projekt Hoffnung*, edited by Dorfentwicklungsgesellschaft Mölbis e.V., 6–7. Mölbis: Pro-Leipzig, 2015.

Kabisch, Sigrun. "Mölbiser Erinnern sich, 1994." In *Mölbis: "Unsere Zukunft hat schon begonnen!"* edited by Gemeinde Mölbis et al., 25. Leipzig: Passage-Verlag, 1995.

Koshar, Rudy. *From Monuments to Traces: Artifacts of German Memory, 1870–1990*. Berkeley: University of California Press, 2000.

Kretzschmar. "Information über einen Eingabevorgang im Zusammenhang mit einem schwerwiegenden lufthygienischen Problem." ["Information about the petition process in connection with a serious air hygiene problem in Mölbis"]. *Mölbis*, January 29, 1988, 33–34. Bundesarchiv, Stasi-Unterlagen-Archiv, Außenstelle Leipzig (BAS) MfS HA XVIII, Nr. 21592.

Krieg, Roswitha. "Spuk– und andere Geschichten." ["Haunted and Other Stories"]. In *Eythra. Am Rande der Großstadt* [*Eythra. On the Outskirts of the Big City*], edited by Werner Klötzer, 247–251. Leipzig: Engelsdorfer Verlag, 2012.

Nabert, Thomas. "Freizeit und Erholung im Südraum Leipzig." ["Leisure and Recreation in the Southern Leipzig Area"]. In *Freizeit- und Erholungslandschaft Südraum Leipzig: Werte, Projekte, Visionen* [*Leisure and Recreational Landscape Southern Leipzig Area: Values, Projects, Visions*], edited by Brigitte Steinbach. Leipzig: Passage Verlag, 1996.

Palmowski, Jan. *Inventing a Socialist Nation: Heimat and the Politics of Everyday Life in the GDR, 1945–1990*. Cambridge: Cambridge University Press, 2009.

Pehlivanian, Christa. "Ein ganzer Betrieb zieht um – ohne Stillstand der Produktion." [An Entire Enterprise Moves – Without Halting Production"]. *Neues Deutschland,* April 25, 1983, 3.

Rampley, Matthew, ed. *Heritage, Ideology, and Identity in Central and Eastern Europe: Contested Pasts, Contested Presents*. Woodbridge: Boydell Press, 2012.

Riedel, Iris. "Mölbis. Im ehemals 'dreckigsten Dorf der DDR'." ["Mölbis. In the Former 'Filthiest Village of the GDR'"]. *Deutschlandfunk*, October 6, 2014. www.deutschlandfunk.de/moelbis-im-ehemals-dreckigsten-dorf-der-ddr-100.html, accessed March 21, 2022.

Schulz, Holger. Foreword. In *Eythra. Am Rande der Großstadt*, edited by Werner Klötzer, 9–10, Leipzig: Engelsdorfer Verlag, 2012.

Sperling, Wolfgang. "Zur Geschichte des Dorfes Mölbis." ["On the History of the Village of Mölbis"]. In *Mölbis: "Unsere Zukunft hat schon begonnen!"* edited by Gemeinde Mölbis et al., 5–21, Leipzig: Passage-Verlag, 1995.

Tetzlaff, Kurt. *Erinnerung an eine Landschaft – Für Manuela*. DEFA, 1983. 1 hr., 24 min.

Thieme, Tobias. "Mölbis – Wandlungen zwischen 'dreckigstem Dorf Europas' und neuer Blüte." ["Changes between the 'Filthiest Village of Europe' and New Prosperity"]. In *Bergbau und Umsiedlungen im Mitteldeutschen Braunkohlenrevier*, edited by Andreas Berkner and Kulturstiftung Hohenmölsen, 231. Beucha: Sax Verlag, 2022.

Vollbach, Ekkehard. "Die Kirche – ein bedeutender Kulturträger." ["The Church – An Important bearer of Culture"]. In *Eythra. Am Rande der Großstadt*, edited by Werner Klötzer, 242–246. Leipzig: Engelsdorfer Verlag, 2012.

Winderlich, Britt, and Roland Kolzenburg, eds. *Zwenkau: gestern und heute*. [*Zwenkau: Yesterday and Today*]. Berlin: Albatros Verlag, 1994.

Part 3

Sharing Energy Landscapes

7 A Post-Industrial 'Adventure Land?'

Challenges for Cultural Tourism Development in the Estonian Oil Shale Region[1]

Saara Mildeberg

> We are standing on the windy side of the local Vesuvius, looking down on another lost town. The 40-metre-high spoil tip emits smoke from cracks in the terrain, self-ignited; the organic mining leftovers are burning inside the hill. Our tour guide warns us that the land beneath our feet could collapse at any moment, sending me and my friend, standing a bit further from the others, into the 300°C hidden flame. A literal tourist trap—how exciting! I look at the stream of smoke surfacing from below the ground, and I can feel its heat when I step closer. It is a rainy day in December, but none of us feels particularly cold; we have been sitting in our cars the whole day, driving from place to place.
>
> Author's notes from the 'The Shocking Ida-Virumaa' tour organised by Avasta Eesti on December 5, 2020.

Ida-Virumaa is the north-easternmost of Estonia's 15 counties, covering an area of 2,972 km^2 and spanning four types of terrain: the coastal plain of the Gulf of Finland, the Viru limestone plateau, the Alutaguse lowland, and the coastal plain of Lake Peipus. Located on the periphery of the European Union (EU) bordering Russia, it is home to approximately 138,000 people living in seven towns, 12 smaller urban settlements, and 175 villages. This Estonian region, which has traditionally relied on carbon-intensive oil shale extraction, is undergoing economic diversification supported by €353 million received via the European Commission's Just Transition Funds. Heavy industrialisation has not only caused environmental damage, but it has also given rise to ongoing social challenges emanating from the gradual transition from socialism to capitalism and the introduction of new technology. At the same time, the region is attracting attention as a tourist destination, branding itself enticingly as a 'land of adventure.'

Tourism is both a process and a context in which mediation occurs at every level, from the international scale to the grassroots. This chapter explores the challenges arising in this context for local cultural institutions. These issues are particularly salient in countries like Estonia, which have numerous post-industrial and rural regions but lack opportunities to generate environmentally, socially, and culturally sustainable economic development.

DOI: 10.4324/9781003473534-11

Several researchers have demonstrated the complexities of incorporating a landscape into a tourism economy. They point to the dissonance created through the representation of a distinct image of a place,[2] whereby the needs and wants of locals are sacrificed to accommodate those of short-term visitors, who in turn reproduce the touristic version of the place, thereby contributing to circular referencing. Mostly, these concerns relate to social or cultural sustainability. For example, Aet Annist, who conducted a case study of rural Setomaa in Southeast Estonia, criticised the state for promoting tourism entrepreneurship to evade responsibility for socio-economic conditions in rural areas.[3] Tiina Peil observed that tourism can detrimentally affect the development and preservation of a local identity, as revealed by her research in Paldiski, in Northern Estonia. This can happen when the tourism provider sacrifices local peculiarities, choosing instead to follow international tourism trends, which are created and maintained by researchers, specialists, and entrepreneurs, who rely on external interpretations.[4] Helen Sooväli found an apparent mismatch between tourism representations and local realities on Estonia's largest island, Saaremaa, in the western part of the country.[5] The present work adds to this literature through a case study of northeastern Estonia, known for its oil shale production and, more recently, branded as an 'adventure land' within tourism circuits. The issue of economic diversification is even more complicated in this region, as both the idea of tourism as an economic field and concern about climate change, fuelling the ecological imperative to transition from extractive industries, have met with scepticism. Several previous studies have documented and evaluated the situation in the region,[6] and while its oil shale heritage and landscapes have received attention, realising their potential in practice has proved challenging.[7] This chapter explores the complexities entailed in reanimating industrial landscapes using a heritage tourism lens from the perspectives of local cultural institutions and tourism entrepreneurs. In doing so, it reveals the challenges entailed in conducting ethnographic research in post-industrial landscapes.

How do the people who work with—and often in—such post-industrial landscapes approach tourism? My ethnographic fieldwork, comprising mainly participant observation and interviews conducted with tourism stakeholders in Ida-Virumaa examined different ways of mediating (post-)industrial energy landscapes in the context of tourism. With the help of the regional tourism coordinator, I compiled a list of representatives of the main cultural tourism organisations in Ida-Virumaa as potential subjects for semi-structured interviews focusing on tourism in the region. To contextualise the results of the interviews, I interviewed the representatives of Visit Estonia, the national tourism board located at Enterprise Estonia. In total, I conducted 14 semi-structured, in-depth interviews in September and December 2021 and one in September 2022. Ten of these interviews were especially relevant for exploring energy landscapes and industrial heritage. They were conducted with an independent tour guide from Sillamäe and representatives of the Kohtla-Järve Museum of Oil Shale, the Estonian Mining Museum, the Sillamäe Museum, Narva Gate, Narva Art Residency, Visit Estonia (Enterprise Estonia), the Alutaguse Parish, the Viru Film Fund, and the Ida-Viru Tourism Cluster, with the last two being projects of the Ida-Viru Enterprise Centre. The remaining four interviews provided necessary background context. Most of the guided tours where I conducted participant observation took place between 2020 and 2022. In the winter of 2020, Ida-Virumaa was Estonia's COVID-19 hotspot. In addition to these guided tours, this research is informed by numerous independent site visits, art and culture events, workshops on cultural tourism, conferences on the Just Transition, and informal discussions that took place between 2020 and 2023.

This chapter is organised in five parts. The first introduces the geography and history of Ida-Viru County and the oil shale, which is the centrepiece of the region's industrial heritage. The second part focuses on industrial landscapes, conceptualising them as heritage sites in the economic transition from mining to tourism, and outlines the challenges of conducting anthropological fieldwork in post-industrial sites. The third part foregrounds Ida-Virumaa, selected as a case study of industrial tourism, juxtaposing the potential that scholars have pointed to in their research projects focusing on policy documents at the municipal and national levels. The fourth part amplifies voices heard on site, drawing attention to the insider perspectives of various tourism stakeholders. The framework for this chapter centres on the idea that tourism potentially offers an environmentally, socially, and culturally sustainable means of diversifying the economy in and of post-industrial areas. This idea is problematised using empirical data gathered from the oil shale region of Ida-Virumaa to present and analyse practical issues faced by people and institutions engaging with industrial landscapes during the transition

Oil Shale (Waste)lands

The Viru limestone plateau in Ida-Virumaa is known for its light-brown layer of deposited kukersite, a type of oil shale found in Estonia and named after the place where it was first discovered. In the Estonian language, the word for oil shale is *põlevkivi*, meaning 'burning rock,' referring to its tendency to burn due to the high proportion of organic matter contained in the sedimentary rock. Oil shale is mined both above and below the ground. In addition to being used as fuel, oil extracted from the rock through a complex process is used in the chemical industry.[8] The environmental problems associated with its production range from vast leftover limestone dumps (which are prone to self-ignition) to the toxic waste from the oil shale chemical industry.[9] Although oil shale is still being produced in large quantities in Estonia, the limited reserves of usable resources and the impact of oil shale mining, chemical industries, and power engineering on the environment have compelled a consideration of alternative energy sources in the country. Moreover, the future of both closed and operational mines, factories, and waste storage facilities is a topic of constant discussion and debate.

Ida-Virumaa is distinguished by the traces of the oil shale industry that mark its environment: hills and holes, waste, and pollution. These human-made landscapes comprise depressions caused by the collapse of underground mines, basins filled with surface water, wavy wastelands and levelled holes, semi-coke landfills near oil shale factories, conical ash piles next to thermal power plants, mine-waste tips at the entrances to underground mines, and so forth.[10] During the Soviet era, north-eastern Estonia supplied the Soviet Union with oil shale products, but the consequences of their production—damaged landscapes and waste and pollution—have remained a contentious local issue.

> Residential buildings made of prefabricated blocks stand like ghost monuments in the middle of abandoned cities, which have been downsized into villages or demoted to distant districts of bigger towns. The freshly renovated promenade of Narva is presented as the only good kind of shock; most of the other places have a story of former glory attached. Even if we saw or met a local dweller, we were advised to keep our distance: Ida-Virumaa is currently the county most affected by the novel coronavirus. And yet, we, as urban explorers on wheels, are able to tour freely in our very own adventure land at a time when most countries on the continent are already in or preparing for a lockdown. The guide uses anecdotal comparisons to introduce the sights on our expedition to highlight the absurdity and contrast of the natural and man-made landscapes: local ghost towns reminiscent of Pripyat, a basin like the Blue Lagoon, a modest version of the Potemkin Stairs, the Nordic Riviera, fjords of gravel, and Estonia's highest cliff where mobs used to settle their scores in the 1990s. …
>
> Author's notes from 'The Shocking Ida-Virumaa' Tour, December 5, 2020

The 'brown gold'[11] has a long history in Estonia, with the first documentation of the discovery of oil shale dating back to 1777. Industrial mining of oil shale in Ida-Virumaa began in 1916 as an initiative aimed at helping Russia to survive the fuel crisis during World War I by developing alternatives to coal.[12] It subsequently expanded during Estonia's first period of independence (1918–1940), and while foreign investors helped to establish factories, local kukersite, used to produce energy, became a source of national pride during this period.[13] This was also the period when systematic research on this resource was initiated. During World War II, the Germans took advantage of the oil shale deposits in Estonia. However, it was during the Soviet occupation that oil shale mining in north-eastern Estonia increased substantially due to the growth of heavy industry in Estonia, especially after Leningrad (now St. Petersburg) in the former Soviet Union and towns in north-eastern Estonia switched from using natural gas to shale gas. It was this new momentum within the oil shale industry that influenced the development of all settlements that were (re-)established in Ida-Virumaa in the aftermath of the destruction wrought by World War II. War-torn villages were replaced with mostly monofunctional industrial towns, which had to accommodate both large-scale industry and the required labour flowing in from other Soviet republics. Compared with urban settlements in capitalist societies, which develop organically according to private interests, these centrally prioritised and rapidly built industrial towns had fundamentally different relations to their surroundings.[14] This situation created an 'urban anomaly'[15] relative to the rest of Estonia, where industrialisation took place on a smaller scale.

Oil shale extraction in Estonia peaked in the 1980s and subsequently declined in the 1990s due to post-Soviet restructuring of the industry. Privatisation of formerly state-owned enterprises, an orientation towards new markets, and the depletion of several mines resulted in a sharp decrease in production and the emigration of many former workers and their families. In the early 1990s, Russophones were impacted not only by neoliberal economic reforms but also by the ethnic policies of Estonia, aimed at restoring the nation-state with a predominantly Estonian population and pushing out most Russophones from state-level politics and, more broadly, from the public sphere.[16] The historical distinctiveness of north-eastern Estonia is reflected in specific socio-economic and cultural challenges. As Estonia's self-image centres on a rural lifestyle, architecture, and nature, Ida-Virumaa, with its Russophone population and post-industrial towns, has experienced cultural marginalisation. However, in recent years, the shrinking towns, (post)industrial landscapes, and marginalised inhabitants of Ida-Virumaa have prompted many scholars, artists, and entrepreneurs to focus their attention on the region.

Oil shale mining and the accompanying industry in Estonia are relatively unique and have evolved over time. However, the resulting landscape resembles those of other areas that are mined (for coal or lignite) in Poland, Ukraine (Donbas), Germany (Ruhrgebiet and the Lusatian and Central German lignite mining areas), France (Nord-Pas-de-Calais), Great Britain, and elsewhere. The Ida-Virumaa region is comparable to rust belt zones, generally comprising monofunctional industrial areas characterised by economic decline and shrinking populations. Although Estonia's population density is nowhere near that of the world's leading industrialised countries, the towns in Ida-Virumaa are densely populated compared with other towns in Estonia. Moreover, they are located close to each other in the oil shale basin and are directly affected by changes in the industry. Nevertheless, it would be unjust to claim that Ida-Virumaa, with all its peculiarities, has become a stagnant periphery.

Industrial Landscapes as Heritage

Open-pit quarries and waste heaps in Estonia are usually reclaimed as forest lands and are seldom used for agriculture. After direct threats have been addressed, afforestation is initiated to reduce dust emissions, prevent erosion, and restore soil productivity, with the end goal of adding economic value to the degraded land and also, more recently, to restore biodiversity.[17] Recultivating land means creating new economic and living spaces, thus requiring additional development of a landscape's social and cultural dimensions. Industry, too, has both an economic and a cultural dimension: it is a practice characterised by investments in production, construction, equipment, and labour and is simultaneously a social order with its own set of symbols, values, and attitudes. Industries may be large scale or small scale, operate in buildings or outdoors, and depend on limited or diverse investments. However, what all industrial enterprises share in common is their inherent temporality relating to influences of demand and supply, monopoly and competition, upsizing and downsizing, and the force fields of mergers and closures.[18] Post-industrial landscapes are defined by objects, interactions, and memories that endure into the present and can serve as agents of change.[19] Framed by the visual legacy of the heavy industry that created them, they form sociocultural scars.[20] Therefore, the relationship between landscape and heritage and collective memory[21] and the reanimation of industrial areas depends on tangible and intangible practices around them and on processes of remembering and forgetting within engaged communities. These communities, comprising daily actors within the landscape as well as the intellectual network that situates the heritage within a larger context, play a major role in the reanimation of industrial landscapes.[22]

The use of a reanimation lens in the turn from the material aspects of industrial landscapes to the history and memories connected to them enables a focus on three sociocultural considerations relating to the treatment of industrial spaces after the withdrawal of the industry that defined and shaped them. These considerations are the dynamics of forgetting and remembering, the continual transformation of heritage through everyday experience, and the transformative potential of scholarship and community engagement.[23] They provide the foundations of the approach, described here, which takes into account the sites, their representations, and the surrounding networks of memories, habits, and traditions, which are constantly changing and being negotiated.

The material remains of an industry that has demanded huge amounts of natural resources, labour, and space over the last century[24] pose a considerable challenge for heritage initiatives. In 2017, the Ida-Viru Tourism Cluster and its 40 partners decided to market the region as an *Adventure Land (*in Estonian: *Seiklusmaa)* in order to tackle the region's longstanding, media-propagated image of stagnation, which continues to be reinforced in certain sections of society. The Cluster, which was established in 2008, is coordinated by the Ida-Viru Enterprise Centre and, as of 2023, had 49 members, eight of which are local municipalities, the rest comprising tourism enterprises and various organisations that are open to visitors. Clustering gives the members opportunities to initiate joint appeals, applications, and projects; participate in workshops, study tours, and competitions; host journalists; and collect regional tourism statistics. Moreover, it has enabled the development of a shared concept of tourism for the region, aimed at presenting Ida-Virumaa as a tourism destination full of business opportunities, attracting new target groups and increasing numbers of visitors. The *Adventure Land* is a concept that aims to bring together several different approaches within one strategy for creating new business opportunities in the region. It entails a blending of nature and culture: sandy beaches, vast forests, and wetlands, as well as industrial landscapes, military history, and educational and recreational opportunities already established on site. The Development Strategy of Ida-Virumaa County, which was formulated by the Union of Municipalities of Ida-Virumaa (IVOL), supports the strategy and anticipates that Ida-Virumaa will become the most preferred tourist destination in Estonia after Tallinn by 2030.[25]

As a researcher from outside the region, I was also a visitor with my own preconceived ideas and a certain tourist gaze influenced by my exposure to different media. However, an embodied exploration with local guides in the lost places of industrial waste constitutes an intervention of norms and standards associated with regular tourism, a way of doing research that can be described as 'scavenger ethnography.'[26] It was as if I was chewing on someone's leftovers, trying to make sense of lost places like post-industrial complexes or industrial wastelands, while striving for a more practical, future-oriented way of engaging with them, which was not based on nostalgia. In studies of subcultures, the term 'scavenger ethnography' has been used to describe personal and interdisciplinary combinations and the autobiographical perspective of a researcher who uses multiple and diverse sources to gather data, often in illegitimate ways, to gain knowledge about those who deviate from the norm.[27] These diverse sources are then juxtaposed through a type of sociological cross-referencing. The aforementioned aspects of data collection were also present when I made every effort to participate in any tourism activities, especially those that disregarded or circumvented the government-imposed restrictions during the COVID-19 lockdown. These efforts included searching for advertised postings on guided tours on less stringently controlled Facebook sites or hiding from the flash of light from the police car that surprised our group of 20 standing together on a hillside at a time when a maximum of ten people were allowed to gather.

Scavenger ethnography involves the upcycling of ideas and materials, and within this conceptual framework, it meant an exploration of both the material and symbolic repurposing of mining waste, thereby empowering the research subject. It encompasses individual and collective ways of adjusting to a phenomenon through a reconsideration of societal values and ideologies. This method of collecting empirical data can be compared with dumpster diving—a recycling of what has been discarded as a response to overproduction and consumption.[28] The aim of such participant observation and the resulting ethnography is to reduce the distance between what is deemed 'deviant' and what is understood as 'normal,' thereby narrowing the gap and allowing for an uneasy recognition of similarities and common ground throughout society. Viewing industrial wastelands as tourism objects enables both the ethnographer and the reader to become critical participants who can engage in both dialogue and the construction of meaning.[29]

Adventure Land: Visions for Sustainable Development

Within the framework of the EU's Just Transition Fund, tourism is viewed as a contributor to economic diversification. Already in 2014, attention had shifted towards the tourism potential of the region, as reflected in the Estonian Regional Development Strategy for the years 2014–2020, which suggested utilising the 'remaining resources of mining and industry, and abandoned mining areas and industrial buildings in tourism and other functions' and the 'tourism potential of Ida-Virumaa more generally through recreational and cultural services and local marketing'[30] to ensure the positive development of Ida-Virumaa. The Regional Development Policy Programme for the years 2021–2024 has a broader focus and promotes different funding and (cross-border) collaboration opportunities, municipal reorganisation, and spatial planning to improve the living and business environment. Although it does not discuss the region's tourism potential, it does include the 'Ida-Virumaa programme,' to support 'sustainable development of urban areas of Ida-Virumaa' and an analysis of the 'restructuring [of] Ida-Virumaa's economy.' In the revised Regional Development Policy Programme for the years 2022–2025, the Just Transition Fund has been added as an additional support measure for Ida-Virumaa.

In the 'Adaptation of Ida-Virumaa County's Economy and Labor Market to the Reduction of Oil Shale Industry' report released in 2020 by the Praxis Centre for Policy Studies, tourism is highlighted as an economic sector that can potentially grow. However, a mapping of resources, including cultural monuments or industrial objects suitable for cultural heritage protection as UNESCO World Heritage, is required.[31] The report states that it is possible to develop the 'oil shale region as a whole into a leisure, sports, tourism, and industrial heritage region' but only if the value of the 'oil shale heritage and landscapes' is assessed by 'a group of specialists from different fields' in consultation with 'local officials and stakeholders.'[32] The final report of the 'Heritage Tourism in Estonia: Potential and Development Perspectives' project published in 2021 by the University of Tartu's Institute of History and Archaeology and supported by the European Regional Development Fund of the Estonian Research Agency also expresses optimism about the industrial heritage of Ida-Virumaa. The report emphasises the importance of thematic clustering of exhibits and the provision of expert explanations.[33] Tourism is also included in the local development plans of the municipalities of Ida-Virumaa, of which eight are members of the Ida-Viru Tourism Cluster.

The *Adventure Land* concept developed by the Ida-Viru Enterprise Centre together with its partners within the Ida-Viru Tourism Cluster is a double-edged sword, bringing industrial remains into the spotlight without revealing their origin story. As previously highlighted, two recent academic studies conducted in Estonia assessed the potential of industrial heritage tourism to reanimate these remains, noting that mapping, assessing, marketing, and cooperation among various experts and interest groups are needed for it to succeed. At the same time, none of the local municipalities list industrial (heritage) tourism as a priority in their development agendas. If mentioned at all, buildings and architecture are referred to as 'cultural' and/or 'historical' heritage, whereas landscapes that have been shaped and formed by the industry are portrayed as adventure tourism attractions. Nevertheless, there are several ways in which industrial objects and landscapes attract visitors. For example, Ida-Virumaa already hosts the Estonian Mining Museum, the Kohtla-Järve Museum of Oil Shale, and the Kiviõli Adventure Centre, while the Aidu Water Sports Centre is currently under development. In Narva, there are plans to develop the former Kreenholm Textile Factory into a creative district much like the Telliskivi Creative Hub in Tallinn, while the power plants on the town's outskirts are slowly being demolished without much ado. As the marketing materials of the Ida-Viru Tourism Cluster reveal, the industrial and post-industrial sites of Kiviõli Adventure Centre, the Estonian Mining Museum in Kohtla-Nõmme, and the Kohtla-Järve Museum of Oil Shale rely heavily on industrial remains and have found their own niche within the tourism market, attracting visitors.

The tourism development specialists at Visit Estonia, the national tourism board of Enterprise Estonia, view the brand *Adventure Land* as an opportunity to turn disadvantages into the region's advantages and strengths. They suggest that the combination of Soviet heritage and the predominance of the Russian language in some parts of the region could draw young adventurous visitors interested in the Soviet Union and/or Russia and help to bridge the generational gap between those who lived in either of those states and their children.[34] According to the European Travel Commission's 2023 report, Russia's full-scale invasion of Ukraine that commenced on February 24, 2022, has had a twofold effect on tourism: on the one hand, it has limited the post-pandemic recovery of tourism in terms of the entry of foreign visitors in countries bordering Russia and Ukraine, and on the other hand, it has led to a decrease in the proportion of Russian tourists in Europe in general.[35] However, Ida-Virumaa had already begun reacting to the gradually worsening relationship between Estonia and Russia before 2022, strongly targeting Finnish tourist groups since 2021.[36] Although exoticism and cultural distance could be disincentives, Visit Estonia representatives are convinced that the unique and distinct topics and services offered are more attractive than repelling. A survey conducted among the mostly domestic visitors of Ida-Virumaa during the fall of 2020 confirmed that those who visited the area did not regret doing so.[37] Moreover, residents of this region ranked post-industrial complexes and landscapes, such as the Kreenholm area, the Kiviõli Adventure Centre, and the Estonian Mining Museum, highest next to historical locations and objects among the places deemed the most attractive tourist destinations in the survey.[38] Events held at these locations have already become established. They include the Station Narva music and urban festival and Narva Opera Days, which have made use of the Kreenholm complex, and the Mägede Hääl music festival, formerly held on the grounds of the Estonian Mining Museum and more recently beside Lake Peipus.

Though they are all part of the region's industrial heritage, these sites emerged out of different initiatives. Plans to develop the industrial waste heap left by the oil shale industry in Kiviõli into an adventure centre were first discussed among a group of friends in 2001. After conducting a round of detailed studies, in 2007, the Ministry of Environment granted permission for its construction, and in 2013, the "hill" received its first skiers. Nowadays, the centre is open for visitors and also welcomes locals, hosting ski lessons for the local school and offering the youth job opportunities in the winter. As a ski hill, the Kiviõli Adventure Centre is an attraction limited to the winter season, but its organisers harbour a strong ambition to develop it into an all-season family leisure centre. The Estonian Mining Museum in Kohtla-Nõmme, established in the former Kohtla oil shale mine (1937–2001) after its closure, is the largest and most popular industrial attraction. Opened to the public in April 2001, the museum-park is run by a dedicated foundation and features industrial artefacts ranging from buildings to underground tunnels. In its first year, the museum received more than 27,000 visitors. In recent years, however, this number has been considerably lower, remaining, on average, at around 20,000 annual visitors. Moreover, attracting repeat visitors has been a constant challenge. The museum plays an important role in the community as well, organising

Christmas programmes for the community and facilitating the local history club in its initiative to collect and publish memories about life in Kohtla-Nõmme.

The Kohtla-Järve Museum of Oil Shale, run by and in the town of Kohtla-Järve, has adopted a different approach. Established in 1966 during the Soviet era, it exhibits photographs and artefacts and provides information boards explaining the history and current state of oil shale as a strategic resource along with professional art inspired by mining. The museum's archives contain more than 30,000 documents and 900 works of art. In addition, its staff has been actively collecting, researching, and preserving information on the town's history.

Industrial heritage also forms an essential part of sports tourism, exemplified by the Kiviõli Adventure Centre; art and education displayed in the Kohtla-Järve Museum of Oil Shale; and edutainment in the Estonian Mining Museum. This multifunctionality is symptomatic of a more general issue concerning industrial heritage in Estonia: it lacks clear definition, even in the current Heritage Conservation Act, which includes historical, archaeological, built, art, and technological monuments as well as sacred natural sites.[39] Of these, the technological monuments are closest to industrial heritage, which is defined as 'a collection of movable objects or things of historical, technological, scientific, cultural or social value, which reflects the processes that took place in the society and exemplifies the developments in economic and social history, industrialisation and aesthetics.'[40] On the one hand, this definition is almost all-encompassing, as long as the object is portable. On the other hand, none of the 35 technical monuments listed in the registry of cultural monuments as of 2023, which include a car, clock, locomotives, and a railroad switch, but also military equipment, such as tanks, cannons, and missiles, are located in Ida-Virumaa; rather, they are distributed elsewhere across the country. Anyone attempting to find an industrial item in the pan-Estonian National Register of Cultural Monuments,[41] which is the Estonian government's official list of cultural buildings, sites, and objects managed by the National Heritage Board, must know what they are looking for and where to look. For example, the list includes the administrative building of the Sompa mine, but it is categorised as built heritage, and no object can be simultaneously listed in two categories. The 'Architecture of the 20th Century' project initiated by the Ministry of Culture and the Heritage Board in 2007 also examined newer architectural heritage. Its inventory includes about 25 places and areas that are associated with oil shale heritage. Among these are the building complex of the Kiviõli Chemical Factory, the oil tower and adjacent limestone buildings of the chemical industrial complex in Kohtla-Järve, the administrative building of the Eesti Põlevkivi Company in Jõhvi, and even a massive residential district in Kohtla-Järve.[42] The overview of industrial heritage written by Henry Kuningas[43] is conceptually the most comprehensive and best illustrated publication available, but similar to the previous lists and inventories, it focuses on individual buildings or clusters, such as the headquarters of the Sompa mine or the factory complex in Kiviõli.

The valorisation of industrial heritage from the Soviet period has also posed challenges for international projects. Between 2017 and 2019, the Kurzeme, Riga, and Vidzeme regions, along with Lääne-Eesti Turism (Western Estonian Tourism), an NGO, and the Valgamaa Arenguagentuur (Valga County Development Agency), a foundation, participated in an Estonian–Latvian cooperation project. This project, titled 'Revival of Industrial Heritage for Tourism Development (Industrial Heritage),' was part of the EU's Interreg Programme.[44] Its goal was to map and systematise industrial heritage for tourism purposes by arranging objects in five categories: mills, rail transport, factories, lighthouses, and water towers. The project's main emphasis was on pre-war foreign influences and interwar Estonian establishments, so the Soviet era was not considered. As its focus was on the western and southern regions of Estonia, Ida-Virumaa County was represented solely by the Estonian Mining Museum and the Kreenholm Textile Factory, which were marked with coordinates but without any explanations provided.

A similar approach can be observed in the 'Small Handbook of Heritage Culture,' a 152-page publication created as an outcome of a Southern Finland and Estonian Interreg programme supported by the European Regional Development Fund. This programme aimed to popularise the notion of heritage among the broader public. Lembitu Tarang and colleagues classified everyday objects of heritage or folk culture across six categories, each with varying numbers of subcategories (5–43).[45] In the above publication, industrial heritage is above all considered the product of locally active, small-scale, and artisanal pre-industrial manufacturing, including small-scale agriculture and construction, relating, for example, to hunting, fishing, and beekeeping sites; gravel and sand quarries; local power plants; rafting sites; dairies; and glasshouses. Fewer than ten objects from the oil shale industry are included in the list.

Revival of Post-Industrial Complexes through Tourism

As per the Local Government Organisation Act, every urban and rural municipality in Estonia must have a development plan and a budget strategy to guide the development of that town or parish. Tourism is included in the development plans of all municipalities of Ida-Virumaa. However, with three exceptions, these plans rarely emphasise industrial heritage. The Mining Museum in Kohtla-Nõmme has been identified as the main visitor centre of the Toila municipality, and a future goal is to promote unspecified 'cultural events' for locals and establish public skiing tracks on the premises of the Mining Museum.[46] However, the main focus of tourism in the municipality's development plan is on manors and spas. The town of Kohtla-Järve, where the Museum of Oil Shale is located, emphasises event tourism, hosting large events that attract visitors.[47] In the development plans of Narva-Jõesuu, Jõhvi, and Sillamäe, industry and tourism remain entirely disconnected. The rural parish of Alutaguse places particular importance on nature tourism, but religious places, manors, spiritual heritage, and architectural objects, including the Tudulinna hydroelectric plant, are also mentioned as tourist attractions.[48]

In Narva's development plan, the former Kreenholm Textile Factory, located on a 300,000 m^2 site, is one of the prioritised tourism sites, with plans to accommodate a visitor centre, which focuses both on historical and modern textile production, a 'cultural quarter' for hosting cultural events, and spaces for private businesses to rent. These plans are, however, contingent on EU support and private investments.[49] Another notable exception that does emphasise and embrace post-industrial landscapes in their development plan is the municipality of Lüganuse. Their development plan for 2018–2028 recognises the unique tourism potential of spaces created by combining artificial landscapes and the natural environment.[50] Among others, Kiviõli Adventure Centre and Aidu Water Land (a water sports centre) are spotlighted for the sports and leisure opportunities that they offer as visitor attractions with international potential.[51] This example reveals that even those municipalities that integrate industrial landscapes within their tourism development plans rarely see them as objects of heritage and culture.

In practice, industrial heritage sites in Ida-Virumaa feature in many collaborative projects. These projects work together with local communities, relying on them as sources of oral history and of tour guides in museums, while also giving back to the communities through the provision of employment and places for expressing and strengthening their identities. One of the best examples of the practice of sociocultural inclusion in tourism is provided by the Estonian Mining Museum, which employs former miners as guides in its underground tunnels and organises a local oral history club where local people can share their stories, which are also presented in the museum (Figure 7.1).

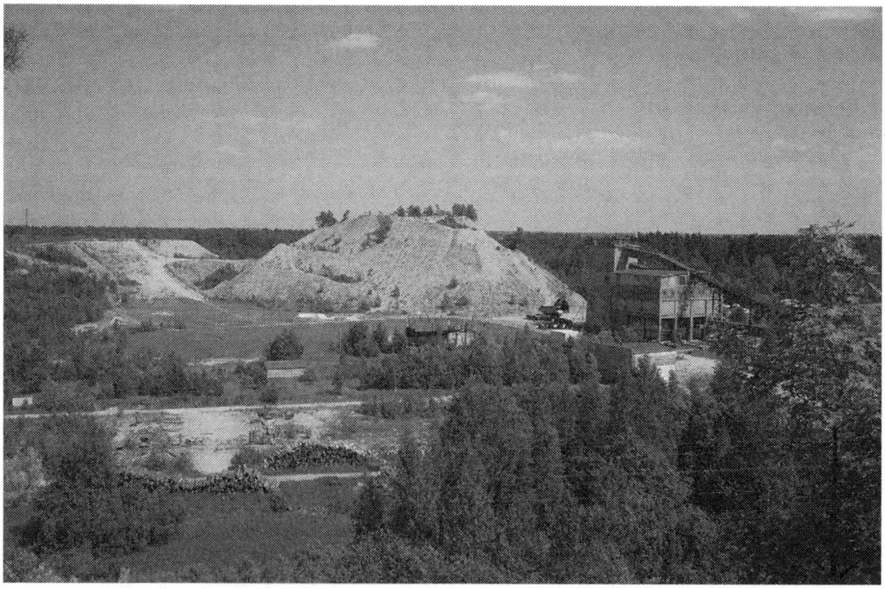

Figure 7.1 The Estonian Mining Museum, which opened to the public in 2001 in the former Kohtla open-pit and underground mine (1937–2001) in Kohtla-Nõmme. Photograph by the author, June 19, 2021.

The Estonian Mining Museum has participated in several collaborative initiatives. For example, from 2017 to 2020, the 'Mägede Hääl' ('Voice of the Mountains') music festival was held on the museum's grounds (the festival was subsequently relocated to Kauksi by Lake Peipus in 2021). Between 2020 and 2022, the 'Seiklusfest' ('Adventure Festival'), which is *Adventure Land's* signature event, was hosted at this venue. During this event, visitors could participate in an orienteering competition, navigating through post-industrial landscapes and completing challenges at various checkpoints set up by local businesses and institutions to introduce their services and products. However, for the 2023 edition, the venue shifted to the Kiviõli Adventure Centre, and in 2024, the event did not take place at all. The director of the Mining Museum also expressed optimism when describing the process and effects of hosting another collaboration, the *Life in Decline* exhibition[52] curated by Francisco Martinez, which was held from June 17, 2021, to October 2, 2021, and featured ten contemporary artists. The director praised the exhibition, which was commissioned especially for the Mining Museum and installed in the former administrative building of the Kohtla mine, signifying a synergy between cultural heritage and contemporary art:

> [T]his exhibition of 10 artists that was commissioned for us—the cultural heritage of those rooms in the 1980s style and with a contemporary art exhibition tailored for it—it is synergy. Absolutely. Similar to a safari on the premises of the quarry, which is fun, right? You drive around, you get adrenaline, and then you're being told what is behind this, what has been mined, and what has happened to the quarry.[53]

Others, however, are concerned that the oil shale industry is either too 'fresh' or too 'stained,' in the sense of being too contemporary and controversial, to become a touristic attraction. An independent tourism entrepreneur and tour guide expressed the view that time and investments are needed to enable what remains of the energy industry to transition into cultural objects. By contrast, the coordinator of the Tourism Cluster held little hope for incorporating oil shale heritage into tourism, as it 'bears the fossil [fuel] footprint,'[54] even though the use of oil shale is not limited to fuel; it also extends to the chemical industry.

For some, the region's state of limbo and the traces that remain of its industrial and Soviet past are what make it attractive (Figure 7.2). The coordinator of the Viru Film Fund echoed feedback from the Film Fund's familiarisation tours:

> Whereas a local person may say: "Lord, how ugly it is!" a filmmaker may say: "Thank God everything hasn't been fixed up here. Thank God you have paint chipping on the walls and a different landscape." We came here with one American film producer who said, "hey, your whole region is like a film studio"—a new era with different architecture every 12 or 20 km. Next town, next era, the distances are so awesomely short—you have the Kuremäe Monastery on one side, you have flying squirrels in the forest, then you have Sillamäe, then Narva, then Kiviõli—different eras, all here at hand.[55]

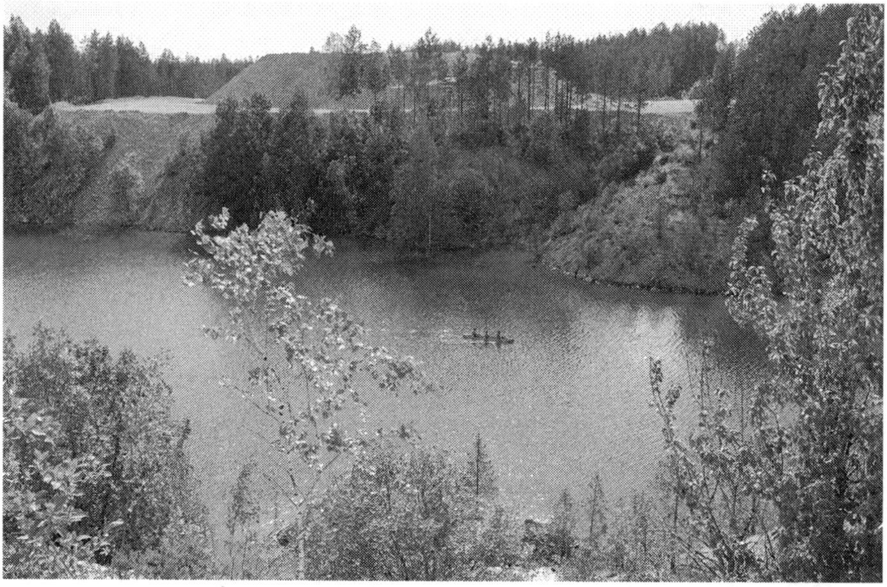

Figure 7.2 Aidu water land, a former quarry (1974–2012). Photographed by the author during the Ida-Viru Tourism Cluster's 'Seiklusfest 2021' on June 19, 2021.

Surprisingly little research has been conducted among the locals to understand their views on the industrial landscapes. An article that reported on the results of research conducted in Ida-Virumaa by the Estonian University of Life Sciences in 2011 and 2012 revealed that the relevance of environmental issues for locals has decreased, most likely due to more stringent regulations and damaged landscapes, including collapsed mine tunnels and buildings. Nevertheless, unavailable or poor-quality drinking water, dust, noxious smells, and other aspects of air pollution are still mentioned as problems.[56] Another article reporting on the same project presented the perspective of representatives of the mining industry, who viewed their landscaping activities and the compensation paid to local municipalities for mining as enriching local life.[57] The experiences of the inhabitants of the former Maidla Parish (now Lüganuse) whose villages were demolished to make way for the October mine (later renamed as Aidu) in the 1970s are only briefly alluded to as 'negative' and 'painful.' They are framed as important reminders for those seeking to take advantage of the mining landscapes to (also) bring benefits to the local community.[58] This brief acknowledgement of the pre-mining communities is more than can be found in many scientific papers on oil shale production in Estonia. The stories of the people who were affected by the mine are instead documented on the web page of the Estonian Green Movement, a non-governmental, non-profit environmental organisation, which is behind the project, 'Estonian stories about environmental changes.'[59]

Even before 2012, when the mine was already depleted and the operation of the pumps that kept water from flooding the quarry ceased, plans for the future of the area had been drawn up. They included a water sports centre, a wind farm, a shooting range for the Estonian Defence Forces, and rock recycling facilities.[60] Although oil shale mining has stopped, the landscape will remain in industrial use through the upcycling of oil shale waste into new products and the elevation of energy production to new spaces occupied by wind turbines.

The region's function as a sacrifice zone for maintaining energy independence has been internalised. The 'green turn' is viewed through a focus on its socio-economic aspects, such as future unemployment[61] or scepticism regarding the ability of the Just Transition Fund to compensate for the financial losses resulting from the closure of the mines.[62] During one of the tours, our local guide explained, 'I consider myself "green," but I wouldn't want to live in a cave,'[63] signifying both her general acknowledgement of environmental issues and also her scepticism about the 'green turn' and the associated transition.

Reflecting on the future of the region's first mine-waste tip, the Kukruse Hill, which is a remnant of the Kukruse mine that operated between 1916 and 1967, the director of the Kohtla-Järve Museum of Oil Shale emphasised the state of flux of the once industrialised landscapes:

> Today's knowledge, technologies, and consumption—I always say this is the baseline. [...] What happens next? Miners tell me that the whole mine that was closed in 1967 could be re-opened today with ease, to work there for a couple more years. With new technologies, we could still produce there. [...] So, this is the reality: on the one hand, the scarcity of resources and on the other hand, the development of technologies enables [us] to recycle and use the hills already. Why start a new mine?[64]

During the two tours to 'Witches' Wells and Devil's Basins,' organised by the Alutaguse Hiking Club on April 4 and 17, 2021, respectively, our guide proposed a different folklorised version of post-industrial landscapes, which challenges the prevailing dichotomy between nature and culture. Among the highlights of the tour are the Ratva Two Witches' wells: two boreholes drilled in Alutaguse Parish next to the Mäetaguse–Kohtla–Järve road. The boreholes were developed collaboratively by Enefit, a subsidiary of the Eesti Energia energy company, and the Alutaguse Parish in 2018, five years after the closure of the Viru mine in 2013. This was done to save nearby fields from floods after the energy company admitted that they did not expect this to happen for a year. Two 40-metre-deep pipes draw water out from the former mining tunnels to a ditch, which leads to a stream opening into a river and ultimately into the Gulf of Finland. During the high-water seasons in spring and autumn, when snow melts or there is heavy rain, the boreholes are transformed into erupting fountains (Figure 7.3). Thanks to the spectacle unfolding in an accessible location near the main road, and despite its seasonal nature—or perhaps fortuitously because of it—the 'two witches' wells' have become a popular tourism spot, equipped with a parking lot, information board, and streetlights.

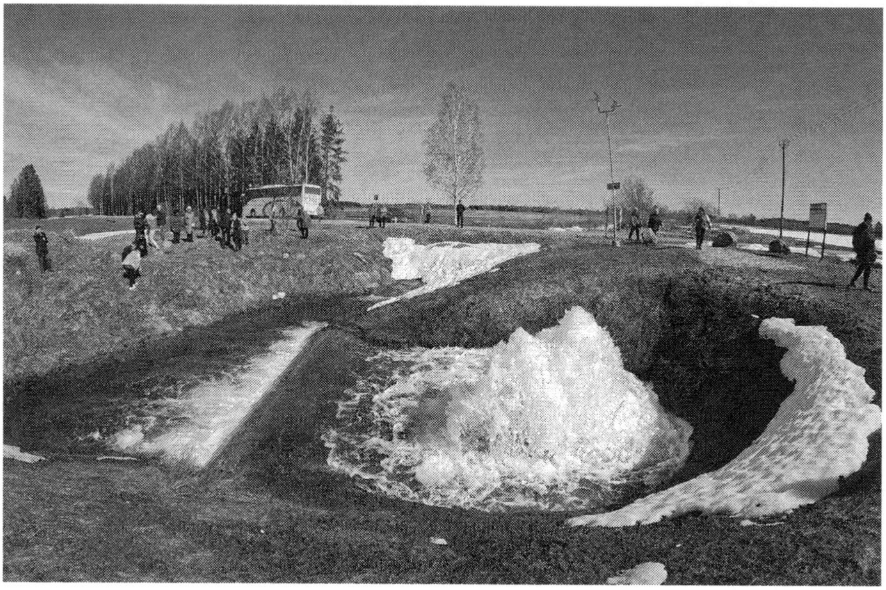

Figure 7.3 Ratva 'Two Witches' Wells' formed from two boreholes drilled on the premises of the former Viru mine (1965–2013) in 2018. Photographed by the author during the 'Bus Tour to Witches' Wells and Devil's Basins,' organised by the Alutaguse Hiking Club on April 4, 2021.

'Who else would make the water overflow [other] than witches, and who else would swim in those basins [other] than devils?'[65] The guide who asked this question was comparing the Ratva Two Witches' wells to the Tuhala Witch's Well, a karst spring that overflows after heavy rains in Kose Parish. Tuhala Witch's Well is a popular tourism spot, and it is also well-known in folklore: the overflowing is said to be caused by witches lashing out at each other underground.

As there is no other space in which to narrate these stories, local guides have taken things into their own hands by coming up with a pseudomythology. Here, the ethos of scavenger ethnography echoes in the tour guide's words to a journalist from Estonian Public Broadcasting:

> It has been said that maybe we shouldn't introduce such places; that this is connected to mining and it's not "pure nature." But then I say that in that case, we shouldn't introduce historical objects at all. This is also our history, our life; it goes with us. Of course, we should show it, talk about it, and talk about its different aspects.[66]

For a visitor, this conflict makes the on-site discovery of the origin of various hills and bodies of water, and their connections to the mines, power plants, and adjacent towns, especially rewarding.

Conclusion

The positioning of oil shale within Estonian culture is ambivalent. On the one hand, it guarantees energy independence and has consequently been a source of national pride. On the other hand, the drastic environmental and social changes that accompanied industrialisation, especially during the Soviet era, have left their mark. The tangible and intangible elements of the vast energy landscapes in Estonia pose a challenge for a nation that takes pride in its rural scenery, pre-twentieth-century architecture (e.g., the medieval Old Town of Tallinn), and pristine nature. The *Adventure Land* brand created by the Ida-Viru Tourism Cluster aims to blend all of this together: the seaside, forests, manor houses and museums, the Narva Castle and Kreenholm complex, and the controversial legacies of the twentieth century, namely the Soviet era and the heavily debated and ongoing oil shale industry. Within the framework of the EU's green transition, tourism is favoured as a means of economic diversification in post-industrial regions, as it can help the communities affected by the transition achieve not only environmental and economic sustainability but also social and cultural sustainability. This idea is endorsed in national-level reports, which acknowledge the tourism potential of industrial heritage and include recommendations for developing industrial heritage tourism. However, a critical policy analysis in Ida-Viru County revealed that it is rarely included in the development plans of the eight local municipalities. Industrial objects are included as unspecified 'cultural' or 'historical' heritage, and municipalities that do make use of post-industrial landscapes define them not as industrial heritage but rather as sports and leisure objects.

The observations of the representatives of enterprises and organisations that have established themselves in and around the industrial complexes in Ida-Virumaa echo the dissonance between general suggestions and actual opportunities for developing oil shale heritage tourism. Their board members, coordinators, marketing specialists, and other representatives admit that domestic tourism is not sufficient to promote industrial tourism, pointing to the need to increase thematic cooperation in the county, develop international networks, and draw attention to industrial tourism among other types of tourism. The different institutions, including museums, sports centres, and the real estate development companies that represent them, have their own priorities and funding schemes, which cannot be easily combined with others. Even though many collaborative events are organised, the potential of industrial tourism is not being fully realised because of several constraining factors. These include the different interests and limited financing of the organisations involved, a lack of interest within the still functioning industry, scepticism about tourism among local residents, who hope that the industry will continue or return to its earlier state, and a lack of visitors, which exacerbates the vicious cycle of low supply and demand. There is optimism among the stakeholders, but a clear definition and the prioritisation of industrial heritage are necessary to achieve visibility, increase thematically based collaboration in the region, develop international networks, and apply for funding available at regional, national, and international levels.

The Just Transition process casts another shadow on the oil shale industry and what remains of it as something undesirable. Ironically, seen from a broader perspective, there is no guarantee that post-industrial landscapes will not change again: as knowledge is gained and technologies are developed, consumption increases, and post-industrial landscapes become resources for upcycling. This is an important reminder that industrial landscapes are, like the industry itself, always in flux. In terms of long-term planning, the current state of oil shale landscapes can be seen as a liminal in-between state between different stages or ways of extraction. The folklorisation of post-industrial landscapes and their incorporation within films and art amidst the EU's green transition, and especially at the grassroots level, could provide an important key to unlocking insight on energy landscapes, past, present, and future, even if at first glance the locals may seem content with living their lives in the sacrifice zone. Tourism, which has been recommended as a means of economic diversification in post-industrial societies, provides not only financial income for the locals but also a cultural platform for sharing and learning from personal life experiences acquired in this industrial environment, which is the backstage of energy production.

Notes

1 Part of this research was conducted within the framework of the 'Social and Innovative Platform on Cultural Tourism and Its Potential Towards Deepening Europeanisation' (SPOT) Project, funded under the European Union's Horizon 2020 programme.
2 Olwig, "This is not a Landscape."
3 Annist, "Heterotoopiast hegemooniaks."
4 Peil, "Turismimaastikud ja ruumilood."
5 Sooväli, "Saaremaa Waltz."
6 See Bogdanov, "Kaevanduspärandi väärtustamise"; Janushkevich, "Põlevkivitööstuspärandi väärtustamine"; Metsaots, "Holistic Development," 373–392; Metsaots et al., "Evaluation of Oil Shale."
7 Metsaots et al., "Public Opinions," 383.
8 Bauert and Soesoo, "Kukersite Oil Shale," 3.
9 Bauert and Soesoo, "Kukersite Oil Shale," 7.
10 Pae et al., "Artificial Mountains."
11 Bauert and Soesoo, "Kukersite Oil Shale," 23.
12 Luud et al., "Landscape Evaluation."
13 Bauert and Soesoo, "Kukersite Oil Shale," 23.
14 See, for example, Printsmann, "Public and Private," for a discussion on the town of Kohtla-Järve.
15 Vseviov, "Kirde-Eesti."
16 Ehala, "The Bronze Soldier," 17–19.
17 Vaht et al., "Assessment of Vegetation Performance," 4.
18 Orange, "Introduction," 13–14.
19 Orange, "Introduction," 15.
20 Storm, *Landscape Scars.*
21 Printsmann, "Maastiku amneesia," 365.
22 Orange, "Introduction," 15.

23 Orange, "Introduction," 15.
24 Sepp, "Piiludes põlevkivitööstuse hälli," 44–47; Sepp et. al, "Piiludes põlevkivitööstuse hälli II," 46–49; Tammiksaar, "Põlevkivitööstuse sünd ja areng,"
25 Union of Municipalities of Ida-Virumaa (IVOL) "2021–2030+," 29.
26 Zhanna Baimukhamedova used this term to describe her fieldwork conducted in the Bavarian Forest National Park during the COVID-19 pandemic at a session of the Environmental Storytelling and Narrative course organised by the Environmental Humanities Research Group of the Norwegian University of Science and Technology in 2021.
27 Blackman, "Scavenger and Bricoleur," 26.
28 Clark, "The Raw and Rotten," 2.
29 Clark, "The Raw and Rotten," 36.
30 Estonian Regional Development Strategy, 30.
31 Michelson et al., *Ida-Virumaa*, 55.
32 Michelson et al., *Ida-Virumaa*, 56.
33 Valk et al., *Päranditurism Eestis*, 43.
34 For an Estonian case study of the use of Soviet heritage in tourism, see Mildeberg and Vider, "Soviet Heritage(scape)."
35 Tourism Economics, "European Tourism," 21.
36 Coordinator of the Ida-Viru Tourism Cluster (Ida-Viru Enterprise Centre), interview with the author, online, December 15, 2021.
37 Statistical data from the 2020 dataset collected for the case studies of the SPOT report. Available at www.spotprojecth2020.eu/reportsandoutcomes.
38 Statistical data from the 2020 dataset collected for the case studies of the SPOT report. Available at www.spotprojecth2020.eu/reportsandoutcomes.
39 "Heritage Conservation Act 2019," paragraph 11, subsection 1.
40 "Heritage Conservation Act 2019," paragraph 11, subsection 1.
41 National Register of Cultural Monuments maintained by the National Heritage Board, 2009. https://register.muinas.ee.
42 Hansar, "Ida-Virumaa."
43 Kuningas, *Eesti tööstuspärand.*
44 Interreg Estonia-Latvia, "Industrial Heritage."
45 Tarang et al., *Väike pärandkultuuri käsiraamat.*
46 Toila Municipal Council, "Toila valla arengukava," 17.
47 Kohtla-Järve City Council, "Kohtla-Järve linna arengukava," 20.
48 Alutaguse Municipal Council, "Alutaguse valla arengukava," 25.
49 Narva City Council, "Narva Linna arengukava," 76.
50 Lüganuse Municipal Council, "Lüganuse valla arengukava," 9.
51 Lüganuse Municipal Council, "Lüganuse valla arengukava," 9.
52 Description of the exhibition on the website of the Estonian Academy of Art: www.artun.ee/en/calendar/exhibition-life-in-decline/.
53 Director of the Estonian Mining Museum, interview with the author, Kohtla-Nõmme, September 13, 2021.
54 Coordinator of the Ida-Viru Tourism Cluster (Ida-Viru Enterprise Centre), interview with the author, online, December 15, 2021.
55 Coordinator of the Viru Film Fund, interview with the author, online, September 14, 2021.
56 Metsaots et al., "What Hinders" 380.
57 Metsaots et al. "Public Opinions," 6.

58 Metsaots et al. "Public Opinions," 6.
59 A collection of 42 personal vignettes, published in March 2017, can be found at: https://roheline.ee/category/eestilood/.
60 Kebbinau, "Põlevkivitööstuse pärand."
61 Representative of the Sillamäe Museum, interview with the author, Sillamäe, September 14, 2021.
62 Director of the Kohtla-Järve Museum of Oil Shale, interview with the author, Kohtla-Järve, September 2, 2021.
63 Alutaguse Hiking Club tour guide, "Car Tour to Witches' Wells and Devil's Basins," Ida-Virumaa, April 17, 2021.
64 Director of the Kohtla-Järve Museum of Oil Shale, interview with the author, Kohtla-Järve, September 2, 2021.
65 Alutaguse Hiking Club tour guide, "Car Tour to Witches' Wells and Devil's Basins," April 17, 2021.
66 Kundla, "Ida-Virumaa's Old Mining Area."

Bibliography

Alutaguse Municipal Council. "Alutaguse valla arengukava 2018–2030." ["Development Plan of the Alutaguse Parish, 2018–2030"]. Riigi Teataja IV, 26. 2018. Accessed August 10, 2024. www.riigiteataja.ee/akt/410072018026.

Annist, Aet. "Heterotoopiast hegemooniaks. Võim ja kultuur Setomaal." ["Heterotopia and Hegemony: Power and Culture in Setomaa"], In *Maastik ja mälu [Landscape and Memory]*, edited by Linda Kaljundi and Helen Sooväli-Sepping, 418–444. Tallinn: Tallinn University Press, 2014.

Bauert, Heikki, and Alvar Soesoo. *Kukersite Oil Shale.* Tallinn: MTÜ GeoGuide Baltoscandia, 2007.

Blackman, Shane. "Scavenger and Bricoleur: A Critical Analysis of Dick Hebdige's Repurposing of Subculture Through the Intersection of Biography and History." In *Hebdige and Subculture in the Twenty-First Century*, edited by Keith Gildart, et al., 29–50. Palgrave, 2020. https://doi.org/10.1007/978-3-030-28475-6_3.

Bogdanov, Risto. "Kaevanduspärandi väärtustamise ja kasutamise võimalused Ida-Virumaal." ["Possibilities of Valuing and Using Mining Heritage in Ida-Virumaa"]. MA thesis, Estonian University of Life Sciences, 2013.

Clark, Dylan. "The Raw and the Rotten: Punk Cuisine." *Ethnology* 43 (2004): 19–31.

Ehala, Martin. "The Bronze Soldier: Identity Threat and Maintenance in Estonia." *Journal of Baltic Studies* 40, no. 1 (2009): 139–158. https://doi.org/10.1080/01629770902722294.

Hansar, Lilian, ed. *Ida-Virumaa: 20. sajandi arhitektuuri inventeerimine [Ida-Virumaa: Inventory of the 20th Century Architecture].* Tallinn: National Heritage Board, 2008. Accessed August 10, 2024. https://register.muinas.ee/ftp/XX_saj._arhitektuur/maakondlikud%20ylevaated/ida-virumaa/Ida-Virumaa.pdf.

"Heritage Conservation Act 2019." *Riigi Teataja*, March 19, 2019, 13. Accessed August 10, 2024. www.riigiteataja.ee/akt/119032019013.

Interreg Estonia-Latvia: "Industrial Heritage for Tourism." Accessed August 10, 2024. https://industrialheritage.travel.

Janushkevich, Viktoria. "Põlevkivitööstuspärandi väärtustamine Ida-Viru maakonnas." ["Valuing the Oil Shale Industrial Heritage in Ida-Viru County"]. MA thesis, Estonian University of Life Sciences, 2016.

Jõhvi Municipal Council. "Jõhvi valla arengukava 2019–2025 koos eelarvestrateegiaga 2019–2022." ["Development Plan of the Jõhvi Parish 2019–2025, with Budget Strategy 2019–2022"]. Riigi Teataja IV, 14. 2018. Accessed August 10, 2024. www.riigiteataja.ee/akt/419102018028.

Kebbinau, Kristiina. "Põlevkivitööstuse pärand." ["The Legacy of Oil Shale Industry"]. *Eesti Geoloog*, May 2, 2017. Accessed August 10, 2024. https://eestigeoloog.ee/kategooriad/geoloogia-ja-keskkond/polevkivitoostuse-parand.

Kohtla-Järve Municipal Council. "Kohtla-Järve linna arengukava 2016–2034." ["Development Plan for the Kohtla-Järve Town 2016–2034"]. Riigi Teataja IV, 18. 2015. Accessed August 10, 2024. www.riigiteataja.ee/akt/420102015018.

Kundla, Rene. "Ida-Virumaa vanast kaevandusalast saab suurvee ajal vaatamisväärsus." ["Ida-Virumaa's Old Mining Area Becomes a Sight Worth Seeing during the Flood"]. *Estonian Public Broadcasting*, April 11, 2022. Accessed August 10, 2024. www.err.ee/1608562072/ida-virumaa-vanast-kaevandusalast-saab-suurvee-ajal-vaatamisvaarsus.

Kuningas, Henry. *Eesti tööstuspärand [Estonian Industrial Heritage].* Tallinn: National Heritage Board, 2015.

Luud, Aarne, Valdo Liblik, and Mait Sepp. "Landscape Evaluation in Industrial Areas." *Oil Shale* 20, no. 1 (2003): 25–32. https://doi.org/10.3176/oil.2003.1.04.

Lüganuse Municipal Council. "Lüganuse valla arengukava 2018–2028." ["Development Plan for the Lüganuse Parish 2018–2028"]. Riigi Teataja IV, 5. 2018. Accessed August 10, 2024. www.riigiteataja.ee/akt/424102018005.

Metsaots, Kaie. "Holistic Development of the Oil Shale Region as an Industrial Heritage, Recreational, Sports and Tourism District." PhD diss., Estonian University of Life Sciences, 2019.

Metsaots, Kaie, Anu Printsmann, Tanel Kaart, and Kalev Sepp. "What Hinders and What Facilitates Good Living in the Oil Shale Mining Region?" *Oil Shale* 32, no. 4 (2015): 373–392.

Metsaots, Kaie, Anu Printsmann, and Kalev Sepp. "Public Opinions on Oil Shale Mining Heritage and Its Tourism Potential." *Scandinavian Journal of Hospitality and Tourism* 15, no. 4 (2015): 380–399.

Metsaots, Kaie, Kalev Sepp, and Antti Roose. "Evaluation of Oil Shale Mining Heritage in Estonia." In *Sustainable Development and Planning V: ECUSUD 2011, Alicante, Spain. 13.-15.04.2011,* edited by C. A. Brebbia and E. Beriatos, 453–468. Southampton: Wit Press, 2011. https://doi.org/10.2495/SDP110381.

Michelson, Aleksandr, Kaupo Koppel, Kirsti Melesk, Kadri Arrak, Merilen Laurimäe, Meeli Murasov, Gerli Paat-Ahi, and Andreas Piwek. *Ida-Virumaa majanduse ja tööturu kohandamine põlevkivitööstuse vähenemisega [Adaptation of Ida-Virumaa County's Economy and Labour Market to the Reduction of Oil Shale Industry].* Tallinn: Praxis Center for Policy Studies, 2020. Accessed August 10, 2024. www.praxis.ee/uploads/2020/12/Lopparuanne_Polevkivitoostuse-kahanemise-mojud.pdf.

Mildeberg, Saara, and Jaanika Vider. "Soviet Heritage(scape) in Sillamäe: Documenting the Potential in an Emerging Tourism Destination." *Societies* 12(5), 127 (2022). https://doi.org/10.3390/soc12050127.

Ministry of Finance. "Estonian Regional Development Policy 2021–2024." Accessed August 10, 2024. www.fin.ee/media/2654/download.

Ministry of Finance. "Estonian Regional Development Policy 2022–2025." Accessed August 10, 2024. www.fin.ee/media/5682/download.

Ministry of the Interior. "Estonian Regional Development Strategy 2014–2020." Accessed August 10, 2024. https://energiatalgud.ee/sites/default/files/images_sala/8/8b/Siseminis teerium._Eesti_regionaalarengu_strateegia_2014-2020._2014.pdf.

Narva City Council. "Narva Linna arengukava 2035." ["Development Plan for Narva Town 2035"]. Riigi Teataja IV, 18. 2022. Accessed August 10, 2024. www.riigiteataja.ee/akt/403102023018.

Narva-Jõesuu City Council. "Narva-Jõesuu linna arengukava kuni 2025 ja eelarvestrateegia 2019–2022 vastuvõtmine ja kinnitamine." ["Development Plan for Narva-Jõesuu up to 2025 and Adoption and Approval of a Budget Strategy for 2019–2022"]. Riigi Teataja IV, 1. 2018. Accessed August 10, 2024. www.riigiteataja.ee/akt/421112018001.

Olwig, Kenneth, "'This is not Landscape': Circulating Reference and Land Shaping." In *European Rural Landscapes: Persistence and Change in a Globalising Environment*, edited by Hannes Palang, Helen Sooväli, Marc Antrop and Gunhild Setten, 41–66. Dordrecht: Kluwer, 2004.

Orange, Hilary. "Introduction." In *Reanimating Industrial Spaces: Conducting Memory Work in Post-Industrial Societies*, edited by Hilary Orange, 13–27. Walnut Creek, CA: Left Coast Press, 2015.

Pae, Taavi, Aarne Luud, and Mait Sepp. "Artificial Mountains in North-East Estonia: Monumental Dumps of Ash and Semi-Coke." *Oil Shale* 22, no. 3 (2005): 333–343.

Peil, Tiina. "Turismimaastikud ja ruumilood." ["Looking at Spaces and Stories of Tourism"]. In *Maastik ja mälu [Landscape and Memory],* edited by Linda Kaljundi and Helen Sooväli-Sepping, 399–417. Tallinn: Tallinn University Press, 2014.

Printsmann, Anu. "Public and Private Shaping of Soviet Mining City: Contested History?" *European Countryside* 3 (2010): 132–150. https://doi.org/10.2478/v10091-010-0010-8.

Printsmann, Anu. "Maastiku amneesia. Põlevkivimaa rekultiveerimine" ["Amnesia and Landscape: Recultivating the Land of Oil Shale"]. In *Maastik ja mälu [Landscape and Memory],* edited by Linda Kaljundi and Helen Sooväli-Sepping, 365–398. Tallinn: Tallinn University Press, 2014.

Sepp, Mait. "Piiludes põlevkivitööstuse hälli." ["Peeking into the Cradle of Oil Shale Industry"]. *Eesti Loodus* 11 (2009): 44–47.

Sepp, Mait, Erki Tammiksaar, and Taavi Pae. "Piiludes põlevkivitööstuse hälli II." ["Peeking into the Cradle of Oil Shale Industry II"]. *Eesti Loodus* 65, no. 12 (2014): 46–49.

Sillamäe City Council. "Sillamäe linna arengukava 2023–2027." ["Development Plan for Sillamäe Town 2023–2027"]. Riigi Teataja IV, 1. 2023. Accessed August 10, 2024. www.riigiteataja.ee/akt/414092023001.

Sooväli, Helen. "Saaremaa Waltz. Landscape Imagery of Saaremaa Island in the 20th Century." PhD diss., University of Tartu, 2004.

Storm, Anna. *Post-Industrial Landscape Scars.* New York: Palgrave Macmillan, 2014.

Tammiksaar, Erki. "Põlevkivitööstuse sünd ja areng Nõukogude Liidus aastail 1918–1945 ning seosed Eestiga. 100 aastat tööstusliku põlevkivikaevandamise algusest Eestis ja Venemaal." ["The Birth and Development of the Oil Shale Industry in the Soviet Union in 1918–1945 and Connections to Estonia. 100 Years from the Beginning of Industrial Oil Shale Mining in Estonia and Russia"]. *Akadeemia* 29, no. 8–9 (2017): 1403–1423.

Tarang, Lembitu, Kristiina Hellström, Rein Kaljuvee, Jürgen Kusmin, Triin Neljandik, and Maire Raid. *Väike pärandkultuuri käsiraamat [Small Handbook of Heritage Culture].* Tartu: Eesti Loodusfoto, 2007. Accessed August 10, 2024. http://files.voog.com/0000/0004/9218/files/V%C3%A4ike%20p%C3%A4randkultuuri%20k%C3%A4siraamat.pdf.

Toila Municipal Council. "Toila valla arengukava aastateks 2018–2030." ["Development Plan for Toila Parish, 2018–2030"]. Riigi Teataja IV, 6. 2020. Accessed August 10, 2024. www.riigiteataja.ee/akt/403122020006.

Tourism Economics. *European Tourism: Trends & Prospects.* Quarterly Report Q4/2023. Brussels: European Travel Commission, 2024. Accessed August 10, 2024. https://etc-corporate.org/reports/european-tourism-2023-trends-prospects-q4-2023/.

Union of Ida-Virumaa County Municipalities. "Ida-Virumaa arengustrateegia 2012–2035+." ["Development Plan for the Ida-Viru County 2021–2030+"]. 2021. Accessed August 10, 2024. https://ivol.kovtp.ee/documents/9867329/32040780/IVAS+avalikustamiseks+september+2021.pdf/1d605201-ed53-41b2-b3fb-b451e7956660.

Vaht, Riina, Margus Pensa, Mait Sepp, Aarne Luud, Helen Karu, and Tiina Elvisto. "Assessment of Vegetation Performance on Semicoke Dumps of Kohtla-Järve Oil Shale Industry, Estonia." *Estonian Journal of Ecology* 59, no. 1 (2010): 3–18. https://doi.org/10.3176/eco.2010.1.01.

Valk, Heiki, Aivar Kriiska, Ave Matsin, Janika Oras, Anu Ormisson-Lahe, Taavi Pae, Aigi Rahi-Tamm, and Mari-Ann Remmel. *Päranditurism Eestis: potentsiaal ja arenguvõimalused [Heritage Tourism in Estonia: Potential and Development Perspectives].* Tartu: University of Tartu Institute of History and Archaeology, 2021. Accessed August 10, 2024. www.etag.ee/wp-content/uploads/2021/03/Paranditurism-Eestis.pdf.

Vseviov, David. "Kirde-Eesti urbaanse anomaalia kujunemine ning struktuur pärast Teist maailmasõda" ["The Development and Structure of the Urban Anomaly of Northeast Estonia after World War 2"]. PhD diss., Tallinn Pedagogical University, 2002.

8 Curating a Future for Coal and Petrochemicals

Ruhrkohle AG's Corporate Influence on the Zeche Zollverein Heritage Site

Marin Kuijt and Gertjan Plets

The Ruhr district in Germany was formerly one of the most significant industrial regions in Europe. At present, its coal mines and steel mills, no longer operational, constitute part of the region's heritage landscape. The Zeche Zollverein industrial complex in Essen epitomises this heritage. The site, comprising a well-preserved coal mine and coking plant, was explicitly conceived as a 'Star of Bethlehem'; a lighthouse that would show Europe how from the ruins of industry a new, green future could emerge.[1] It thus serves as a reminder of the region's industrial past and as a symbol of its transformation.

Previous research on Zollverein has mainly foregrounded the state and civil society in shaping this industrial heritage site and the message it projects.[2] Listed as a UNESCO World Heritage site since 2001, it often features in discussions to illustrate how industrial heritage is created and curated within local history workshops, labour unions, and political parties.[3] Politicians, former industrial workers, and trade unionists certainly played an important role in the creation of heritage in the region. However, as Berger and others have observed, the corporation that operated the mines and industrial site was itself 'vitally important for the victory of industrial heritage in the Ruhr.'[4] Nevertheless, the role of this mining and fossil-fuel conglomerate, named the Ruhrkohle AG (hereinafter referred to as RAG), has not been the focus of scholarship so far. Our examination of the role of RAG in this chapter complements earlier histories of the Zeche Zollverein, foregrounding the roles of the state and civil society. RAG is not a typical privately owned company; throughout its history, it was controlled by the state, trade unions, and private capital, making it a corporatist corporation.

This chapter delves into the increasingly acknowledged role of corporations in curating landscapes and extraction sites. Exhibitions sponsored by corporations are mostly seen as attempts to 'art-wash' their image.[5] We emphasise the idea that corporate contributions to museums and heritage sites are strategic manoeuvres that 'attempt to rewrite social histories.'[6] Such manoeuvres are particularly salient for the Ruhr district, where social history is the dominant mode of history writing. Building on the work of Rodney Harrison and others, we contend that corporate

DOI: 10.4324/9781003473534-12

investments in heritage are not merely about historic rehabilitation; rather, they are future-oriented and revolve around the creation of imaginaries.[7] Our own recent research has shown that corporate involvement extends beyond mere image management, encompassing the shaping of societal visions and behaviours concerning sustainability, extractive industries, and energy, thus influencing political-economic norms and perceptions.[8]

In this chapter, we focus on heritagisation efforts in the late 1990s and specifically on the abandoned Zollverein industrial site located at the heart of the Ruhrgebiet (Ruhr district) by the Emscher River in Essen. At the time, this site was being transformed into a heritage site under the Internationale Bauaustellung Emscherpark (IBA), a programme to redevelop this coal production region. We base our analysis on sources extracted from the IBA archive located in Bochum in the Haus der Geschichte des Ruhrgebiets (House of the History of the Ruhr). Using material from this archive, we trace RAG's role in shaping the Zollverein heritage site. The archive, which is the repository of the documents and records of the organisation that transformed Zollverein into a heritage site, contains correspondence and contracts with RAG as well as the minutes of its meetings and exhibition drafts and concepts. We combine this archival material with a reading of the site and exhibition to show how internal politics and deliberations animated the heritage site and exhibition.

We ask: how did the RAG shape the heritage site established at the *Kokerei* (coke plant) within the Zollverein complex? To elucidate how and why RAG operated in the ways that it did, we first introduce the heritage site itself and outline its history. Next, we look at the political economy of energy in Germany and assess RAG's balance sheet. We subsequently read the heritage site and RAG's involvement in its development against the backdrop of RAG's corporate interests, showing how RAG's actions relating to Zollverein were guided largely by its interests in the fossil economy. We reveal how RAG inserted a fossil fuel-intensive vision of the future into the heritage infrastructure at the site. Around the time of the opening of Zollverein, the Ruhr was undergoing a transformation. In the face of deindustrialisation, widespread job losses, and economic restructuring, the region looked toward new economic horizons.[9] While different groups in Germany saw mining as a dead end and pushed for a green energy transition, RAG questioned this path. At the launch of the Zollverein exhibition, the conglomerate wrote fossil fuels into this transition narrative, and, more recently, RAG has inserted itself into a widely shared narrative that celebrates that industrial past and imagines a green future by constructing its headquarters at the Zollverein site. However, this narrative hides the fact that the present energy regime, defined by fossils and powerful multinationals, such as RAG, is deeply unsustainable.[10] To borrow the words of Donna Harraway, in this chapter, we are 'staying with the trouble' of the fossil economy to show how, despite talk of a green transition, a corporation's desire to reproduce fossil capital (re)shaped a heritage site.[11] The unsustainable interests of RAG have undergirded and shaped Zollverein from the 1990s till the present.

Energy and Political Economy in West Germany

The Ruhr district in Germany has a rich industrial history. Coal mining and steel production were traditionally the region's main industrial sectors. Starting in the nineteenth century, the region was rapidly industrialised. By the first half of the twentieth century, a dense network of mines, factories, and railways had been established, making it a vital economic industrial hub region for Germany. Coal processing comprised two sides: the 'black side' of coal, namely the mining of coal and its direct uses, and the 'white side' of coal, that is, downstream processing of coal into other products, such as coke, required for the production of steel and chemical products.

The Zeche Zollverein complex, positioned between Essen and Gelsenkirchen in the Ruhr region, embodies this industrial past and the two faces of coal. The complex, covering an area of 100 hectares, is divided into two parts. Mining shafts that had been active since 1851 make up the eastern part. The mines are connected to the *Kokerei*, built in 1961. Together the mines and the *Kokerei* make up the 'black side' of Zollverein. The white side, a petrochemical complex, was attached to the *Kokerei*. Zollverein's mines were active until 1986, while the *Kokerei* and adjacent white side were operational until 1993. Their closure was part of the deindustrialisation of the region. Most of the heavy industry around Essen has been closed down, whereas the 'white side' of coal remains active in the region through the operations of large petrochemical plants. After industrial activity came to a halt at Zollverein, a process of preservation, musealisation, and heritagisation was initiated. Initially, RAG, which owned the Zollverein complex, planned to demolish the mining infrastructure and develop properties on the eastern side of Zollverein.[12] However, the German state listed the mines as heritage, and the North Rhine-Westphalia (NRW) government acquired the eastern side of Zollverein in 1986 to protect and preserve the iconic landmark.

As a former landscape of extraction and production, the site symbolically encapsulates how the extraction of fossil fuels not only fuelled the industrial revolution but also reshaped labour relations and modernity. The transformation of the mining and coking complex was part of the IBA redevelopment programme. From 1989 to 1999, the German state spent DM 5 billion to provide the region with a future beyond the ailing coal and steel industries.[13] Led by geographer and urban planner Karl Ganser (1937–2022), IBA was very ambitious in scope. In the face of mine closures, the programme was aimed at preserving the past and opening up a new future.[14] The *Kokerei* of Zollverein had special significance for IBA. Not only was it one of the most iconic examples of industrial heritage in the region, and thus important to preserve, but the complex was also the site of one of IBA's main attractions.[15] In 1999, the *Kokerei* hosted the final celebrations and exhibition of IBA titled 'Sonne, Mond, und Sterne' ('Sun, Moon, and Stars'). The exhibition revolved around the supposed transition of the region away from coal and towards a future dominated by solar energy (Figure 8.1).

Coal powered Germany throughout the twentieth century.[16] Until the mid-twentieth century, it was the dominant fuel within Germany's energy mix. It was

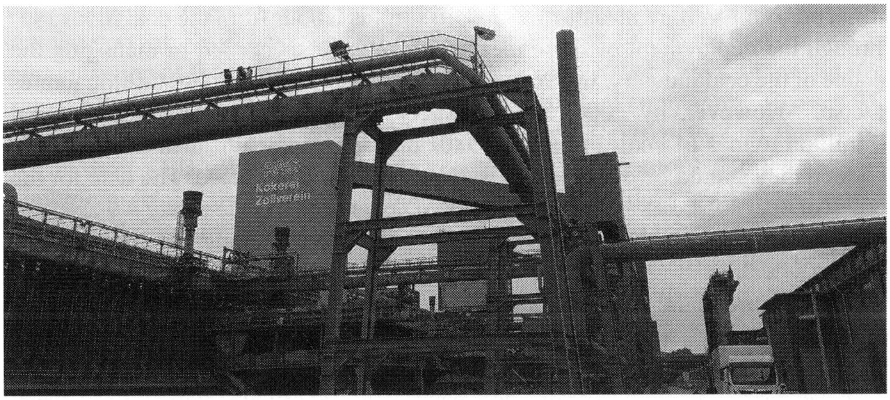

Figure 8.1 The Zollverein *Kokerei* with the RAG logo. Photograph by the authors, July 21, 2021.

also a crucial raw material for the steel and chemical industry. After the Second World War, an ever-increasing range of fuels penetrated the German energy mix. First, the proportion of oil rose steadily, overtaking coal as the most important fuel source around 1970. The discovery of gas in the Netherlands and in the Soviet Union subsequently brought this new energy source into the mix.[17] In the 1980s and 1990s, the Christian Democratic Union (CDU), the largest political faction within the West German parliament, supported the expansion of nuclear power and lignite coal mining. At the end of the century, the Red-Green coalition, which assumed office in a reunited Germany, aimed to initiate a 'green energy transition.'[18] As Stephen Gross argues, this policy meant that Germany became a global outlier. Most countries witnessed an increase in fossil fuel consumption during the same period, often attributed to neoliberal governance of the energy sector.[19] In Germany, the new government pushed for the expansion of solar energy and created a market for renewables through new legislation. An increase in the supply of renewable electricity posed a threat to the coal sector because the demand for electricity was projected to remain stable.[20] Solar energy could drown out coal and nuclear energy, and this provoked opposition from the coal industry and other involved parties. As political scientists Jacobsson and Lauber explained, 'Coal and nuclear interests [were] thus fighting the law with new vigour—probably because there [was] now a real possibility that they might be displaced.'[21]

Coal mining in West Germany had been under pressure for decades. In 1958, imported coal was cheaper than domestically mined coal.[22] To relieve some of this pressure and enable a gradual winding down of the coal industry, the state provided enormous subsidies to the mining industry. Between 1968 and 2018, the German state spent approximately 165 billion euros to close the price gap between domestic and imported coal.[23] Furthermore, mining companies could count on tax breaks and measures to stimulate the demand for coal, such as the creation of a national coal to sell and ship the coke plant to China.[41] However, after that deal fell through, the *Kokerei* became a target for preservation. RAG promised to transfer the plant, together with a dozen other industrial assets in the Ruhr, to the Stiftung Industriedenkmalpflege und Geschichtskultur (Foundation for Industrial Heritage Care and Maintenance and Historical Culture) in 1995.[42] Established by RAG and the NRW state, this foundation would protect and preserve the sites it owned, open them up to the public, and conduct research on them.[43] By transferring the *Kokerei* to the Foundation, RAG was able to avoid the cost of demolishing the plant while also meeting the public and state demand for heritage preservation.[44]

Apart from bringing in assets, RAG provided DM 3.4 million in cash to the Foundation and labour costs amounting to DM 15 million by seconding 30 employees to the Foundation for a period of five years. Notably, the NRW government provided DM 9.75 million to subsidise these costs.[45] Over several years, the NRW government also provided the Foundation with DM 40 million in liquidity to fund the transition to industrial heritage.[46] The difference of DM 21.7 million between the contributions of RAG and the NRW government was compensated for through the assets transferred by RAG to the Foundation. The costs of establishing the industrial heritage sites were configured in such a way that the contributions of RAG and the NRW government to the Foundation were equal. Therefore, both had equal stakes in the Foundation. Cost saving was the reason behind the NRW government's decision not to buy all of RAG's properties.[47] RAG's involvement and support for the Foundation meant that the corporation would carry at least part of the financial burden of the transformation. Consequently, RAG would have a controlling stake in the Foundation and could influence the process of heritagisation. This was also true for IBA as a whole. As Pinch and Adams observed, for IBA 'project delivery remained dependent upon the region's large players and interests.'[48]

Prior to its transfer to the Foundation, which was partly controlled by RAG, the plant was first decommissioned and renovated by RAG. Before it was opened to the public in 1999, physical access to the *Kokerei* could only be obtained via RAG. It was still a polluted industrial plant with hidden dangers. The decommissioning division of RAG oversaw and executed the clean-up of the site. Well into the autumn of 1998, half a year before the opening of the *Kokerei* to the public, the company was still removing soil polluted with polycyclic aromatic hydrocarbons.[49] This decommissioning work marked the inception of the process of reshaping the *Kokerei* into a heritage site. RAG was involved in all the steps relating to both the industrial and cultural work before the opening of the site to the public.

Planning the Exhibition

During the decommissioning work, IBA's preparations were well underway. In 1997, IBA's director, Karl Ganser, wrote to RAG's manager in charge of the *Kokerei* to explain his plans.[50] 'The *Kokerei* will be the biggest and most beautiful industrial monument in the area,' Ganser wrote. The monument would pay tribute to the proud past and point towards the future.[51] Ganser had two specific plans. Firstly, the *Kokerei* would host SMS, an exhibition centring on energy. Secondly, a solar power plant would be installed at the *Kokerei*. Together, these two initiatives would constitute a *Gesamtkunstwerk* (total work of art) juxtaposing 'the mining industry of the past and the solar industry of the future.'[52] Ganser envisioned a clear transition from the old mining and coke industry to a solar-powered future.[53] From the industrial ruins, a new future would emerge.

Three weeks later Ganser discussed the plans for the solar panels with executives at RAG's corporate headquarters in Essen.[54] Given its stake in the Stiftung Industriedenkmalpflege und Geschichtskultur, RAG was among the core stakeholders involved in the organisation of the SMS exhibition, and its interests and preferences mattered for the exhibition's organisers. However, the Foundation signed the contracts and paid for the exhibition.[55] A document drawn up by IBA in which the plans were discussed noted that they were only feasible if the stakeholders agreed to the plans and were willing to cooperate.[56] The first three stakeholders on the list were RAG, the Stiftung Industriedenkmalpflege und Geschichtskultur, and the RAG Fund, which owned the white side adjacent to the *Kokerei*. Planning an exhibition at the site was evidently next to impossible unless the owners of the building and the surrounding property were on board. Other involved parties were the Essenhe Municipality and former workers at Zollverein who 'keep the history of the place alive.'[57]

The initial plan for the exhibition at the *Kokerei* was developed by Karl Ganser and Feuer und Flamme, an exhibition planning agency.[58] Its stated objective was to 'symbolise the structural transformation of the region.'[59] Accordingly, Ganser asked a group of historians from various German universities and museum professionals to help him conceptualise how 'the theme of "energy" should be developed.'[60] However, this team did not develop the exhibition in a vacuum. Ganser and his team knew they had to come up with an exhibition concept that would be broadly acceptable to the key players, including RAG. Anticipating the positions of the different stakeholders, they came up with a draft exhibition concept and a statement on the message of the exhibition as follows:

> The end of coal subsidies is nearing. That is the reason for looking at the history and present state of the Ruhr through the lens of energy. [...] Globally, the use of fossil energy will not end [...], and not in the Ruhrgebiet either. It will simply become —in the Ruhr—invisible. Global CO_2 emissions are rising and will continue to do so because we do not want to prohibit developing countries from achieving our levels of prosperity. That is why industrialised countries must [...] make the effort to develop energy sources that are not as dangerous for the climate as coal is.[61]

A series of conversations with the aforementioned stakeholders would follow to make this project "*konsensfähig*" (agreeable/capable of being consented to).[62] One of these conversations took place in 1998 at a session of the *Beirat* (Advisory Board) attended by academics, officials from different state institutions, and IBA personnel, who met to discuss the message of the upcoming exhibition.[63] RAG also had one of the 16 seats in the *Beirat*. During the meeting, a discussion took place between a representative of the environmental ministry of NRW, Thomas Neiss; one of the academics who contributed to the initial message, historian Franz-Josef Brüggemeier; and the RAG representative, Karl Kleineberg. Kleineberg had a long career in mining, but in the late 1990s and early 2000s, he became involved in heritage preservation.[64] He was a proponent of public–private partnerships in this area.[65] In his view, the state (local and federal) should collaborate with private industry to preserve mining heritage.

Neiss argued that the *Kokerei* should be presented as a 'fossil of the industry's history.'[66] The specifics of the history of technology should not be foregrounded. Kleineberg countered this suggestion, noting that coke was still irreplaceable in steelmaking. Hence the *Kokerei* was not a fossil; rather, it was an example of an industry that would continue to be necessary. Brüggemeier emphasised that the focus of the exhibition should be on the industrial period in the Ruhr. Anticipating the concept of the Anthropocene that would be coined several years later, he said: 'For the first time, humans [have] remade the planet from the ground up.'[67] He continued, 'how we will represent this and what carries the most weight in our exhibition will need to [be] decided, also with an eye on the "black side" of industrialisation.'[68] He concluded that the *Kokerei* could provide an opportunity to reflect on the environmental aspects of industrialisation.

Neiss picked up on Brüggemeier's last remark. 'Energy,' he noted, 'is the central question of civilisation. It gives us the chance to develop and progress, yet we also overexploit nature.'[69] Kleineberg warned that the exhibition should not be too moralising in this regard, stating that it was easy for those with abundantly available energy to condemn the use of fossil energy. Other countries did not, however, have this luxury. Agreeing with Kleineberg, Gottfried Korf, one of the historians involved in planning the exhibition, chimed in: 'The exhibition should not have an easy message or clear moral point of view, rather it should be open.' Korf's intervention was indicative of the attitude of the exhibition designers. They wanted the exhibition to be ambivalent in its messaging rather than conveying a clear message or linear narrative.[70] This emphasis on openness and ambivalence may well reflect contemporary curatorial norms and standard practices to keep the messages of exhibitions ambiguous. It is this attitude that allowed conflicting visions of the future of energy to coexist within the same exhibition.

In the *Beirat*, the exhibition team emphasised its willingness to listen to the views of stakeholders. One of these stakeholders was RAG. The corporation agreed with the overall idea and concept of the SMS. It agreed that solar panels could have their place at the *Kokerei* as long as solar energy did not pose a direct threat to coal. In their view, the exhibition should not depict the Ruhr's transformation as a straightforward energy transition from coal to solar energy: fossil fuels, especially coal and its derivatives, would be central to the future of the region and the world.

A Semi-Sustainable Transition

In May 1999, the *Kokerei* was opened to the public with the launch of the SMS. Visitors followed 'the path of coal' through the *Kokerei*.[71] They arrived just as coal used to arrive, conveyed through the transport system that connected the coal mines in the Zollverein complex to the *Kokerei*. Once inside, the visitors passed through the building from the mixing stations to the ovens, following the route that coal had taken. The tour along the path of coal in the *Kokerei* led to a final deck.[72] Here, the visitors would firstly take a ride in a Ferris wheel named the *Sonnenrad* (sun-wheel). The motion of the wheel mimicked the history of the Ruhr. First, the wheel plunged the visitors back down into the darkness of the *Kokerei*, as miners were once submerged in the coal mines, and then it took the visitors back into the light, showing them the rooftop of the *Kokerei* with the solar panels installed on it (Figure 8.2).

After the ride, the visitors arrived at a part of the deck that showcased sunflowers, coal, and solar panels as three forms through which the sun transmits energy to the earth. This juxtaposition of plant-based energy, fossil fuel, and renewable energy sources is embodied in the exhibition's motto: 'coal goes, and the sun comes.'[73] Yet, as was intended by the exhibition's designers, the last room through which the visitors passed added ambivalence to the story of the energy transition. The visitors' final passage was through the 'avenue of fire,' where coal, oil, and natural gas were displayed and an installation demonstrated how these fossil fuels burn. As noted in the catalogue: 'The fossil-fuel economy of the Ruhrgebiet comes nearer to its end, but the end will remain open.'[74] To dispel any illusions that the exhibition was advocating a total transition away from coal, or any other fossil fuels, the catalogue ends with a chart produced by Deutsche Shell, of Germany's energy mix until 2060. The chart predicts that the relative share of fossil fuels in the energy mix will decline because of the rise of solar and wind energy. However, it also predicts that the use of coal, oil, and natural gas will remain stable throughout this period.

The exhibition had the potential to provide a critical perspective on the German energy economy and show the possibility of a sustainable transition, as advocated by the national government at the time. However, it ultimately presented a contradictory view of the energy transition. On the one hand, the solar panels embodied the quest for a renewable future and an end to mining. On the other hand, the exhibition did not speak of a general energy transition; fossil fuels would not be completely replaced by solar energy, and coal would remain an important fuel source. This contradiction in the exhibition reflected opposing visions of the future of energy in the German energy sector.

Notably, the corporatist interests of RAG lay in the fossil economy and not in the solar sector. The labour union of RAG employees also advocated the continuation of hard coal mining to preserve jobs. In 1997, the trade union IG Bergbau und Energie organised a human chain extending across the Ruhr district to demonstrate solidarity with the miners. A total of 220,000 people stood together, hand in hand, forming a 'band of solidarity.'[75] Despite all the mine closures in Germany, RAG was still heavily invested in hard coal mining. In Germany, the company, through a subsidiary, continued to operate mines until 2018, employing tens of thousands of workers. Furthermore, RAG was not slowly winding down its stakes in mining. Around the time of the exhibition, the company was actively expanding its global mining activities.[76] In 1999, the corporation bought an American mining

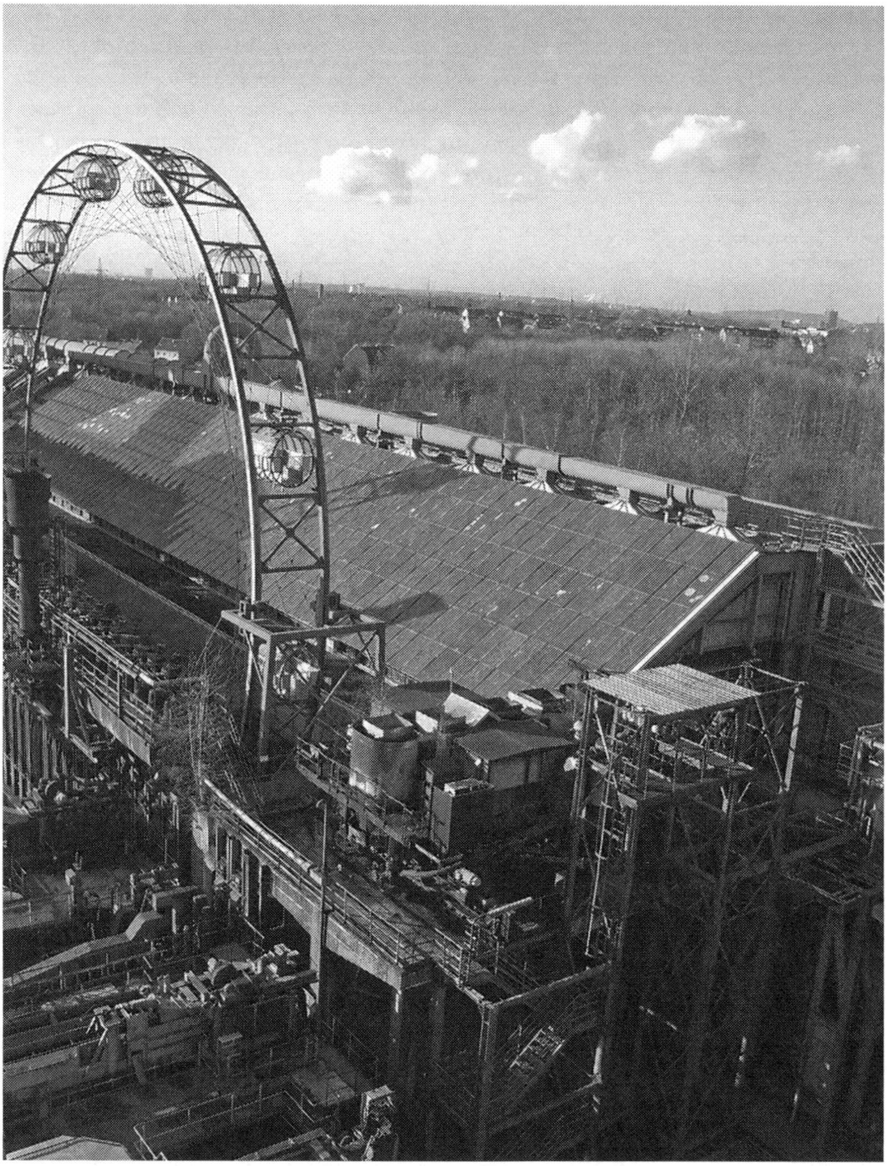

Figure 8.2 The solar panels and Ferris wheel at the 'Sonne, Mond, und Sterne' exhibition at Zollverein. Rainer Halama/CC-BY-Sa-2.5.

concern that operated 14 mines, becoming the fifth-largest coal-mining company in the United States. RAG also operated mines in Venezuela and Australia, and it had the third-largest coal-trading division in the world. RAG's vested interests in coal went beyond operating mines. The company also exported mining technology and expertise to mining corporations worldwide. By showing the visitors that coal would remain in use for the foreseeable future, SMS produced a vision of the future that would be profitable for RAG, which provided access to the site and partly controlled the foundation that paid for the exhibition. The gravitational pull of RAG thus steered the exhibition towards its own goals and preferences.

At the same time, the new federal Red-Green coalition was pushing for the expansion of solar power. In 1999, the government introduced the '100,000 roofs' programme, which subsidised the installation of solar panels.[77] Ganser saw the *Kokerei* as a means to promote and expand enthusiasm for solar energy. In the introductory section of the SMS catalogue, Ganser wrote that the solar panels installed on the *Kokerei* marked an important moment in 'energy history.'[78] The solar panels had not been financed by big energy companies because these companies were conservative when it came to energy sources of the future. Rather, Ganser and his exhibition team advertised solar modules for adoption in order to donate money to the project. This was intended to be a response to the pro-coal human chain of 1997, demonstrating that 'the people of the Ruhr are also willing to show their commitment to the future.'[79] Of the 50,000 invitees, only 813 people donated money for the solar panels.[80] Popular support lacked lustre, but the solar agenda was an important component of the exhibition.

Fossil Present

Today RAG strategically presents itself in association with the *Kokerei* as a sustainable company that is proud of its industrial past. An introductory video on its website explains that by financing *Ewigkeitslasten*, it protects water-dependent 'ecosystems' from pollution by closed mines.[81] At this point, the images of a bee and a flower appear, capturing the positive, eco-friendly image that the company wants to convey about itself. When the headquarters come into focus, the narrator states that RAG presents 'a sustainable model, admired around the world' that demonstrates how to deal with the legacy of a bygone industry. RAG's current asset portfolio is also briefly mentioned, but no mention is made of a key part of this portfolio that is not sustainable at all.

Visitors to the *Kokerei* area today, just like those who visit RAG's website, are presented with a selective image of RAG. In this area, the company associates itself with its illustrious industrial past. Visitors see, for instance, the RAG logo on the *Kokerei* (Figure 8.1). This logo dates back to the time when the *Kokerei* was still in operation. In the present, it suggests to visitors that the past industrial period of coal and steel production and RAG are synonymous. This is also the message conveyed by RAG's recently constructed corporate headquarters perched at the edge of the site (Figure 8.3). The building embodies IBA's message concerning the company's transformation of industry into heritage: proud of the past, it is eagerly looking ahead to the future.[82] The colour combination used for the exterior of the headquarters is ochre and dark brown, resembling rusted steel. Through this colour-based allusion, RAG has embedded itself in the industrial history of the site. As Chloe Taft has argued in the case of casinos located in former steel-industry buildings in the United States, through interventions in the built environment, companies seek to 'blend past, present, [and] future,' presenting themselves as exponents of a 'morally legitimate form of capitalism.'[83] RAG's headquarters also reference sustainability through its rooftop garden. As the architects proudly claim, the building's construction adhered to the highest sustainability standards.[84] The corporate headquarters overlooking the heritage site thus marries the industrial past to a supposedly sustainable present.

The marriage between the industrial past and sustainability signalled by the heritage site occludes the fossil present that sustains the heritage site. If visitors riding the *Sonnenrad* (Ferris wheel) at SMS had turned their gaze from the *Kokerei's* solar panel-covered rooftop in the opposite direction, they would have seen a petrochemical complex attached to it but closed to the public. This was the only way to see the 'white side' of coal at the *Kokerei*. While preservation efforts have focused on maintaining and restoring the buildings, machinery, and infrastructure relating to the black side of coal mining and coking operations, the white side has received less attention. What remains of this complex is still owned by RAG and is destined to be demolished and redeveloped using the company's property funds. Today visitors can see remnants of the complex that once stood there—pipes that transported gases, storage kettles, and the steel skeletons of two large chimneys. Some of the buildings in this complex have been transformed into event venues, shops, and offices. If visitors walk a little further from the *Kokerei* towards the edge of the terrain, they will see an underground parking lot covered with grass. It is a quiet corner of the Zollverein complex. To visitors, petrochemicals are not a central part of the narrative or visitor experience—there is a relative silence surrounding

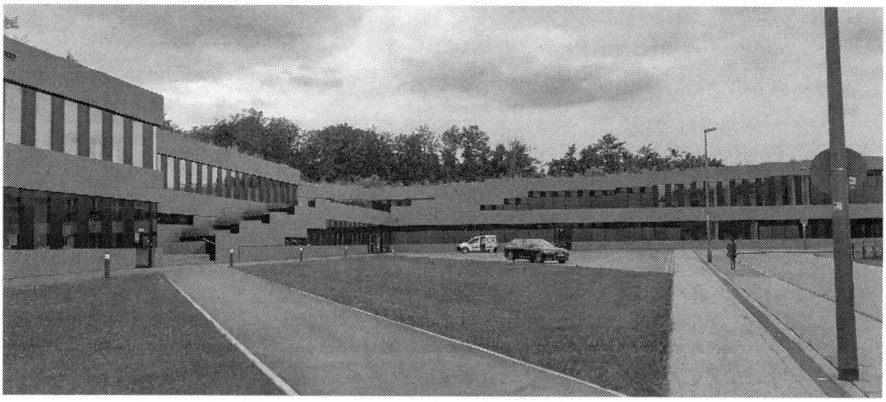

Figure 8.3 RAG's headquarters at Zollverein. Photograph by the authors.

petrochemicals, thus supporting a narrative that associates the fossil side of RAG only with coal and places it squarely in the past.

Whereas petrochemicals are marginally present at the *Kokerei*, they occupy a central place in RAG's balance sheet. In 2021, Evonik was the second-largest chemical company in Germany and the nineteenth-largest chemical company in the world, selling 15 billion euros worth of products and generating 2.38 billion euros in profits.[85] After selling some of the shares in Evonik to reinvest in private equity, the RAG Foundation still held 56.8% of Evonik's shares in 2021, which yielded a dividend of 305 million euro.[86] This dividend alone suffices to cover the *Ewigkeitslasten* of that year, which came down to 264.5 million euros. In 2021, other assets of RAG yielded an additional 347 million euros, and RAG accumulated an additional 479.5 million euros. Through the sale of some Evonik shares that year, RAG's total profit was 665 million euros.

The closing off of the 'white side' during the SMS and the demolition of large parts of the petrochemical infrastructure near the *Kokerei* rendered this portion of RAG's balance sheet largely invisible at the heritage site. It is not apparent from a visit to the heritage site that petrochemicals took the place of coal, and the paradoxical invisibility of petrochemicals to ordinary citizens is thus reinforced at Zollverein. Because petrochemicals are so ubiquitous—think, for example, of plastics—we often lose sight of them.[87]

RAG's stake in Evonik is what Andreas Malm calls 'fossil capital,' or assets that generate profits through the use of fossil fuels.[88] The chemical industry needs to be fed with natural gas, oil, and coal to manufacture its products.[89] This results in unsustainably high CO_2 emissions; in 2021, the global chemical industry was the third-largest industrial emitter of greenhouse gases.[90] Apart from contributing to global warming, the chemical industry is also a heavy polluter of the ecosystems near its plants, causing harm to communities and the environment.[91] For the past two decades, to maintain the *Ewigkeitslasten*, RAG depended structurally on its stake in Evonik.

RAG funds the closure and heritagisation of coal with the proceeds of petrochemicals, an unsustainable industry. This link between the two fossil industries is not, however, apparent to visitors to the heritage site because RAG did not preserve the petrochemical section of the complex in the same way as the coal part. This allows the conglomerate to present itself through its corporate architecture and public relations videos as greener and more sustainable than it actually is. RAG suggests to the *Kokerei*'s visitors that fossil fuels are a thing of the past and that the present is sustainable. This is meant to put visitors at ease and legitimise the company. In reality, a lot of hard political work and struggle against entrenched fossil interests is still needed to achieve a sustainable transition.

Conclusion

In this chapter, we examined the *Kokerei* at the Zeche Zollverein heritage site, calling attention to the involvement and role of the corporatist fossil fuel conglomerate, RAG, in shaping the heritage site. As a key stakeholder alongside the state and civil society in the variegated heritage landscape, RAG was able to inscribe its fossil interests in the heritage site, starting with the opening of the *Kokerei* to the public and continuing up to the present. Beneath the veneer of sustainability, fossil interests are crucial to the site. In 1999, RAG curated a future for coal in an exhibition that was meant to portray a sustainable future powered by solar energy. Its pro-coal message resonated with the company's interests and investments in coal mining in Germany and beyond. At the time of writing, the conglomerate is still heavily invested in the fossil economy through its substantial stake in the petrochemical company, Evonik. The fossil profits of Evonik contribute to the upkeep of heritage in the region, including Zeche Zollverein. In this chapter, we have argued that RAG hides the centrality of hydrocarbons to its business at the heritage site. Yet, despite the importance of petrochemicals to RAG, the former petrochemical facility at the Zollverein site received little attention and was partly demolished. Furthermore, the company moved its headquarters to the heritage site in the 2010s, choosing an architectural design that references the industrial past and a sustainable present, thereby occluding its investments in the fossil economy.

The heritagisation of Zollverein and the specific exhibition discussed in this chapter illustrate a difference between civic and corporate uses of industrial heritage in the Anthropocene. On the one hand, industrial heritage sites can stimulate critical engagement with anthropogenic climate change. As public institutions, heritage sites and associated museums can 'collaborate in building communities able to *engage with* and *respond to* climate change.'[92] This possibility was apparent at the SMS exhibition through its emphasis on solar power and the energy transition. However, SMS also presented an unsustainable vision of the future by emphasising the continued use of coal. This can be attributed to RAG's corporate interests and its central role in the exhibition's design and creation. The corporate instrumentalisation of heritage undermined the potential for civic use of heritage to reflect on climate change and energy transitions. At present, RAG continues to undermine civic engagement with climate change by presenting a façade of sustainability at the *Kokerei* site. More broadly, this case study demonstrates that to understand the history of the Zollverein heritage site, it is important to study RAG.

We therefore call for more research on how RAG and other corporations instrumentalise heritage and museums to further their fossil interests in and beyond the Ruhr. The involvement of fossil capital in museums and heritage sites raises further themes for research. As some high-profile examples of climate activism in museums show, activists and citizens demand that museums facilitate the civic use of heritage vis-à-vis climate change.[93] Corporations, such as RAG, undermine this use of heritage. It is important to gain a better understanding of the tension between civic and corporate uses of (industrial) heritage. Furthermore, as this chapter illustrates, corporations are able to use museums to curate business and fossil-friendly futures. How and why are they able to do this? The answers can be found by studying corporate actors within the field of heritage and museum studies.

Notes

1 Wagner, *Vergangenheit als Zukunft?*, 240–241.
2 Raines, "Wandel durch (Industrie)Kultur," 191–193; Dorstewitz, "Planning and Experimental Knowledge," 431–449; Berger, Wicke, and Golombek, "Burdens of Eternity?" 21–43; Zweger, *Cultural Heritage in Transition.*
3 Pinch and Adams, "The German Internationale Bauausstellung (IBA)," 233; Slach and Rumpel, "Transferable Impulses of IBA," 206.
4 Berger, Wicke, and Golombek, "Burdens of Eternity?" 26.
5 Evans, *Artwash,* 13.
6 Rectanus, *Culture Incorporated,* 18.
7 Harrison et al., *Heritage Futures.*
8 Plets and Kuijt, "Gas, Oil and Heritage," 50–77; Plets, *Heritage Statecraft and Corporate Power.*
9 Hospers, "Restructuring Europe's Rustbelt."
10 For a critique of the current fossil fuel energy regime, see Boyer, *No More Fossils.*
11 Harraway, *Staying with the Trouble.*
12 Raines, "Wandel durch (Industrie)Kultur," 191–193.
13 In 1999, the exchange rate between the euro and the DM was set at 1 euro = 1,95583 DM; Oei, Brauers, and Herpich, "Lessons," 969.
14 Wagner, *Vergangenheit als Zukunft,* 237–278.
15 Wagner, *Vergangenheit als Zukunft,* 242.
16 For the energy history of Germany, see: Gross, *Energy and Power.*
17 Högselius, *Red Gas.*
18 Gross, *Energy and Power*, 270.
19 Garavini, "Thatcher's North Sea," 38.
20 Jacobsson and Lauber, "The Politics and Policy of Energy," 269.
21 Jacobsson and Lauber, "The Politics and Policy of Energy," 269.
22 Storchmann, "The Rise and Fall," 1470.
23 Oei, Brauers, and Herpich, "Lessons," 969.
24 Storchmann, "The Rise and Fall."
25 Oei, Brauers, and Herpich, "Lessons," 969.
26 Storchmann, "The Rise and Fall," 1471.
27 Katzenstein, *Policy and Politics,* 7–10.
28 Katzenstein, *Policy and Politics*, 23.
29 Haipeter, "Works Councils," 679–695.
30 Katzenstein, *Policy and Politics*, 23; Buch-Hansen and Grau Larsen, "The Chemical Brother," 157–180.
31 Deeg, *Finance Capitalism Unveiled.*
32 Katzenstein, *Policy and Politics;* Streeck, *Re-Forming Capitalism*; Deeg, *Finance Capitalism Unveiled*; Katzenstein, "Industry in a Changing West Germany," 2–29.
33 Jenßen, "Ziele und Aufgaben," 393.
34 Deeg, *Finance Capitalism Unveiled*, 255.
35 Jenßen, "Ziele und Aufgaben," 397.
36 Handelsblatt, "RAG gibt Kokerei-Pläne auf."
37 Deeg, *Finance Capitalism Unveiled*, 255.
38 Handelsblatt, "RAG gibt Kokerei-Pläne auf."
39 RAG Foundation, *Geschäftsbericht*, 64.

40 Wagner, *Vergangenheit als Zukunft,* 237–242.
41 Zollverein, "Chronik der 80er und 90er Jahre."
42 Archive of the House of the History of the Ruhr, Bochum, 02.01, Internationale Bauausstellung Emscher Park (henceforth: IBA), 116B, Letter from the Ministerium für Stadtentwicklung, Kultur und Sport NRW to Landschaftsverband Rheinland, rheinisches Amt für Denkmalpflege und Landschaftsverband Westfalen-Lippe, Westfälisches Amt für Denkmalpflege, Sepember 5, 1995.
43 IBA 116B, "Stiftung Industriedenkmalpflege Grundsatzvermerk," 1.
44 IBA 116B, "Letter," March 22, 1995, 2.
45 IBA 116B, "Letter," March 22, 1995, 2.
46 IBA 116B, "Letter," March 22, 1995, 2.
47 IBA 116B, "Kabinettvorlage, Handlungsrahmen," July 2, 1992, 42.
48 Pinch and Adam, "The German Internationale Bauausstellung," 234.
49 IBA 827B "Ergebnisniederschrift der Steuerungsgruppe," September 25, 1998, 1.
50 IBA 1032B, "Letter Karl Ganser to Herr Simsch."
51 Wagner, *Vergangenheit als Zukunft,* 240.
52 IBA 1032B, "Letter Karl Ganser to Herr Simsch."
53 See also Wagner, *Vergangenheit als Zukunft,* 237–242.
54 IBA 1032B, "Letter Karl Ganser to Herr Müller."
55 IBA 827B, "Steuergruppe," 4.
56 IBA 1032b "Die Machtbarkeit," 12.
57 IBA 1032B "Die Machtbarkeit," 12.
58 IBA 1032B, "Letter Ganser to Borsdorff."
59 IBA 1032B, "Letter Ganser to Borsdorff."
60 IBA 1032B, "Letter Ganser to Borsdorff."
61 IBA 1032B, "Die 'Botschaft' der Ausstellung," 24/4/x.
62 IBA 1032B, "Letter Ganser to Borsdorff."
63 IBA 827B 1, "Sitzung des Ausstellungsbeirats," 1
64 Mazzoni, "Utopia," 13.
65 Kleineberg, "Stillegen – was dann?" 64–69.
66 IBA 827B 1. "Sitzung des Ausstellungsbeirats," 2–3.
67 IBA 827B 1. "Sitzung des Ausstellungsbeirats," 2–3.
68 IBA 827B 1. "Sitzung des Ausstellungsbeirats," 2–3.
69 IBA 827B 1. "Sitzung des Ausstellungsbeirats," 2–3.
70 Borsdorf, *Sonne, Mond und Sterne* 15.
71 Borsdorf, *Sonne, Mond und Sterne,* 22.
72 Borsdorf, *Sonne, Mond und Sterne,* 324–327.
73 Borsdorf, *Sonne, Mond und Sterne,* 326.
74 Borsdorf, *Sonne, Mond und Sterne,* 252.
75 Winkelman, "IGBE, Band der Solidarität."
76 Jakob, "Modern Mining Technology," 40.
77 Gross, *Energy and Power,* 294.
78 Ganser, "Die Reinaissance einer Kokerei," 10.
79 Ganser, "Die Reinaissance einer Kokerei," 10–11.
80 IBA, 1036A, "Register of donations."
81 Ruhrkohle AG, "RAG-Stiftung – A Unique Model."
82 Wagner, *Vergangheit als Zukunft*, 237–238.
83 Taft, *From Steel to Slots,* 88–89.
84 Kadawittfeldarchitektur, "RAG-Stiftung und RAG AG Zollverein."

85 Evonik, "Financial Data, 2022." https://corporate.evonik.com/en/company/facts-figures/financial-data
86 RAG Foundation, *Geschäftsbericht* (2021), 64.
87 Hanieh, "Petrochemical Empire," 25–51.
88 Malm, *Fossil Capital,* 316.
89 Hanieh, "Petrochemical Empire," 25–51.
90 International Energy Agency, *Chemicals*, https://www.iea.org/reports/chemicals.
91 Feltrin, Mah, and Brown, "Noxious Deindustrialization"; Davies, "Slow Violence."
92 Newell, Robin, and Wehner, "Introduction: Curating Connections."
93 Evans, *Artwash*.

Bibliography

Berger, Stefan, Christian Wicke, and Jana Golombek. "Burdens of Eternity? Heritage, Identity, and the 'Great Transition' in the Ruhr." *The Public Historian* 4, no. 39 (2017): 21–43.

Borsdorf, Ulrich, ed. *Sonne, Mond und Sterne: Kultur und Natur der Energie*. [*Sun, Moon and Stars: Culture and Nature of Energy*]. Essen: Peter Pomp Verlag, 1999.

Boyer, Dominic. *No More Fossils*. Minneapolis: Minnesota University Press, 2023.

Buch-Hansen, Hubert, and Anton Grau Larsen. "The Chemical Brothers: Competition and the Evolution of the Board Interlock Network in the German Chemical Industry, 1950–2015." *Business History* 1, no. 66 (2024): 675–695. https://doi.org/10.1080/00076791.2021.1923696.

Davies, Thom. "Slow Violence and Toxic Geographies: 'Out of Sight' to Whom?" *Environment and Planning C: Politics and Space* 40, no. 2 (2022): 409–427.

Deeg, Richard Edward. *Finance Capitalism Unveiled: Banks and the German Political Economy*. Ann Arbor: University of Michigan Press, 2010. https://doi.org/10.3998/mpub.15451.

Dorstewitz, Philipp. "Planning and Experimental Knowledge Production: Zeche Zollverein as an Urban Laboratory." *International Journal of Urban and Regional Research* 38, no. 2 (2014): 431–449.

Evans, Mel. *Artwash: Big Oil and the Arts*. London: Pluto Press, 2015.

Evonik. Financial Data," 2022. Accessed May 15, 2024. https://corporate.evonik.com/en/company/facts-figures/financial-data.

Feltrin, Lorenzo, Alice Mah, and David Brown. "Noxious Deindustrialization: Experiences of Precarity and Pollution in Scotland's Petrochemical Capital." *Environment and Planning C: Politics and Space* 40, no. 4 (2022): 950–969. https://doi.org/10.1177/23996544211056328.

Ganser, Karl. "Die Renaissance einer Kokerei." ["The Renaissance of a Coking Factory."] In *Sonne, Mond Und Sterne: Kultur Und Natur Der Energie*, edited by Ulrich Borsdorf, 8–17. Essen: Peter Pomp Verlag, 1999.

Garavini, Giuliano. "Thatcher's North Sea: The Return of Cheap Oil and the 'Neo-liberalisation' of European Energy." *Contemporary European History* 33, no. 1 (2024): 37–52.

Gross, Stephen. *Energy and Power: Germany in the Age of Oil, Atoms, and Climate Change*. Oxford: Oxford University Press, 2023.

Haipeter, Thomas. "Works Councils as Actors in Collective Bargaining: Derogations and the Development of Codetermination in the German Chemical and Metalworking Industries."

Economic and Industrial Democracy 32, no. 4 (November 1, 2011): 679–695. https://doi.org/10.1177/0143831X10393039.

Handelsblatt. "RAG gibt Kokerei-Pläne auf." ["RAG Gives up Coking Factory Plans."] May 31, 2005.

Hanieh, Adam. "Petrochemical Empire: The Geo-Politics of Fossil Fuel Production." *New Left Review* 130 (August 19, 2021): 25–51.

Harraway, Donna. *Staying with the Trouble: Making Kin in the Chthulucene.* Durham: Duke University Press, 2016.

Harrison, Rodney, et al. *Heritage Futures: Comparative Approaches to Natural and Cultural Heritage Practices.* London: UCL Press, 2020.

Högselius, Per. *Red Gas: Russia and the Origins of European Energy Dependence.* New York: Palgrave Macmillan, 2013.

Hospers, Gert-Jan. "Restructuring Europe's Rustbelt: The Case of the German Ruhrgebiet." *Intereconomics* 39, No. 3 (2004): 147–156.

International Energy Agency. *Chemicals.* 2024. Accessed April 15, 2024. www.iea.org/reports/chemicals.

Jacobsson, Staffan, and Volkmar Lauber. "The Politics and Policy of Energy System Transformation—Explaining the German Diffusion of Renewable Energy Technology." *Energy Policy* 34, no. 3 (2006): 256–276.

Jakob, K. F. "Modern Mining Technology and Its Application within RAG Coal International." In *IMCET 2001: Proceedings of the 17th International Mining Congress & Exhibition of Turkey*, 39–43. Ankara: The Chamber of Mining Engineers of Turkey, 2001.

Jenßen, Jens. "Ziele und Aufgaben bei der Neuordnung des Deutschen Steinkohlenbergbaus: Eine Retrospektive aus der Sicht der Ruhrkohle AG." ["Objectives and Tasks in the Reorganization of the German Hard Coal Mining Industry: A Retrospective from the Perspective of Ruhrkohle AG."] *Zeitschrift Für Öffentliche Und Gemeinwirtschaftliche Unternehmen: ZögU* [*Journal for Public and Nonprofit Services*] 10, no. 4 (1987): 387–402.

Kadawittfeldarchitektur, "RAG-Stiftung und RAG AG Zollverein." Accessed August 9, 2024. www.kadawittfeldarchitektur.de/en/projekt/rag-stiftung-und-rag-ag-zollverein/

Katzenstein, Peter J. *Policy and Politics in West Germany: The Growth of a Semisovereign State.* Philadelphia: Temple University Press, 1987.

Katzenstein, Peter J. "Industry in a Changing West Germany." In *Industry and Politics in West Germany: Toward the Third Republic*, edited by Peter J. Katzenstein, 2–29. Ithaca: Cornell University Press, 2018.

Kleineberg, Karl. "Stillegen – was dann? Ansätze der Ruhrkohle AG." ["Shutdown – What then? Approaches of Ruhrkohle AG."] *ICOMOS* 21 (1996): 64–69.

Malm, Andreas. *Fossil Capital: The Rise of Steam Power and the Roots of Global Warming.* London: Verso, 2016.

Mazzoni, Ira. "Utopia hinter den sieben Bergen." ["Utopia behind the Seven Mountains"]. *Tageszeitung* (August 15, 2001) Accessed June, 27, 2024. https://taz.de/Utopia-hinter-den-sieben-Bergen/!1156693/.

Newell, Jennifer, Libby Robin, and Kirsten Wehner. "Introduction: Curating Connections in a Climate-Changed World." In *Curating the Future: Museums, Communities and Climate Change*, edited by Jennifer Newell, Libby Robin, and Kirsten Wehner, 1–16. London: Routledge, 2016.

Oei, Pao-Yu, Hanna Brauers, and Philipp Herpich. "Lessons from Germany's Hard Coal Mining Phase-Out: Policies and Transition from 1950 to 2018." *Climate Policy* 20, no. 8 (2020): 963–979.

Pinch, Philip, and Neil Adams. "The German Internationale Bauausstellung (IBA) and Urban Regeneration: Lessons from the IBA Emscher Park." In *The Routledge Companion to Urban Regeneration*, edited by Michael E. Leary and John McCarthy, 230–241. London: Routledge, 2013.

Plets, Gertjan. *Heritage Statecraft and Corporate Power: Altai Activism in Post-Soviet Siberia.* London: Routledge, 2024.

Plets, Gertjan, and Marin Kuijt. "Gas, Oil and Heritage: Well-oiled Histories and Corporate Sponsorship in Dutch Museums (1990–2021)." *BMGN-Low Countries Historical Review* 1, no. 137 (2022): 50–77. https://doi.org/10.51769/bmgn-lchr.7028.

RAG Foundation. "RAG-Stiftung – A Unique Model," Video, Essen, Accessed July 11, 2024, n.d., www.rag-stiftung.de/en/.

RAG Foundation. *Geschäftsbericht.* 2021. Accessed August 9, 2024. www.rag-stiftung.de/fileadmin/user_upload/Publikationen/Jahresabschluesse/RAG-Stiftung_GB21_Gesamt_DE_220530.pdf .

Raines, Anne Brownley. "Wandel durch (Industrie)Kultur: Conservation and Renewal in the Ruhrgebiet." *Planning Perspectives* 26, no. 2 (2011): 183–207.

Rectanus, Mark. *Culture Incorporated: Museums, Artists, and Corporate Sponsorships.* Minneapolis: University of Minnesota Press, 2002.

Slach, Ondrej, and Peter Rumpel. "Transferable Impulses of IBA Emscher Park – Czech Perspective." In *Internationale Bauausstellung Emscher Park: Impulse: Lokal – Regional – National – International*, edited by Christa Reicher, Niemann Lars, and Angela Uttke, 210–220. Münster: Klartext, 2011.

Storchmann, Karl. "The Rise and Fall of German Hard Coal Subsidies." *Energy Policy* 33, no. 11 (July 1, 2005): 1469–1492. https://doi.org/10.1016/j.enpol.2004.01.006.

Streeck, Wolfgang. *Re-Forming Capitalism Institutional Change in the German Political Economy*. Oxford: University Press, 2009.

Taft, Chloe. *From Steel to Slots: Casino Capitalism in the Postindustrial City.* Cambridge, MA: Harvard University Press, 2016.

Wagner, Helen. *Vergangenheit als Zukunft? Geschichtskultur und Strukturwandel im Ruhrgebiet.* [*Past as Future? Historical Culture and Structural Change in the Ruhr District.*] Köln. Böhlau 2022.

Winkelman, Jürgen. "IGBE, Band der Solidarität, Menschenkette für Bergleute." ["IGBE, Band of Solidarity, Human Chain for Miners."] *NRWision*, June 11, 2018. Accessed April 15, 2024. www.nrwision.de/mediathek/emschertal-movie-camera-igbe-band-der-solidaritaet-menschenkette-fuer-bergleute-181106/.

Zollverein. "Chronik der 80er und 90er Jahre." ["Chronicle of the 80s and 90s."] Accessed June 25, 2024. www.zollverein.de/ueber-zollverein/vom-bergwerk-zum-unesco-welterbe/.

Zweger, Bart. *Cultural Heritage in Transition: A Multi-Level Perspective on World Heritage in Germany and the United Kingdom, 1970–2020.* Cham: Springer, 2022.

9 Representing What Has Been Destroyed

The Sunken Island of Ada Kaleh in Museums of the Iron Gates Region in Romania[1]

Merve Neziroğlu

> Seven years and eight months of pure construction time were necessary. Nature and man have created a work of unique beauty here, at the legendary Iron Gates. [...] But the human being has refined this landscape with his work. He has created an uplifting monument [testifying] to his abilities, his skills. A massive spillway dam spans the stream, built in a hard struggle with unleashed forces of nature, and its meaning being known even to the non-expert, the outsider.[2]

With its distinctly poetic touch, the above quote reflects the keen anticipation surrounding the inauguration of the run-of-river hydropower plant, Iron Gate I (Porțile de Fier in Romanian and Ђердап in Serbian)[3] in 1972. Named after the kilometre-long valley at the rocky foothills of the Carpathian Mountains along what is now the Romanian–Serbian border, the plant on the Danube effortlessly became a political symbol of power, similar to many other hydraulic infrastructures.[4] For Josip Broz Tito and Nicolae Ceaușescu, respectively, the former presidents of (then) Yugoslavia and Romania, who were both present at the inauguration ceremony, the construction of this plant was a technical masterpiece. It proved to the world the ability of these leaders to tame the unpredictable and ever-dangerous rapids of the Iron Gates breakthrough valley. At the same time, the power plant was celebrated as a visible marker of the socialist friendship between the two states that continues today, decades after the demise of state socialism.

The Iron Gate I power plant joins a long line of hydropower plants built during the boom in dam construction in the twentieth century. This boom was not linked to any particular political system, such as socialism; dams were and still are constructed all around the globe.[5] Building a hydropower plant is a huge technological undertaking involving thousands of workers, as Rune Frandsen shows in Chapter 5 of this volume, and with far-reaching consequences for local environments. The drastic interventions in the landscape associated with dam development entail the destruction of homes and livelihoods through flooding, raising the question of how to handle both the displaced people and cultural artefacts from the destroyed sites. In some cases, for example, during the construction of the Aswan High Dam in Egypt, which began in 1960, the temples of Abu Simbel and other cultural assets were relocated with international assistance.[6] In other instances, historical artefacts were submerged and lost under the rising water levels. A very

DOI: 10.4324/9781003473534-13

recent case is the Southeastern Anatolia Project in Turkey, where the Ilısu Dam was planned and eventually built despite national and international protests during the previous decade. Although the protests drew attention to the serious ecological and social consequences of the dam's construction on the Tigris, the project was realised, resulting in the resettlement of tens of thousands of locals, many of whom are Kurds, and the destruction of the millennia-old city of Hasankeyf as well as countless villages. The Turkish government made an attempt to save a few artefacts, including a mausoleum and a mosque. However, countless artefacts that have shaped the region were gradually submerged by 2020.[7] This last and recent example reveals that certain issues relating to the impacts of dam construction have hardly changed over the decades. Although the devastating environmental impacts of dam construction are well known, they continue to be built. People who live in a region affected by such huge projects are often unable to contest such projects, even when there is international pressure against their construction. The question remains as to whether cultural assets, at least, can be saved, and to what extent.

The Danube, Europe's second largest river, has been subjected to numerous technical interventions over several centuries, which have shaped the contemporary riparian landscape.[8] Following the construction of Iron Gate I, the water level of the Danube rose by over 33 m, necessitating the relocation of people living on both banks of the river, whose homes were destroyed. Besides several villages located along the riverbank, the historic centre of Orşova, a Romanian port town, and the river island of Ada Kaleh[9] (from the Ottoman-Turkish name meaning 'Fortress Island') were flooded. Ada Kaleh was populated by less than 1,000 people, most of whom were Muslim and belonged to Romania's Turkic minority. The years leading up to the destruction of the island in the late 1960s were overshadowed by the question of what should be done with the island's inhabitants and its artefacts. The Romanian government made an attempt to save the citadel and the cemetery by translocating them around 20 km downstream to Şimian, an uninhabited islet directly facing the city of Drobeta-Turnu Severin. The idea was to rebuild the island of Ada Kaleh there, but the construction works were never fully completed. The fate of the inhabitants soon became the subject of international diplomacy, as the issue was discussed in the Turkish parliament in 1964, when news was received about the dam construction and subsequent evacuation. The governments of Turkey and Romania came to an agreement, allowing Turkey to invite the islanders to move there 'in order to prevent them from vanishing.'[10] Some families stayed in Romania, but several dozen decided to take up the invitation and emigrate. Eventually, the last inhabitants of Ada Kaleh left, and the island disappeared, submerged under water.

To this day, Ada Kaleh is remembered both nationally and beyond the borders of Romania. Focussing on the Iron Gates region in Romania, in which the sunken island was located, this chapter examines how Ada Kaleh is remembered in the exhibitions of several museums in this region that showcase artefacts from the island. It explores the romantic representation of the island, in which the tradition of depicting the 'oriental other' continues to prevail, leaving no room for unheard stories. Following a brief synopsis of the island's history and the period of Romania's transformation, the chapter provides an overview of the four museums and exhibitions in the Iron Gates region. Two of these museums are located in the

Romanian city of Drobeta-Turnu Severin, one at the power plant itself and one in the partially rebuilt town of Orşova. The fieldwork for this study was conducted in 2022, a couple of weeks after the Iron Gate I power plant celebrated its 50th anniversary.

A Sunken Island

Commencing in the nineteenth century, the increasing technical advancement of infrastructure and the commercial use of travel routes along the Danube prompted a rise in the number of travellers, and with them, depictions of the local environment and people.[11] With a long history as a popular travel destination on the Danube, Ada Kaleh was located throughout its existence at an 'intersection of borders.'[12] In the nineteenth century, amidst conflicts over the 'Oriental question,' the borders between the Habsburg and Ottoman empires, where the island was located, were contested. Numerous nationalist independence movements emerged, and the empires gradually disappeared as newly emerging nation states took shape during and following violent conflicts. The new nation states in Southeastern Europe were founded on a belief of belonging to a national community,[13] and the re-drawing of political borders created new minorities, in particular, Turkic communities, in the post-imperial states. As an exclave, in 1878, Ada Kaleh remained the Ottoman Empire's last bastion in the Danube region until it was annexed by Austria-Hungary in 1913. A few years later, in 1923, the newly founded Turkish Republic officially ceded the island to the Kingdom of Romania, but the islanders remained connected to the cultural heritage of the Ottoman Empire through language and religion.

Being located on the Danube, Ada Kaleh was a frequently visited travel destination on the itineraries of river cruises. The goods produced on the island, such as cigars and Turkish Delight, were exported, and the local bazaar was a trading hub. However, the island's fate was about to change with the initiation, in 1953, of discussions between the governments of the Socialist Federal Republic of Yugoslavia and the Romanian People's Republic over optimal utilisation of the Iron Gates valley. The gorge's topography was ideal for the construction of a dam, and the official ground-breaking ceremony was held in 1964.[14] Over the following years, the islanders prepared to say goodbye to their home, which they left by 1969. Soon after, the island's full submergence began. In May 1972, the Iron Gate plant was inaugurated and became Europe's second largest hydraulic power station.[15] By then, Ada Kaleh's former inhabitants had been resettled elsewhere, and the island itself was submerged under the water.

A sunken island attracts interest, and in Ada Kaleh's case, its Ottoman-Turkish heritage has added another dimension. Ada Kaleh's position on the Danube, combined with the flooding, led to its idealisation in the narratives and descriptions of the island within the popular media, which had already been shaped by Orientalism linked to increased tourism.[16] Orientalism refers to stereotypical assumptions about the East and actions based on these stereotypes. At the same time, as noted by Edward Said, it only exists through its relation to the West, which includes 'dealing with it by making statements about it' or 'describing it.'[17] Discursive attributes were assigned to Ada Kaleh's environment, such as 'wild' or 'untamed,' not just in the reports of Western travellers but also in the impressions of Ottoman and Turkish travellers. These attributes contrasted with those of the infrastructurally more advanced upper reaches of both the Danube and the Iron Gates, considered the gateway between East and West, civilisation and wilderness, and Occidental and Oriental.

This view of the island was associated with a political context in which it was considered the last bastion or even the last stretch of homeland on the Danube.[18] Because of the Ottoman Empire's forced retreat in the nineteenth century, the Danube symbolised the Empire's loss, whereas Ada Kaleh, by contrast, symbolised its existence. The fact that Ada Kaleh was conquered and reconquered several times and that it was ultimately considered an island inhabited by a cultural and religious minority through its final incorporation into the Kingdom of Romania are factors that shaped perceptions of the island. In recent Turkish literature as well as in newspaper articles and online media, Ada Kaleh is usually framed as a lost homeland,[19] and while the Danube itself might not be a central element in the Turkish remembrance culture, the island is clearly acknowledged as an important part of the cultural heritage of the Ottoman Empire. This can be seen in poems, songs, and even marches, for example, the 'Tuna nehri akmam diyor' March (Pleven March or Osman Pasha March), in which the Danube features prominently.[20]

The displacement and loss of homeland affected not only the islanders inhabiting Ada Kaleh but also many Romanians and Yugoslavians living in the Iron Gates region. A multi-generational oral history study of displaced people and their descendants from places that were destroyed because of the power plant, which was published in 2018, identified four key moments in the interviewees' memories. These were the official announcements of displacement, the destruction of villages, the flooding, and their relocation. 'People kept the harsh memories connected to demolishing their own houses; they were forced to leave the idyllic landscapes of their childhood, the sense of place they loved.'[21] When the villages around the hydropower plant were destroyed, several thousand people had to be resettled. The study's findings revealed that for those who were affected by this caesura, this region constituted a *traumascape*, that is, a landscape characterised by the traumatic events that took place there: the way the impending destruction of their home was announced, the helplessness of those who lived there, and ultimately the loss of home.[22] A similar situation applied to the inhabitants of Ada Kaleh, as revealed by contemporary witnesses in an anthropological dissertation from 2015.[23]

The traumascape of the Iron Gates (Figure 9.1) has been inserted into Romania's contested memory culture in the present. The year 1989 marked a turning point for many Eastern European countries, and many studies, some adopting a comparative perspective, have analysed post-communist memory and heritage.[24] In Romania, the regime ended with the violent Romanian Revolution in December 1989 following a ten-day period of demonstrations and unrest that culminated in the fall of the system and the execution of Ceaușescu along with the death of approximately 1,000 civilians.[25] Immediately after the Revolution, the so-called

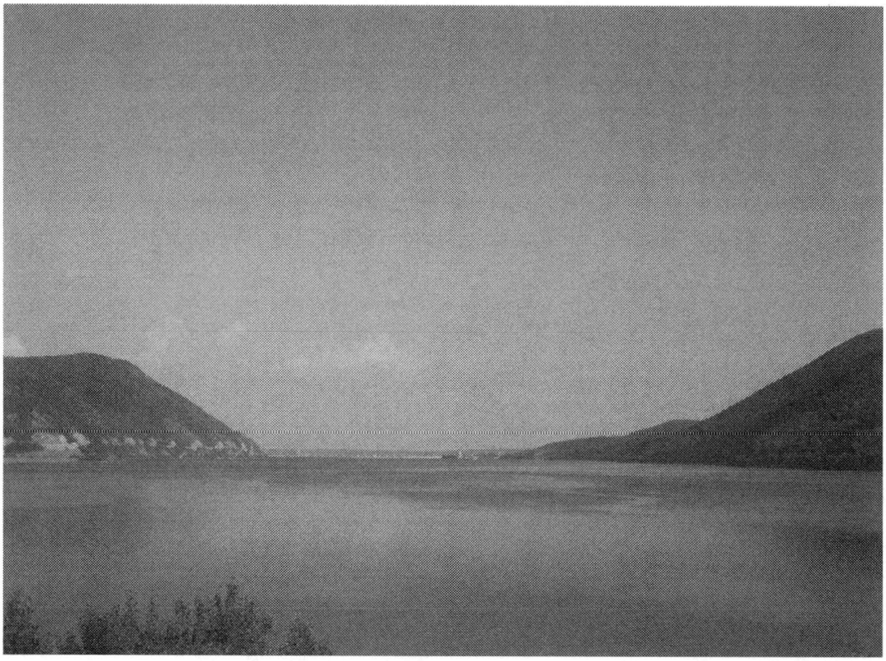

Figure 9.1 Iron Gate I with the breakthrough valley in the background. The view follows the Danube down to the hydroelectric power station, revealing the dam that blocks the river. Photograph by the author, July 15 2022.

neo-communists who rose to power were keen to forget the recent past and thus glossed over the regime's past crimes.[26] It was the international sphere, and, in particular, aspirations of joining the European Union (EU) that introduced a new dimension into the memory of communism. In 2006, one year before Romania joined the EU, the Council of Europe adopted a resolution condemning the abuses inflicted by communist regimes. Traian Băsescu, who was president at the time, ordered a commission to investigate the crimes committed under communism and, in light of the report's findings, condemned the system as 'criminal and illegitimate.'[27]

The transformation of the Iron Gates, which was perceived as traumatic for the residents, is just one of many stories in Romania's communist history. Less than two decades separate the end of communism and Romania's integration into the EU. During this period of transformation, one of the measures that helped Romania and other candidate countries from Eastern Europe on their way to EU membership was the Phare programme implemented between 1989 and 2006, which included financial aid to support the development of those countries so that they met EU membership requirements.[28] This specific funding is explicitly mentioned on several panels about Ada Kaleh in the exhibition at the Iron Gate I Hydropower Plant Museum.

Against this backdrop, this chapter examines how Ada Kaleh is represented in museums in the Iron Gates region. The unresolved issues relating to a socialist past become apparent in the memory of the island represented and preserved in museums and in the public space in this region. Museums can be places where history is remembered, sometimes even serving as designated places for discussing or critically examining a traumatic past and finding a way to come to terms with it. Stories of (forced) displacement can be found in museums, but like any other topic, these stories are interpretations offered to the visitor by the museums.[29] Hence, there needs to be consideration of the political framework in which these interpretations of the past are made. While conducting my field research, I visited four museums and exhibitions in the Romanian cities of Drobeta-Turnu Severin, Gura Văii, and Orşova. All of them are state-funded and charge a small entrance fee.[30] The distance between Drobeta-Turnu Severin and Orşova is approximately 30 min by car, and the village of Gura Văii, where the hydroelectric power station is located, is situated more or less midway along this route. All four museums include sections representing the history of the sunken island of Ada Kaleh. The locations of these exhibitions, where extensive interventions in the landscape took place, from the power plant to towns close by, shape their representations.

Beginning in Drobeta-Turnu Severin and following an imagined upstream movement, I will trace the locations of the museums and briefly describe the content of the exhibitions here. Muzeul Regiunii Porţilor de Fier (the Museum of the Iron Gates Region), which is located in the Drobeta-Turnu Severin city centre, is the main branch of the Regional Museum, comprising three branches, and was founded in 1912. The languages used in the exhibitions are English and Romanian. With its focus on the local flora and fauna as well as the region's history of settlement, the museum may remind the visitor of typical socialist regional museums.[31] It comprises several rooms in which the displayed objects and texts narrate the broader settlement history in the Iron Gates region. They contain archaeological finds from the excavation performed during the construction of the power plant, a small replica of Trajan's Bridge, some folkloric objects, and information on fabrics. Ada Kaleh is featured through a handful of objects that are everyday items as well as a few images of the island and its inhabitants. These exhibits of objects and images, which are on a relatively small scale, are displayed in a corridor close to the museum's exit. Part of the museum consists of an aquarium in the basement of the building, which contains not only fish tanks but also information on local and endemic flora and fauna.

The second branch of the Iron Gates Regional Museum is the Muzeul Hidrocentralei Porţile de Fier I-Gura Văii (Iron Gate I Hydropower Plant Museum, hereinafter referred to as the Hydroelectric Museum). Founded in 1976, this museum hosts the largest exhibition on Ada Kaleh in the region. It is located a few kilometres upstream of the city of Gura Văii next to the Power Plant Iron Gate I. The museum has been conceptualised as a single large room in which several glass cabinets and large panels are exhibited. Mostly displaying photographs and texts about Ada Kaleh, it also showcases a living room with typical objects found in a home on the island. Furthermore, the exhibition features information about the local flora and fauna and the earliest settlements in the Iron Gates region. There are many photographs of the construction of the power plant displayed in a corridor leading towards the turbine room. The language used for most of the exhibition is Romanian, although some information is also presented in English.

The third branch, Muzeul Regiunii Porților de Fier – Filiala Orșova (the Museum of the Iron Gates Region, Orșova branch), is located in Orșova, a town that had to be partially rebuilt because of the construction work around the dam. Founded in 2017, this is the newest branch and arguably the smallest facility. A few objects, notably a replica of a house front, are displayed here, and a few texts provide information on Ada Kaleh. A strong focus of this museum's exhibits is on Schela Cladovei, a Stone Age site located in the Iron Gates and dating back several millennia. Like the other museum branches, this museum too provides information on flora, fauna, and folklore. The information is presented mostly in Romanian. Further information on all three museums can be found on the website of the Museum of the Iron Gates Region.[32]

In addition to these museums, there is another exhibition located in a historical landmark site that uses technology, namely an augmented reality app, to bring Ada Kaleh to life with images of people moving about in their daily lives. Located in the city centre of Drobeta-Turnu Severin, the *Castelul de Apă* (Water Castle) is a former reservoir built in the early twentieth century and currently serving as a cultural centre. It hosts a small exhibition focussing on how the reservoir worked and how it was used. The interactive exhibition of the island is thus embedded in a longer history of water infrastructure along the Danube. The exhibition titled 'Foto Kaleh. Bazarul din Insulă o expositie interactivă' (Bazaar on the Island – An Interactive Exhibition) has only been running for a few years and is exclusively in Romanian. A poster on the wall lets the visitor know that this project is co-funded by the National Cultural Fund Administration, with the Local Council and City Hall of Drobeta-Turnu Severin and the Department of Historical Heritage and Cultural Tourism as co-organisers. In sharp contrast with the other exhibitions, this exhibition uses modern technology to present the exhibition content in an exciting and somewhat unusual way. The visitor has to use a smartphone to get the full experience of the exhibition's interactive elements.

The degree of the prominence of Ada Kaleh varies from museum to museum. What they do have in common is an overly strong emphasis on a romanticised portrayal of the island, with only a superficial reference, if any, to the historical and political context of the island's destruction. This superficial, yet overly romantic, representation is deeply connected to the inherent appeal of the motif of a sunken island, which happens to be a formerly Ottoman (read: 'oriental') island. The uglier or more controversial sides of the island's history remain hidden. This is because the underlying objective of these exhibitions is not to reflect upon the islanders' destroyed environment (and that of many other people in the region) in the context of the country's socialist past but rather to emphasise the value of the island as a special tourist destination.

Resurfacing of a Destroyed Past

When I travelled to the Iron Gates region in May 2022, the 'Iron Gate I' power plant had just celebrated its 50th anniversary. A large poster was still hanging on a wall outside the entrance of the Hydroelectric Museum as a reminder of the celebrations attended by politicians from both Serbia and Romania. Depicting the dam, the poster read '50 years of energy! 50 years of Iron Gates 1972–2022' scripted in large, golden letters.[33] It reminded me of the photographs I had seen that were taken half a century ago at the same spot. In these images, Tito and Ceaușescu celebrated the project's inauguration along with thousands of people from Yugoslavia and Romania. That day, a plaque was unveiled, emphasising the importance and success of the cooperation of the two socialist states in both the Serbo-Croatian and Romanian languages:

> Today, 16 May 1972, in the presence of the President of the Council of the Romanian Socialist Republic, Nicolae Ceaușescu, and the President of the Socialist Federal Republic of Yugoslavia, losip Broz Tito, the "Iron Gates" hydro-energy and navigation system was officially inaugurated; a work of the joint efforts of the working people of the Socialist Republic of Romania and the Socialist Federal Republic of Yugoslavia, a symbol of friendship and fruitful socialist cooperation and good neighbourliness.[34]

The socialist periods for Romania and Serbia, one of the successor states to Yugoslavia, have been over for a while now. However, the curious visitor to the hydropower plant can catch a glimpse of their traces on the opposite side of the Danube bank, in Serbia. In the middle of the hilly landscape is a huge flag of Yugoslavia, made of concrete, its colours slowly fading. Those with particularly keen vision (or who consult Google Maps)[35] can decipher the letters T I T O above the flag. This relic from the past frames the power plant, which is also a visible monument of the past, still resonating in the present.

On the Romanian side, the power plant encompasses the Hydroelectric Museum, providing information for those interested. Given its name and location, visitors might assume that it focuses on technology, as do other power plant museums elsewhere in the world. For example, the Norwegian Museum of Hydro Power and Industry in Tyssedal, located at a fjord, features guided tours, films, and exhibits on the role of hydropower in Norway's industry.[36] The historical Museum of Hydroelectric Energy in the Italian town of Cedegolo elucidates how water becomes electric power using visual aids such as videos and projections.[37] The Fala Hydro Power Plant in Slovenia even offers a virtual visit to its museum, enabling the digital visitor to tour the turbine and engine rooms.[38] Compared with these other museums that have foregrounded energy production in their exhibitions as their *raison d'être*, the Hydroelectric Museum in the Iron Gates region has no such technical focus. This may seem surprising for someone who would like to learn more about the technical details of the plant. Some basic information on the power plant's technology is available, at least for the Romanian-speaking visitor, who can obtain some information on the technical details during a short (included) tour of the turbine room or by purchasing the brochures available in the museum shop. Additionally, a collection of photographs produced during the construction period is displayed in a corridor leading to the turbine room.

Surprisingly, beside information on the endemic species of flora and fauna and a bit of folklore, the largest focus of the museum is on Ada Kaleh. While the destruction of settlements may only appear as an afterthought, if at all, in other museums that focus on energy production, the Hydroelectric Museum showcases the destroyed island of Ada Kaleh not only at the starting point of the exhibition but also very prominently. Upon entering the exhibition room, visitors immediately see a large photograph of the dam, taken from the Romanian side. A few paces later, the visitor reaches nine larger displays (estimated to be 1 × 2 m) that present plenty of information on Ada Kaleh. Those displays are either solely in Romanian or in both Romanian and English. The bilingual displays have been financed by the previously described EU instrument, the Phare programme. Those displays are simultaneously part of the EU-funded 'Danube – a European corridor' project supported through the 2005 Romania–Serbia Neighbourhood Programme. The displays in the Romanian language provide brief text or sentences on the island's location, its contested history, the subsequent (multiple) name changes for the island, its main attractions, for example, the mosque, and a couple of images depicting life on Ada Kaleh. Only two displays mention the effects of the hydroelectric power plant's construction and the rising water level, which necessitated the relocation of the island's population. The places to which the islanders moved are briefly listed. The bilingual panels also provide plenty of images and a little bit of context, partially repeating what has been already said on the other panels. Overall, the already brief narration of the island's history ends with the departure of the islanders.

The strong focus on Ada Kaleh might appear even more baffling when we look at *how* the island is presented. The Hydroelectric Museum conveys the nostalgic and romanticised portrayal of the island fostered by its destruction. This attribution can be seen in all four exhibitions, partially in the words used to describe the island and partially in the objects that are displayed. All exhibitions have objects on display that could be described as 'typically oriental.' The Regional Museum in Drobeta-Turnu Severin displays a porcelain basin and a soap dish, used 'for the practice of the purification ritual by washing,' candlesticks, and the Koran, which according to the contributing text are 'vestiges of the Mosque from Ada Kaleh,' a coffee grinder, tray and kettles, and lastly, two waterpipes. A handful of objects are also on display in the Water Castle, such as a coffee grinder for Turkish coffee and a tray, said to be original household items. Both the objects associated with consumption and those associated with hygiene and faith reflect the stereotypical image of 'the Muslim,' who spends his day alternating between smoking, drinking coffee, and praying. This stereotypical image was already present five decades ago, shortly after the island's destruction, as demonstrated by an article in a German-language Romanian newspaper published in December 1969. The article discusses the island's reconstruction on Şimian Island as follows: 'The Turkish houses are also being rebuilt with a ground floor and storey, a slightly sloping roof, and a protruding porch on which the Oriental is accustomed to spending his hours of relaxation.'[39]

Replicas of such houses and living spaces, in the tradition of nineteenth-century Oriental chambers,[40] can be found in the Hydroelectric Museum and in the Orşova branch of the Regional Museum. The latter displays a replica of a house front with two showcases set into it, displaying objects such as candlesticks or carafes, conveying to visitors what a typical house on Ada Kaleh looked like. Circling back to the Hydroelectric Museum, a replica of a living room can be found here. A showroom occupying an estimated six square metres contains Ottoman furniture, one regular and one prayer carpet, tea and coffee dishes, a mannequin with typical Ottoman-style clothing for men, and framed Koran verses on the wall.[41] Some brief text provides context:

> Interior of Ada Kaleh Island: When the Turks stopped being nomads, they transposed the old concept of the tent into the organisation of the interior, keeping the carpet as the central element of the living space. This is how the Turkish houses on the island of Ada Kaleh were almost completely devoid of any furniture, apart from divans or sofas or small tables, the floors being covered with carpets, and old paintings with sentences from the sacred writings hung on the walls.[42]

The seemingly direct link between the islanders and the former nomadic lifestyle of Turks with which Ada Kaleh's inhabitants were identified and self-identified implies a certain temporality and a sense of otherness. This otherness of Turks and Muslims (the terms are used interchangeably) is a well-known motif throughout Southeastern Europe and beyond, and therefore not surprising (Figure 9.2).

Beyond the displayed objects, the way in which language is used in the museums is also striking. Regardless of the time period, in the literature on Ada Kaleh as well as in contemporary newspaper articles, certain key words appear repeatedly. Words like 'garden,' 'pearl,' 'Atlantis'—and therefore, mythical adjectives—*legendary*, and *magical* are used in the museums to describe the island's attributes.[43] This depiction, per se, does not imply that this image was absent during the time of the island's existence. As the writings of travellers from both the East and the West indicate, they must have found the island intriguing, given that it was a river island located at the end of the Iron Gates breakthrough valley, a stretch of the river that was particularly dangerous to navigate. However, this portrayal of the island is overly romanticised and seems to leave no room for counter-narratives, as the following example shows.

> *In the Hydroelectrical Museum, where Romanian and English are the languages of communication, one of the panels on Ada Kaleh stands out. It depicts an aerial view of the island accompanied by a quote from an islander, Bob Gheorghe (1942–2012), translated into twelve languages.* Floating garden, I would say as old as the Danube. A place for meditation, oblivion and dream[ing] for all those who have a chance, at least once in their lifetime, to walk on the sacred land and to enchant themselves with the beauties of this pearl of the Danube, not polished enough, that was the island of Ada-Kaleh—the small Atlantis of Romania.[44]

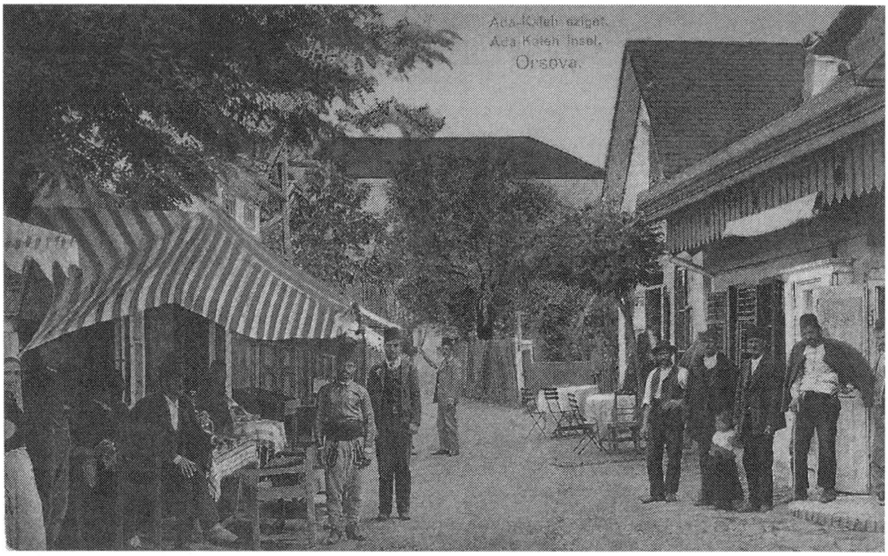

Figure 9.2 On this postcard dating back to 1911, the islanders of Ada Kaleh pose on a street. Most of the men are wearing the *fez*, the customary headgear worn in the Ottoman Empire.

Source: Magyar Nemzeti Digitális Archívum.

Of all of the four exhibitions, this is the only exhibit in which an islander and eyewitness is given an active voice via a quotation. Gheorghe was one of the most prominent islanders because he was collecting data and also strongly promoting and sharing memories of the island.[45] While his quote reflects the special nature of the island and nostalgia for this lost place (and homeland), the perspective that it conveys is highly selective. It reveals who is authorised to interpret the events of the time, given that Bob Gheorghe was of Romanian ethnicity and moved to the island when he was a child. This consideration is in no way intended to devalue his perspective or the choice of quotation. However, the question of why the active voice selected in this exhibition was his and not that of a Turkish islander remains to be answered.

While Gheorghe's quote and the Oriental chambers are embedded in a longer tradition of representing Ada Kaleh in Romania, the exhibition at the Water Castle conveys a more contemporary representation. This exhibition presents a contrasting portrayal of Ada Kaleh as a trading hub, taking viewers on a journey through time via technology. Differing from the exhibitions at other museums, its aim is openly stated, namely to provide insight into everyday life in the island's bazaar through an interactive experience. Visitors can download Artivive, a free software application that enables art to be presented as augmented reality on their smartphones. They can then scan selected images using their smartphones. For this exhibition, which is of a manageable size, approximately 50 glass jars filled with water and containing either an image or brief text are distributed on nine shelves. All of them depict life on Ada Kaleh and its bazaar. Some of these jars have a sticker placed on them; when these particular jars are scanned, the images within them appear on the smartphone screen and begin to move. For example, one interactive image shows an ice cream cart labelled as such in Romanian (*înghețată*). When it is scanned, the same image appears on the phone, and the cart is pushed aside by a person who is running into the picture, revealing another person sitting behind the cart. Among the non-immersive images, some are photographs showing everyday life on the island, while others are inscribed and appear to be reproductions of postcards. The jars without pictures contain pieces of paper with some paragraphs written exclusively in Romanian. They provide information on tourism on the island in days gone by or illustrate how the islanders handled occurrences of flooding from the Danube from time to time.

The exhibition conveys a bittersweet touch of irony, given that the pictures of a sunken island are immersed in jars filled with water. It entails a creative juxtaposition of a contemporary software app with a past way of life, which can now *resurface* and be re-lived. This seems to be the intended goal of the exhibition: people in the present can glimpse how life on the island must have been with the help of modern technology and pictures taken during the time of the island's existence. However, a visitor whose knowledge of the island is limited may not get the full picture of Ada Kaleh's fate. The focus is on the island's appeal as a tourist destination—one that is special because of its 'oriental' character—not on the island that for centuries was the place of residence and the centre of life for a very close-knit community, whose members were eventually deracinated. This observation, which also applies to all of the other museums, will be discussed in the following section.

Untold Stories and Unfinished Business

The foregrounding of Ada Kaleh in the context of the transformation of Iron Gate's landscape could be attributed to its appeal as an 'oriental' island, which, from a museum curator's perspective, could provide more attractive content for visitors. Relative to the overemphasis on Ada Kaleh's destruction, the destruction of Orșova, a nearby town, which has been partially relocated and is known throughout Romania as the country's youngest town, is at least acknowledged. In the Hydroelectric Museum, two displays show images of the town's transformation. One shows four photographs of the 'old' (*veche*) Orșova, while the second display shows three images of the 'new' (*nouă*) town. On one of Ada Kaleh's panels, an inconspicuous sentence mentions that the island 'disappeared under the waters of the reservoir of the Iron Gates, along with Orșova *veche* and other settlements on the Danube plain.'[46] These other settlements, mentioned here in passing, are (partially) mentioned by name in the museum's brochure. However, their stories remain untold, at least in the museums in the Iron Gates region.

Apart from the above-mentioned exception of Bob Gheorghe, none of the exhibitions feature the voices of actual islanders who lived during the time when the dam was constructed and the island destroyed. It is important to keep in mind that the island's close-knit community was torn apart when they had to move. As one panel exhibited in the Hydroelectric Museum notes, about 70 families moved to Turkey. A similar number of families left for Constanţa on the Black Sea coast, and only 14 families stayed in places in close proximity to their former island home, such as Orşova and Drobeta-Turnu Severin. The voices of islanders have only begun to be heard in the last two decades, when Ada Kaleh started to attract increasing interest. More recent sources, such as interviews, which contain the memories that the displaced islanders have themselves shared,[47] also acknowledge the now advanced ages of contemporary witnesses.[48]

Yet, the voices of the islanders are key for grasping the full picture when it comes to Ada Kaleh. However, this is only possible as long as they feel comfortable speaking about their experiences, given their painful memories of loss. An oral history of islanders who stayed in Romania, conducted by anthropologist Maria Iancu, strikingly reveals the extent to which the official narrative in the context of the island's submerging deviates from the experiences and memories of the inhabitants, with frequent omissions of the uglier sides of displacement. Iancu points out that when the islanders moved into the spotlight and were interviewed by journalists or curators, the expectations of both groups clashed. Whereas the islanders may have wanted to talk more critically about aspects of their experiences, the interviewers were more interested in the 'before state' of the island. Some islanders even found that their narrative style was not sufficiently attractive for the interviewers.[49] One islander with whom Iancu spoke mentioned his manifold ideas on how to preserve the memory of the island. One of his ideas was to establish an Ada Kaleh museum providing broad information on the island's history and social life, including information about the last inhabitants who witnessed its destruction and experienced displacement. Moreover, this islander envisioned not only dedicating a special section of this museum to the victims of the Communist regime but also going one step further and exposing the culprits and injustices against individual islanders and the community itself during the socialist period.[50] In general, the more critical statements of islanders revealed that their experiences relating to the destruction of the island and their displacement were profoundly shaped by the communist regime.

The uninhabited island of Şimian, which faces Drobeta-Turnu Severin, offered the closest venue for recreating Ada Kaleh as an open-air museum. Some of the relics of Ada Kaleh, the most important one being the citadel, still exist to this day, preserved on Şimian. In the Hydroelectric Museum, one of the displays under the Phare-funded project is dedicated to Şimian and shows two photographs of the decaying fortress; in one of them, the structure is partly covered by overgrown foliage. Four short paragraphs written in Romanian and English present a brief history of the island and the citadel, informing readers that archaeological research was carried out between 1966 and 1968 to enable the planned preservation of the fortress on Şimian. The Romanian government did indeed attempt to save at least some of Ada Kaleh's treasures in the 1960s. Initially, there was optimism around the Şimian Project. The newspaper, *Neuer Weg*, provided updates on the case: 'Ada

Kaleh was left by its last inhabitants a few weeks ago. The flood that the dam at the Iron Gates will impound is swallowing up the island. [...] And yet, Ada Kaleh will not disappear. [...] On Şimian, besides the "new" Ada Kaleh, a modern tourist complex is also being built.'[51] Years later, in 1980, the same newspaper announced that Şimian had now been handed over to the local tourism authorities and that 'several hundred visitors'[52] were visiting the island every day. According to the paper, the island was still only open to residents of the County of Mehedinţi, to which the island belonged and would only be opened to the general public after catering facilities had been established. The brief paragraphs on Şimian offered in the museums emphasise the efforts made to save Ada Kaleh's artefacts. However, they neither mention the planned tourist complex nor the fact that these plans—only partially realised—were eventually abandoned.

While the newspaper article from 1980 shows that for a short period, visits to Şimian were possible, the island is no longer open to the public.[53] There is no available infrastructure that would make it possible to gain access to the island legally, and from the shore, foliage overgrowth on the island is evident. Discussions on the island's potential and future have reached the international level, with a bilateral agreement signed in 2008 between Romania and Turkey on a 'Cooperation Programme in the Fields of Education, Culture and Arts, Press and Broadcasting, Youth and Sports.' Point 6 in Article 22 of the agreement, which regulates cooperation in the fields of archaeology, museology, and cultural heritage, states the following:

> The Parties, in accordance with their national legislation, shall analyse the possibility of setting up the "Ada-Kaleh" Foundation, meant to protect, preserve and render to the tourist circuit the architectonic complex of Ada-Kaleh on the Shimian Island, with the support of the Turkish community removed from the island and residing in Romania and in the Republic of Turkey.[54]

To date, however, no such foundation has been established. The press has discussed, and continues to discuss, the possibility of turning the island into a touristic hotspot in the border region of Serbia, Romania, and Bulgaria.[55] Yet, the island of Ada Kaleh is still awaiting its physical resurrection, while the museums in the region provide visual and textual representations of carefully selected artefacts.

Conclusion

The construction of the Iron Gate I hydroelectric power plant brought about a drastic transformation of the landscape of this region, which impacted the lives of thousands of people in both (the former) Yugoslavia and Romania. While Ada Kaleh no longer exists in a physical form, memories of this Danubian island live on. One means of remembrance is through the exhibitions on Ada Kaleh in the museums of the Iron Gates region, which have served as the empirical content for the analysis presented in this chapter. In these exhibitions, Ada Kaleh's story strikingly ends with its sinking. The island is depicted as it (apparently) was, and the presentations build heavily on the romanticised image linked to its Ottoman-ness and contested history, which was already present before the imminent flooding. Many aspects of the island and the islanders are addressed only marginally or not at all in the exhibitions, especially those relating to the period of the dam's construction, the flooding, and the question of how to preserve the island's cultural heritage. The reason for the strong focus on Ada Kaleh and little to no mention of other places that were also destroyed could lie in the inherent appeal of that place, being an island with a strongly 'oriental' character.

In the four exhibitions analysed in this chapter, Ada Kaleh is presented in the state in which it had been known to exist decades before the flooding: a small Ottoman-Turkish island on the Danube, where Muslims lived, prayed, and produced and sold goods, such as sweet treats, cigars, or carpets. Given its unique position on the river, Ada Kaleh is assigned legendary or mythical attributes, for example, fostering the image of a wonderful place and even a paradise now lost. Time stands still in the exhibitions, and a certain melancholia hangs in the air. While walking through the museums, I had the impression that life on Ada Kaleh, as it is presented in the exhibitions I visited, must have been peaceful. The island was seemingly a secluded and safe place, where the inhabitants would drink coffee and relax under the roofs of their homes when they were not busy selling their goods. Although the geopolitical peculiarity of the island, which was repeatedly conquered and reconquered, is mentioned in the exhibition at the Hydroelectric Museum, what this meant in the lives of the islanders has, in my view, remained somewhat in the background.

Considering the multi-layered aspects not just of Ada Kaleh's history but also of the Danube and its anthropogenic transformation, it becomes apparent that the museums analysed in this chapter operate in a specific temporal context shaped by political decisions. Evidently, museums are not places where everything can be narrated, especially not the one and only truth, and museum exhibitions are dynamic. It would seem that Ada Kaleh, as *pars pro toto* for the transformation of the landscape, is especially appealing. This is because it was not just an island—albeit an inherently romanticised geographic entity—but also an Ottoman island, whereas the other places that were destroyed were framed more clearly as culturally and ethnically belonging to Romania. Moreover, the strong focus on romanticisation leaves no room for critical voices. In the case of Ada Kaleh, this means that a visit to those museums will stimulate the curious visitor's imagination to visualise what the island must have been like before its destruction. The impact of the destruction, however, is not visible, and this gap can only be filled through excavation from other sources.

Notes

1 The research for and work on this article were made possible through the 'Contested Waterway. Governance and Ecology on the Lower Danube, 1800–2018' project, funded by the Leibniz Association (2020–2023). I would like to thank Mihaela from the Iron Gates I Hydropower Plant Museum for her time and patience and for the many insights

that she offered; Adriana in Drobeta-Turnu Severin, where I stayed, who not only provided useful tips but also made me feel at home; and my Romanian teacher, Diana, in Leipzig, who patiently corrected my grammatical errors.

2 "Sieben Jahre und acht Monate reiner Bauzeit waren nötig. Natur und Mensch haben hier, am sagenumwobenen Eisernen Tor, ein Werk einzigartiger Schönheit geschaffen. [...] Doch diese Landschaft hat der Mensch mit seiner Arbeit veredelt. Er hat seinen Fähigkeiten, seinem Können ein erhebendes Denkmal gesetzt. Über dem Strom spannt sich wuchtig eine Überfallstaumauer, die im harten Ringen mit entfesselten Naturkräften entstanden ist und deren Bestimmung auch dem Nichtfachmann, dem Uneingeweihten bekannt ist." Mesch, "Stern erster Grösse." This and all following translations by the author.

3 Twelve years after the plant's inauguration, a second power plant named Iron Gate II was opened further downstream. However, in this chapter, references to the power plant are only to Iron Gate I, unless explicitly stated otherwise.

4 Bauch et al., *Wasserinfrastrukturen und Macht.*

5 Brendel, *Konvergente Konstruktionen*; Gestwa, *Die Stalinschen Großbauten des Kommunismus.*

6 Hassan, "The Aswan High Dam."

7 Gottschlich, "Wasser als Waffe."

8 For interventions along the Lower Danube, see Gătejel, *Engineering the Lower Danube.*

9 There are different ways of writing the island's name, for example, Ada Kaleh, Ada-Kaleh, Ada-Kale, and Adakale. In this chapter, the name Ada Kaleh is used unless sources are being quoted, in which case the name in the source text is retained.

10 "Doğru ise; bu soydaşlarımızın eriyip yok olmalarını önlemek icin memleketimize getirirmeleri hususunda Bakanlıgımız Romanya Hükümeti nezinde tesebbüse geçmiş midir?" Meclis Tutanakları, 534.

11 Gătejel, "Overcoming the Iron Gates"; Király, "Die Donau ist die Form"; Riedler, "Integrating the Danube."

12 Király, "Am Schnittpunkt der Grenzen."

13 Anderson, *Imagined Communities.*

14 Štanzel, *Wasserträume und Wasserräume,* 194–198.

15 Štanzel, *Wasserträume und Wasserräume,* 198.

16 Blasen, "Mustafa Bego," 8.

17 Said, *Orientalism*, 3.

18 Mende, *"Heutiger Nachbar – gestriger Untertan",* 335.

19 Ağanoğlu, *Tuna nehrinde*; Demir, "Rumeli'nde veda"; "Sular altında."

20 Dursun, *Tuna Güzellemesi*, 275.

21 Cretan and Varan, "Place and the Spatial Politics," 515.

22 Cretan and Varan, "Place and the Spatial Politics," 509–519.

23 Iancu, *Between Past, Present and Future.*

24 Bernhard et al., *Twenty Years after Communism*; Todorova et al., *Remembering Communism*; Pakier et al., *Memory and Change in Europe.*

25 Pop-Eleches, "Romania Twenty Years after 1989," 87.

26 Dobre, "Uses and Misuses of Memory," 301–302.

27 Dobre, "Uses and Misuses of Memory," 310.

28 Phare Programme.

29 Völkering, *Flucht und Vertreibung im Museum*, 17.

30 I paid the entrance fee everywhere, except at the Hydroelectric Museum, where the staff refused to allow me to pay for my visit.

31 Gavrilova, *Russia's Regional Museums*.
32 "Muzeul Regiunii Porţilor de Fier." The website is only available in Romanian.
33 "50 de ani de energie! 50 ani Porţile de Fier 1972-2022." Field research, July 2022.
34 "PORŢILE DE FIER," 1.
35 Google Maps, "Tito Memorial Park."
36 Visit Norway, "Norwegian Museum of Hydro Power."
37 European Route of Industrial Heritage, "Museum of Hydroelectric Energy."
38 Dravske elektrarne Maribor, "History of the Fala Hydro Power Plant."
39 "Baubeginn an Ada Kaleh," 2.
40 Çelik, *Displaying the Orient*.
41 A photograph of the interior can be found on the museum's website: Muzeul Regiunii Porţilor der Fier, "Muzeul Hidrocentralei."
42 Hydroelectric Museum, field research, July 2022.
43 All of these attributes can be found in the accompanying texts exhibited at the Hydroelectric Museum. Field research, July 2022.
44 Bob Gheorghe (1942–2012) was one of the last former inhabitants of the island. The quote is provided in English and in 11 other languages. Hydroelectric Museum, field research, July 2022.
45 Iancu, *Between Past, Present and Future*, 143.
46 "[…], a dispărut sub apele lacului de acumulare de la Porţile-de-Fier, alături de Orșova Veche si alte așezări de pe Clisura Dunării." Hydroelectric Museum, field research, July 2022.
47 Michalache and Andreescu, *Adakale-li*.
48 Bercovici, "La deuxième vie d'Ada-Kaleh." The author identified the study as an 'emergency ethnology' given the advanced ages of the islanders.
49 Iancu, *Between Past, Present and Future*, 86–88.
50 Iancu, *Between Past, Present and Future*, 160–161.
51 "Ada Kaleh ist vor wenigen Wochen von seinen letzten Bewohnern verlassen worden. Die Flut, die der Damm am Eisernen Tor stauen wird verschlingt die Insel. […] Und trotzdem: Ada Kaleh wird nicht verschwinden. […] Auf Şimian entsteht neben dem 'neuen' Ada Kaleh auch ein moderner Touristenkomplex." "Baubeginn an Ada Kaleh," 2.
52 "Wieder ein Ada Kaleh," 1.
53 However, the chats I had with several people I encountered indicated that young people occasionally party there, reaching the island by boat.
54 Government of Turkey, *Türkiye Cumhuriyeti Hükümeti ile Romanya Hükümeti,* Art. 22, no. 6.
55 "Parc tematic Ada-Kaleh construit"; Georgescu, "Insula Şimian."

Bibliography

Ağanoğlu, Yıldırım, H. *Tuna nehrinde bir yitik vatan Adakale* [*Ada Kaleh, a Lost Homeland on the Danube*]. Istanbul: İz Yayıncılık, 2015.

Anderson, Benedict, *Imagined Communities: Reflections on the Origin and Spread of Nationalism*. London/New York: Verso, 1991 (1983).

"Baubeginn an Ada Kaleh. Die romantische Insel entsteht von neuem." ["Construction Begins on Ada Kaleh. The Romantic Island Emerges Anew."]. *Neuer Weg*, December 17, 1969.

Bauch, Martin, and Birte Förster. *Wasserinfrastrukturen und Macht von der Antike bis zur Gegenwart* [*Water Infrastructures and Power from Antiquity to the Present*]. Berlin: De Gruyter, 2015.

Bercovici, Monica B. "La deuxième vie d'Ada-Kaleh ou Le potentiel culturel de la mémoire d'une île disparue." ["The Second Life of Ada Kaleh or The Cultural Potential of the Memory of an Island That Has Disappeared"]. *Ethnologia Balkanica* 13 (2009): 169–180.

Bernhard, Michael, and Jan Kubik. *Twenty Years After Communism. The Politics of Memory and Commemoration*. New York: Oxford University Press, 2014.

Blasen, Philippe H. "Mustafa Bego, türkischer Nargileh-Raucher und ungarischer Nationalheld. Nationale Aneignung und internationale Vermarktung der Insel Ada-Kaleh," ["Mustafa Bego, Turkish Nargileh Smoker and Hungarian National Hero. National Appropriation and International Marketing of the Island of Ada Kaleh"]. *Spiegelungen. Zeitschrift für Deutsche Kultur und Geschichte Südosteuropas* 9, no. 63 (2014): 7–29.

Brendel, Benjamin. *Konvergente Konstruktionen: Eine Globalgeschichte des Staudammbaus* [*Convergent Constructions: A Global History of Dam Construction*]. Frankfurt: Campus Verlag, 2019.

Çelik, Zeynep. *Displaying the Orient. Architecture of Islam at Nineteenth-Century World's Fairs*. Berkeley: University of California Press, 1992.

Cretan, Remus, and Claudia Varan. "Place and the Spatial Politics of Intergenerational Remembrance of the Iron Gates Displacements in Romania, 1966–1972." *Area* 50 no. 50 (2018): 509–519.

Demir, Ayhan. "Rumeli'nde veda edilen son toprağımız: Adakale" ["Our Last Farewelled Land in Rumelia"]. *Nihayet*, August 14, 2020. www.gzt.com/nihayet/rumelinde-veda-edilen-son-topragimiz-adakale-3549089.

Dobre, Claudia-Florentina. "Uses and Misuses of Memory. Dealing with the Communist Past in Postcommunist Bulgaria and Romania." In *Memory and Change in Europe. Eastern Perspectives*, edited by Malgorzata Pakier and Joanna Wawrzyniak, 299–316. New York: Berghahn Books, 2016.

Dravske elektrarne Maribor. "History of the Fala Hydro Power Plant." Accessed September 4, 2024. www.dem.si/en/museum-of-the-fala-hydro-power-plant/history-of-the-fala-hydro-power-plant/.

Dursun, Haluk, *Tuna Güzellemesi* [*The Beauty of the Danube*]. İstanbul: Kapı Yayınları, 2015.

European Route of Industrial Heritage. "Museum of Hydroelectric Energy." Accessed September 4, 2024. www.erih.net/i-want-to-go-there/site/museum-of-hydroelectric-energy.

Gătejel, Luminiţa. *Engineering the Lower Danube. Technology and International Cooperation in an Imperial Borderland*. Budapest: CEU Press, 2022.

Gătejel, Luminiţa. "Overcoming the Iron Gates: Austrian Transport and River Regulation on the Lower Danube, 1830s–1840s". *Central European History* 49 (2016): 162–180.

Gavrilova, Sofia. *Russia's Regional Museums. Representing and Misrepresenting Knowledge about Nature, History and Society*. London: Routledge, 2022.

Georgescu, Alexandra. "Insula Şimian ar putea deveni perla turismului mehedinţean" ["Şimian Island Could Become the Pearl of Tourism in Mehedinţi"]. *Adevărul*, July 22, 2013. https://adevarul.ro/stiri-locale/turnu-severin/insula-simian-ar-putea-deveni-perla-turismului-1458057.html.

Gestwa, Klaus. *Die Stalinschen Großbauten des Kommunismus. Sowjetische Technik- und Umweltgeschichte, 1948–1967*. [*Stalin's Large-Scale Buildings of Communism. Soviet History of Technology and Environmental History, 1948–1967*]. Berlin: De Gruyter, 2010.

Google Maps. "Tito Memorial Park." Accessed September 4, 2024. www.google.com/maps/place/Tito+Memorial+Park/@44.6676182,22.516883,15z/data=!4m6!3m5!1s0x47522928f1d68ccf:0x3f44f4dab41f7e8c!8m2!3d44.6679952!4d22.5191758!16s%2Fg%2F11gbzcrv49?entry=ttu.

Gottschlich, Jürgen. "Wasser als Waffe. Der türkische Ilisu-Staudamm." ["Water as a Weapon: The Turkish Ilısu Dam"]. *Die Tageszeitung*, June 28, 2022. https://taz.de/Der-tuerkische-Ilisu-Staudamm/!5859730.

Government of Turkey. Türkiye Cumhuriyeti Hükümeti ile Romanya Hükümeti Arasında İmzalanan Eğitim, Kültür ve Sanat, Basın-Yayın, Gençlik ve Spor Alanlarında İşbirliği Programı. [*Cooperation Programme Between the Government of the Republic of Turkey and the Government of Romania in the Fields of Education, Culture and Arts, Media, Youth and Sports*]. March 3, 2008. http://ua.mfa.gov.tr/detay.aspx?7522.

Hassan, Fekri A. "The Aswan High Dam and the International Rescue Nubia Campaign." *The African Archeological Review* 24, no. 3/4 (2007): 73–94.

Iancu, Maria. *Between Past, Present and Future: The Displaced Islanders of Ada Kaleh.* PhD diss., University College London, 2015.

Király, Edit. "*Die Donau ist die Form." Strom-Diskurse in Texten und Bildern des 19. Jahrhunderts. ["The Danube is the Form." Discourses on Current in Texts and Images of the 19th Century*]. Vienna: Böhlau Verlag, 2017.

Király, Edit. "Am Schnittpunkt der Grenzen. Topik und Topografie der Schwelle am Beispiel des Eisernen Tores und der Insel Ada Kaleh." ["At the Intersection of Borders. Topic and Topography of the Threshold Using the Example of the Iron Gates and the Island of Ada Kaleh"]. In *Die Donau und ihre Grenzen. Literarische und filmische Einblicke in den Donauraum* [*The Danube and Its Borders. Literary and Cinematic Insights into the Danube Region*], edited by Olivia Spiridon, 203–216. Bielefeld: transcript, 2019.

Meclis Tutanakları, Millet Meclisi, 1. Dönem, 31. Cilt, 119. Birleşim. [Assembly Minutes, National Assembly, 1st Period, 31st Volume, 119th Meeting]. www5.tbmm.gov.tr/develop/owa/td_v2.sayfa_getir?sayfa=534:535&v_meclis=4&v_donem=1&v_yasama_yili=3&v_cilt=31&v_birlesim=119.

Mesch, Erich. "Stern erster Grösse." ["Star of First Magnitude"]. *Neuer Weg*, May 17, 1972.

Michalache, Carmen, and Magda Andreescu. *Adakale-li. Patria din buzunarul de la piept [Adakale-li. The Homeland in Your Breast Pocket]*. Bucharest: Editura Martor, 2012.

Muzeul Regiunii Porților de Fier. Accessed April 27, 2023. https://muzeuldrobetaturnuseverin.ro/.

Muzeul Regiunii Porților de Fier. "Muzeul Hidrocentralei Porțile de Fier I." ["The Hydroelectric Museum Iron Gates I"]. Accessed April 27, 2023. https://muzeuldrobetaturnuseverin.ro/muzeul-hidrocentralei-portile-de-fier-i-gura-vaii/.

Pakier, Malgorzata, and Joanna Wawrzyniak. *Memory and Change in Europe. Eastern Perspectives*. New York: Berghahn Books, 2016.

"Parc tematic Ada-Kaleh construit de turci pe insula Simian" ["Ada-Kaleh Theme Park built by Turks on Simian Island"]. *Adevărul*, June 21, 2005. https://adevarul.ro/stil-de-viata/parc-tematic-ada-kaleh-construit-de-turci-pe-1318728.html.

"Phare Programme." *EUR-Lex. Access to European Union Law*. Accessed September 30, 2024. https://eur-lex.europa.eu/EN/legal-content/summary/phare-programme.html.

Pop-Elches, Grigore. "Romania Twenty Years after 1989. The Bizarre Echoes of a Contested Revolution." In *Twenty Years After Communism*, edited by Michael Bernhard and Jan Kubik, 85–103. New York: Oxford University Press, 2014.

"PORȚILE DE FIER AU INTRAT OFICIAL ÎN ISTORIA ENERGETICII" ["The Iron Gates Have Officially Made Energy History"]. *România liberă*. May 17, 1972.

Riedler, Florian. “Integrating the Danube into Modern Networks of Infrastructure: The Ottoman Contribution.” *Journal of Balkan and Black Sea Studies* 3, no. 5 (2020): 97–120.

Said, Edward. *Orientalism*. New York: Vintage Books, 1979.

Štanzel, Arnošt. *Wasserträume und Wasserräume im Staatssozialismus. Ein umwelthistorischer Vergleich anhand der tschechoslowakischen und rumänischen Wasserwirtschaft 1948–1989.* [*Water Dreams and Water Spaces under State Socialism. A Comparative Environmental History Based on the Czechoslovakian and Romanian Water Management 1948–1989*]. Göttingen: Vandenhoeck & Ruprecht Verlag, 2017.

“Sular altında kalan minyatür Osmanlı” [“Ottoman Miniature Flooded”]. *Yeni Şafak*. July 5, 2015. www.yenisafak.com/yenisafakpazar/sular-altinda-kalan-minyatur-osmanli-2175606.

Todorova, Maria, Augusta Dimou, and Stefan Troebst. *Remembering Communism. Private and Public Recollections of Lived Experience in Southeast Europe*. Budapest: CEU Press, 2014.

Visit Norway. “Norwegian Museum of Hydro Power and Industry.” Accessed September 4, 2024. www.visitnorway.nl/bestemmingen-in-noorwegen/regio-west/hardangerfjord-regio/listings-hardanger-fjord/norwegian-museum-of-hydro-power-and-industry/51663/.

Völkering, Tim. *Flucht und Vertreibung im Museum. Zwei aktuelle Ausstellungen und ihre geschichtskulturellen Hintergründe im Vergleich.* [*Flight and Displacement in the Museum. Two Current Exhibitions and Their Historical Cultural Backgrounds in Comparison*]. Münster: LIT Verlag, 2008.

von Mende, Leyla. *“Heutiger Nachbar – gestriger Untertan.” Impressionen osmanischer und türkischer Südosteuropa-Reisender (1890–1940). [“Today’s Neighbour – Yesterday’s Subject.” Impressions of Ottoman and Turkish Travellers to Southeast Europe*]. Baden-Baden: Ergon Verlag, 2021.

“Wieder ein Ada Kaleh.” [“Another Ada Kaleh”]. *Neuer Weg*, July 18, 1980.

10 Integrating Minority Perspectives into the Heritagisation of Post-Mining Landscapes in Lusatia, Germany

Jenny Hagemann, Fabian Jacobs, and Lutz Laschewski

In the Eastern German region of Lusatia, there is a place called 'Wolkenberg,' which can be translated as 'cloudy hill.' To reach this place, we must walk to the top of a gently rising hill. The path is lined with young plants, including slender locust trees scattered here and there. Each and every plant appears to have been intentionally placed. We immediately sense that we are walking through a carefully designed landscape. At the top of the hill, we enjoy a wide-angled view of a recently established vineyard that serves the wine made there in the summer, revealing why most of the visitors are here. We sit down to enjoy a glass of wine and let our eyes wander further. Close by, forests and fields stretch out across the landscape, but further away to the right, a huge machine is working its way through the landscape, thoroughly churning up the soil. In the distance, and to the left of the colossal machine are two power plants spitting out clouds of steam into the sky. We now finally realise that everything around us—from the hill, to the vineyard, to the forests and fields—is a landscape of energy, shaped by the generation of electricity and therefore constantly changing. The machine is an excavator that is digging for lignite, which is fed to the power plants to generate warmth and electricity.

Behind us, the history of the devastated village of Wolkenberg is presented at a memorial site. Until 1992, Wolkenberg was situated right where we are standing. We learn that people lived here from at least the thirteenth century and that during most of the village's history, they spoke not only German but also Sorbian, a West Slavic language, and named the place 'Klěšnik.'[1]

The site evokes a certain feeling of dissonance within the visitor: The 'Sorbianness' of the landscape, that is, its interweavings with Sorbian cultural practices throughout the region's past and present, is fully positioned in the pre-industrial past and remains within the confines of the memorial site. However, the area's recultivation as a vineyard—originally supported by the former mining operator, Vattenfall, in the mid-2000s—and its cultural and economic as well as touristic value support a narrative of a successfully reappropriated post-mining landscape. This dichotomy between the 'Sorbian' and 'industrial' aspects of the landscape's past has strongly characterised both regional post-mining planning and Lusatia's self-image to this day.[2]

DOI: 10.4324/9781003473534-14

Grounded in these observations, this chapter discusses the heritagisation of post-mining landscapes with a focus on the Sorbian minority. Today, the Sorbs are one of the four legally recognised minorities in Germany (after the Sinti and Roma, the Danish minority, and the West-Frisian minority). At the close of the nineteenth century, approximately two-thirds of their historical and current areas of settlement overlapped with the Lusatian open-cast mining area, and more than half of the population of this area self-identified as Sorbian and spoke this language.[3] The landscape is deeply marked by ruptures, continuities, and imbrications of 'Sorbian' and 'industrial' dimensions. Therefore, our aim is to reflect on how these aspects can be integrated into the process of valuing the post-mining landscape. We also ask how, in this context, Lusatia's regional identity is being re-negotiated and how it could ultimately be made more diverse and inclusive through reorientation towards the minority specificity. These questions have become especially important, as the German government decided to phase out lignite mining by 2038 at the latest, leading to a profoundly transformative process in the region.

Our insights stem from a literature review of historical research on minority rights, traditional land use systems, recultivation, mining law, resettlement, and memory culture within Lusatia; site visits; and interviews conducted between 2020 and 2024. We conducted issue-focused expert interviews with local historians as well as non-structured and problem-centred expert interviews with Sorbian stakeholders/representatives at different levels. The framework for our study is an interdisciplinary research initiative, which supports and reflects on current approaches to achieve UNESCO World Heritage status for Lusatia's post-mining landscape. The Sorbian Institute/Serbski institut (SI) is one of the five partners engaged in this endeavour.[4] The initiative was initially developed by representatives from the Brandenburg Ministry of Science, Research and Culture (Ministerium für Wissenschaft, Forschung und Kultur, MWFK), the Brandenburg University of Technology (Brandenburgisch-Technische Universität Cottbus-Senftenberg, BTU), Branitz Park (a garden in Cottbus, created by the local prince of Pückler-Muskau during the second half of the nineteenth century), the Humboldt Forum Berlin, and the Berlin Biennale for Contemporary Art.

Towards this end, a tentative application was submitted at the national level in 2021 by the MWFK jointly with the Saxonian Ministry for Regional Development (Sächsisches Staatsministerium für Regionalentwicklung, SMR). However, at the German Conference of Ministers of Education and Cultural Affairs (Kultusministerkonferenz) held in December 2022, it was decided that the post-mining landscape of Lusatia would not be included in Germany's tentative list of future UNESCO World Heritage nominations. The main arguments leading to its rejection were inadequate protection mechanisms for the landscape, the incomplete demarcation of the proposed area, and insufficiently clear communication of its value-bearing features in their context. A key issue focused on how the increasing usage of renewable energies within the landscape itself or within its sightlines would be dealt with.[5]

This feedback and its consequences for the entire initiative provide a productive opportunity for reflection on our shifting roles as researchers, heritage makers, and part-time lobbyists for Sorbian interests, as well as the opportunities and limitations of official heritagisation processes. Few case studies within the field of heritage studies have focused on sites that are becoming or failing to become UNESCO World Heritage, especially in contexts involving minority groups. Drawing on Susemihl's work on the Tr'ondëk Klondike area in Canada,[6] we argue that this could be especially beneficial from a minority group's perspective, as the 'heritagisation of landscapes'[7] provides a field for negotiating local and often regional identities. '[P]laces are "produced" by cultural practices [such as heritagisation], and identity is constructed or produced by places. [...] So, heritage in the sense of place helps people to position themselves as a nation, community or individual and in defining their "place" in our physical world.'[8] It is therefore important to question how heritage can also serve as a guiding orientation for Sorbs to position themselves within Lusatia's post-mining developmental process.

To answer these questions, we briefly describe the history and terrain of our study site, namely the region of Lusatia—including Lausitzer Revier (its mining area) and its specificities—considering the overlap with the previously discussed Sorbian settlement area. We present an overview of changing approaches for engaging with aspects specific to minorities in relation to lignite mining in the region as well as their political rights in this context, the state of the (national) energy policy systems, and demographic changes. In the following section, we outline the broader framework of our research embedded within the heritage initiative, followed by a deeper analysis of the possibilities and limitations relating to the integration of a minority-adjusted perspective within the process of heritagising landscapes of energy. We demonstrate that historically evolved dynamics have structured the ways in which the 'minority issue' has been discussed and dealt with, revealing how these dynamics were also powerful in the earlier process of economic restructuring in the region. A particular focus on 'participation' forms the framework for our discussion, followed by a concluding section, which provides an outlook on future steps.

Lusatia, the *Lausitzer Revier*, and the Sorbian Settlement Area

The region of Lausitz (in German) or Łužyca/Łužica (in Lower and Upper Sorbian) is located in southeastern Germany approximately 150 km south of Berlin. Within present-day Germany, it is partly located in the state of Brandenburg (Lower Lusatia) and partly in the state of Saxony (Upper Lusatia). Historically, Lusatia also included areas east of the River Oder, which now fall under Polish territory following the redrawing of borders after World War II. As we are focussing specifically on Lausitzer Revier (the Lusatian lignite mining district), we will only refer to the areas within German territory in this chapter. The mining area is bounded to the east by the River Oder (marking the border with Poland) and by the Czech Republic to the south. While the German federal government decided to phase out lignite mining in 2019, economic and political actions have been aimed at maintaining the region's role as a centre for renewable energy.[9] A total of 40 billion euros has been earmarked to support all German lignite mining regions, with 17.2 billion euros allocated for Lusatia.[10]

Profound transformations such as the one noted above are never solely economic processes. They also challenge existing regionalisms and lead to the redefinition and reorientation of a region's identity. Regions and their identities are always also imagined communities based on shared ideas of particular spaces, their characteristics, symbols, and inhabitants.[11] Regional planning ultimately institutionalises these ideas but is simultaneously shaped by the conscious and unconscious ideas of the respective actors involved in this process.

In the case of Lusatia, these processes were particularly complex given the attempts to implement a minority-adjusted perspective, which both challenges and enriches current understandings of the region and its post-mining landscape. As we noted when introducing Wolkenberg/Klěšnik, Lusatia is also an area of settlement of Sorbs and has been so since the seventh century. Nevertheless, since the time of the integration of this area into Saxony in the tenth century, Lusatian rulers have been German-speaking.[12] Even though Sorbs made up to 35% of the urban population, as, for example, in late-medieval Bautzen/Budyšin, the German-speaking population dominated, especially in areas where urban centres developed.[13] In rural areas, however, Sorbian culture and languages remained highly influential in the historical process of settlement. In many communities, the majority–minority ratio of the population did not change until the turn of the twentieth century.

On the one hand, the increasing dominance of the German language at the time is linked to the emerging romantic idea of the nation, leading to German political representatives claiming its superiority even before the foundation of the German Reich in 1871.[14] The introduction of compulsory education led to a standardisation process, with the German language being preferred over local dialects and the native languages of other population groups, such as the Sorbs. On the other hand, the inception of open-cast mining stimulated industrial growth and fundamentally changed the relationship with the land.[15]

Most of the associated challenges inside as well as outside the Sorbian area stem from the evolving legal framework of mining planning in Germany. For example, the current Federal Mining Act emerged from the Allgemeines Berggesetz für die Preussischen Staaten (the Prussian Mining Act) introduced in 1865. It follows the principle of mining freedom and separates non-mining natural resources, such as minerals, oil, natural gas, and coal, from land ownership. Until the 1930s, the state's right over mineral resources still meant that landowners were free to negotiate the sale of land to companies privately and to make their own decisions. This also applied to the large-scale extraction of lignite, which became possible from the beginning of the twentieth century through the availability of industrial technology.

In the early twentieth century, the loss of hard coal regions after World War I was a crucial factor contributing to the growing economic relevance of lignite for Germany. In Lusatia, this led to mergers, the development of new technologies for overburden removal, the rerouting of rivers, and increased devastation of settlement structures. Neu-Laubusch/Nowy Lubuš was the first settlement to be completely devastated, with the destruction of five villages up to World War II to

Table 10.1 Summary data on numbers of destroyed villages and resettled inhabitants sourced from Berkner and Gesprächskreis Braunkohlenplanung, *Braunkohlenplanung und Umsiedlungsproblematik*; Förster, *Verschwundene Dörfer*; Erler and Stein, *Dokumentation bergbaubedingter Umsiedlung*

Period	*Destroyed villages*	*Resettled inhabitants (registered)*
1924–1944	5	1,285
1945–1989	119	24,964
1990–2024	13	2,729
Total	137	28,978

enable the establishment of the Laubusch/Erika mine in 1924 (Table 10.1). Under the banner of economic autarky, and with the National Socialist regime making preparations for war, this trend was perpetuated and consolidated.[16] At the level of mining planning, the freedom of private landowners, and thus of a large proportion of the rural Sorbian-speaking population, to negotiate the terms of sale of private land with the representatives of the mining companies was lost. In 1934, a *Rohstoffsicherungsklausel* (raw material security clause) was enshrined within mining law, which created the possibility of compulsory expropriation as *bergrechtliche Enteignung* ('cession of land under mining law') to ensure unimpeded extraction of mineral resources. This meant that the ownership of the natural resources awarded in the mining planning proceedings was accorded higher priority than the property rights of the landowner.[17]

After World War II and the partitioning of Germany in 1949, not only the political situation but also the demographic situation in Lusatia changed radically due to the influx of displaced persons.[18] At the economic level, the production rate of the district peaked. In 1957, the Cottbus District was declared the 'Coal and Energy Centre of the GDR.'[19] Most of the 137 Lusatian villages were destroyed to enable lignite extraction between 1949 and 1989 (Table 10.1). Of these villages, 73% were located directly in what Arnošt Muka has termed the 'Sorbian core' and the 'transition area.'[20] Despite being recognised as a national minority in 1949, Sorbs were neither specifically nor systematically considered in resettlement schemes and in the re-appropriation of the post-landscape. Only selective, collectively relevant measures were implemented to compensate for and mitigate negative consequences.

In the following decades, and especially since the onset of the oil crises of the 1970s, the extraction of regional lignite increased massively, and numerous large power plants were built. These developments were accompanied by drastic landscape interventions. By the time of German reunification in 1990, the devastated area had expanded to 68.671 km^2, of which 53.6% was under recultivation (Figure 10.1).[21]

These dramatic changes were also reflected at the level of regional identity. As in other mining regions, open-cast mining and miners became the dominant and unifying force in the forging of the region's identity within most social

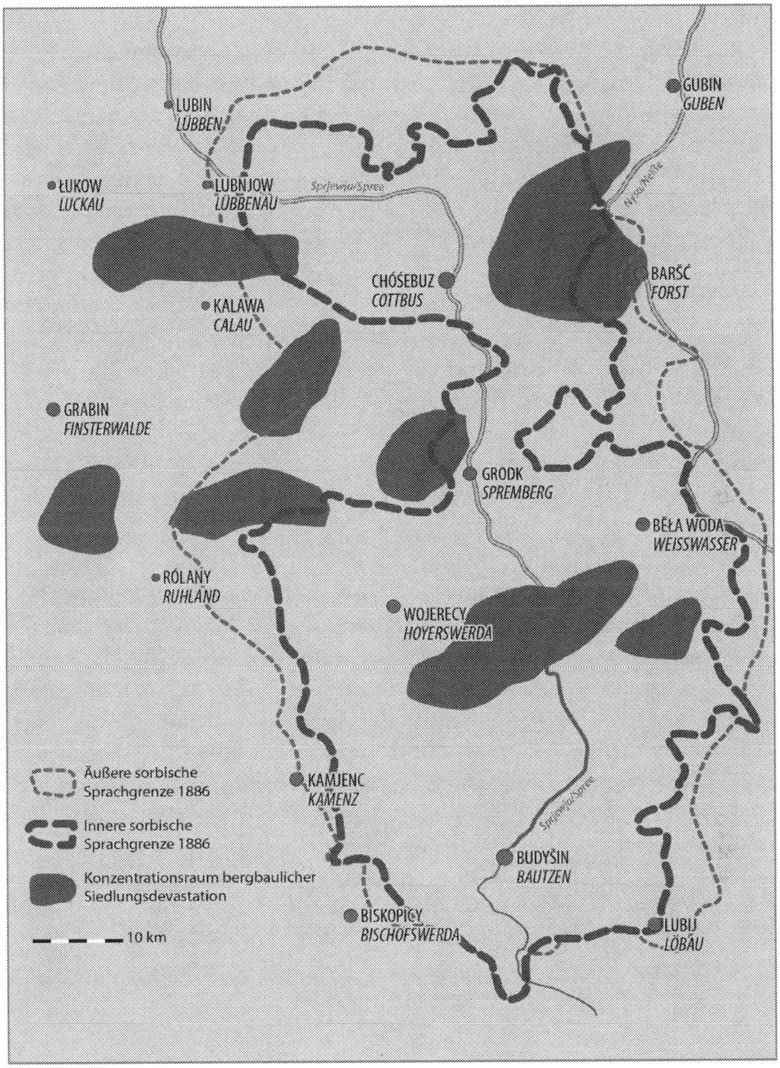

Figure 10.1 Sorbian core and transition areas overlapping with core mining areas.

Source: Frank Förster, *Verschwundene Dörfer*, 26.

discourses,[22] combining varying self-conceptions of new workers who moved to the area. However, this was not a linear process. Even during the twentieth century, workers in older mining areas, such as Lohsa/Łaz, maintained their rural lifestyle, whereas workers in newer areas often moved into housing estates created specifically for this purpose.[23]

During the post-reunification period, lignite planning was reorganised. After 1990, individual concerns relating to various protected goods (*Schutzgüter*) were incorporated within the framework of environmental and social impact assessments.[24] National minority rights legislation was implemented in Lusatia under the Framework Convention for the Protection of National Minorities (FCNM), 1998, and the European Charter for Regional or Minority Languages (Language Charter) introduced in 1999, each being adapted at the federal state level in Saxony and Brandenburg.[25] Consequently, Sorbian political interest groups are also now consulted during the lignite planning process. However, a veto against the utilisation of settlement areas is not possible, and there is no consideration of Sorbian aspects within recultivation practices. Moreover, there are no binding rules, let alone implementation of standards for compensation measures. In light of these gaps, Sorbs have negotiated compensation measures bilaterally with the mining company outside of the planning process via DOMOWINA, their political interest group.

German reunification marked a socio-economic watershed in the formation of Lusatia as a region. The number of employees in the lignite and refining industries fell sharply, and unemployment rates above 20% were the norm in the 2000s. For example, whereas approximately 80,000 people were working in the lignite industry at the end of the 1980s, currently only around 8,700 are still employed in this and associated industries, illustrating the rapid disruption of the socio-economic structure that has taken place.[26] Especially since the introduction of the planned lignite phase-out, various stakeholders have been seeking a new self-understanding of the region that not only transports its mining past but also functions as a guiding principle for a post-fossil future. Thus, the aim of the tentative World Heritage application was also to explore a future-orientated identity of the region as an outstanding example of innovation by raising awareness of technological and scientific achievements in dealing with the post-mining landscape.

The 'Minority Issue' in Lusatian's Post-Mining Heritage

In light of his work in Indigenous contexts, George Nicholas identified seven challenges entailed in the protection of minoritised objects, places, and values,[27] which can be easily adapted to characterise the consequences of mining for Sorbian cultural practices. The heritagisation of the post mining landscape does not and cannot respond to all of these challenges or mitigate the consequences of mining. Even though the tentative World Heritage application has been rejected, these two sources of Lusatian regional identity—Sorbian and industrial—can be linked together within the heritage initiative, and their transformation can be thematised. In other recent regional planning processes, the Sorbian language is apparently being gradually established as an integral, living, and unique part of Lusatia.[28] This development enables Sorbian stakeholders to implement at least some of the measures suggested by Nicholas for handling identified challenges, of which 'safeguarding [...] intellectual property [...] [and] knowledge licencing and sharing [...]'[29] are currently priorities.

Our initiative is part of a transdisciplinary network funded by the Bundesministerium für Bildung und Forschung (Federal Ministry of Education and Research, BMBF) to develop ecologically sustainable solutions for challenges related to climate change within and beyond the region. As one of the few projects associated with the 'cultural landscape' thematic complex, the World Heritage application thus functioned here as a reaction to structural change relating to the de-industrialisation process. The funding structure itself is not remarkable, as governments usually take a leading role in the preservation of heritage or other cultural goods.[30] Nevertheless, as Nguyen and Phan concluded on the basis of their work in Vietnam, it is usually minoritised heritage which 'does not receive adequate conservation resources as equally as other kinds of [nationally marked] heritage do.'[31] In our case, this applies not only to heritage as such but also to the research associated with it. It is because of the SI's engagement in advance of the initiative's commencement that it has been able to act as an equal research partner, receiving the same amount of financial support as our other partners.

As researchers engaged in heritage initiatives, we find ourselves in the complex position of being what Helmut Groschwitz describes as 'cultural brokers.'[32] While our research on the ruptures, continuities, and interconnections of Sorbian and industrial cultures first and foremost follows a scientific approach, as official heritage makers, we simultaneously defined and selected Sorbian elements to include in the application. We also engaged in communication and cooperation with Sorbian interest groups. Therefore, we constantly switch between our roles as critical humanities scholars, official heritage makers, and lobbyists for Sorbian interests.

While reflecting on this position, we implemented various steps to include minority-adjusted perspectives in the process. The application itself focused on the post-mining landscape. Its core argument was that a very large number of technological innovations were developed for renaturalisation and recultivation in Lusatia, and, importantly, they have remained visible over a long period—over 120 years—and formed the world's largest contiguous post-mining landscape.[33] The main parts of the application focused on the re-appropriation of the landscape as a nature conservation area, a recreational area, or a laboratory for new agricultural practices. Considered from a purely technical perspective, Lusatia is a globally significant example of an innovative design for a post-mining landscape. The term 'radicality' was repeated to express the land-grabbing nature of lignite mining and its consequences for the region's population, and especially for those in devastated and/or resettled places, and associated consequences for Sorbian cultural practices.

This narrative is part of the Outstanding Universal Value (OUV) that applicants are required to elaborate on and communicate to UNESCO. However, the post-mining landscape is also characterised, above all, by movement, change, and loss of settlement structures, landscape elements, and cultural practices. We therefore focused on what was fundamentally intangible, prioritising knowledge over materiality.

During the application phase, the challenge for the overall project lay in incorporating these intangible aspects, especially the negotiation of minority rights during the resettlement process and establishing links with pre-industrial, agricultural, and/or Sorbian cultural practices into the selection and planning of the site. This process of what Tuuli Lähdesmäki refers to as 'placing heritage' entails practices of typologising and mapping.[34] In the case of Lusatia, this process resulted in a contradictory situation whereby we first had to differentiate between Sorbian and non-Sorbian land use. Considering that the vast majority of devastated villages were predominantly Sorbian-speaking until well into the nineteenth century, it was almost impossible to distinguish between traditional Sorbian and more general historical forms of agricultural land use.

In their recent work on the heritagisation of Māori culture in Aotearoa, New Zealand, Passey and Burns emphasised what they have termed 'bicultural heritage,' which entails thinking and talking about 'heritage in terms of people and place rather than previous Western-dominated hegemony of heritage objects and buildings.'[35] While generally endorsing their approach, rather than making a sharp distinction between ethnic groups, we tried to include Sorbian aspects as a cross-cutting theme in the UNESCO application. These aspects only became a minority heritage, and thus antithetical to the majority concept of cultural heritage, according to Györgi,[36] as a result of industrialisation, associated immigration, and the development of the German nation-state.

To establish Sorbian aspects as a cross-cutting theme, we extended the period under investigation to include pre-industrial phases, thereby incorporating knowledge about how traditional or pre-industrial forms of land use characterise the post-mining landscape. This knowledge encompasses visible signs of continuity as preserved landscape elements, such as agricultural buildings or settlement structures, within the post-mining landscape. As a result of the discursive dichotomisation of 'Sorbian–pre-modern–rural' and 'industrial–modern–urban,' reappropriations of the post-mining landscape reference the pre-industrial past in ways that are sometimes non-random but more often random. In 1996, for example, the former mining company, LAUBAG, restored the historic Tschuga River near Neu-Petershain/Nowe Wiki. Even if the goal was to drain open-cast mining water,[37] this initiative randomly restored the Tschuga as a very meaningful landmark for the Sorbian population of this area. Rarely does information presented at recultivated sites—especially at post-mining lakes—convey any knowledge about the site's societal aspects, notably about their Sorbian population. Moreover, when they do, as we discovered during field visits to, for example, the aforementioned Wolkenberg/Klěšnik, Sorbian culture is mostly historicised as something that 'vanished' following the onset of lignite mining.

To change this perception, we aimed to incorporate knowledge about the continuities, ruptures, and interweavings of industrial and Sorbian history into the heritage narrative, thereby diversifying the perspective on the landscape. For example, we used terminology such as 'ongoing learning process' within the application text to describe the processuality of the inclusion of minority and population rights within the resettlement process. In this way, we sought to clarify that the more socially acceptable resettlement processes currently being implemented continue to be criticised and debated. Another possibility at this textual level is to use Sorbian languages in the official heritage-making process, for example, in the application text. In all documents and related publications, the German and Sorbian place names of all the places and regional authorities are used. In addition, the official five-page application was submitted not only in English, as is customary, but also in Upper and Lower Sorbian.

To visualise our findings on negotiation processes relating to minority rights within resettlement planning, we selected tangible signs of this ongoing learning process in the area under study. This applied especially to the settlement structures of resettled villages but also to preserved historical buildings or places of remembrance that combine the industrial and Sorbian aspects of their past.

The new village centre of Trebendorf/Trjebin constructed for Hinterberg/Zagora District, which was devastated in 2010, provides a good example of how a relatively minority-adjusted appropriation of a landscape can be achieved. Compared with other resettled villages where Sorbian traditions are maintained, traditions continued to be cultivated in this case, even after the resettlement process was complete, contributing to coexistence at the village and regional levels. From a planning perspective, this is apparent in the construction of the 'Ludki' kindergarten, the site for performing village festivals and traditions, and a sports field for club life. At the core is the 'Schusterhof ensemble,' comprising a timber frame house, a barn, and a bakery with a farm garden, which were relocated from Hinterberg/Zagora by Vattenfall and are now used by the local DOMOWINA group for public readings, theatre, and guided tours arranged for tourists. These showcase traditional farming methods and Sorbian traditions as well as the women's club of Trebendorf/Trjebin (Figure 10.2).

The entire village centre therefore serves not only as a testimony to historical land use and living village cultures but also to effective integration of minority and population rights during the resettlement process. In addition to this ensemble-orientated approach, another factor influencing minority adjustment was the resettlement of Hinterberg/Zagora within its historically grown socio-cultural networks in the Schleife sub-region. This example demonstrates that documentation, transplantation to a new area, and the reconstruction of what has been destroyed there are not sufficient. Rather, the affected place and its inhabitants—whether or not they are Sorbs—are closely interwoven with the surrounding cultural landscape and can only respond with resilience to resettlement if these entanglements are considered within the resettlement process as part of their past, present, and future.

The above example also illustrates how Sorbs not only manage a historicised cultural heritage in the form of traditional costumes and customs but also actively shape the region. There are still tens of thousands of Lusatian inhabitants who self-identify as Sorbian and for whom Sorbian customs and culture are simply part of their everyday lives. This factor can be taken into account in the World Heritage application by extending the period analysed not only into the past but also into the present and future. Therefore, places that are currently being relocated, such as the last village of Mühlrose/Miłoraz and the associated new village centre, Neu-Mühlrose, can be included in the area under study.

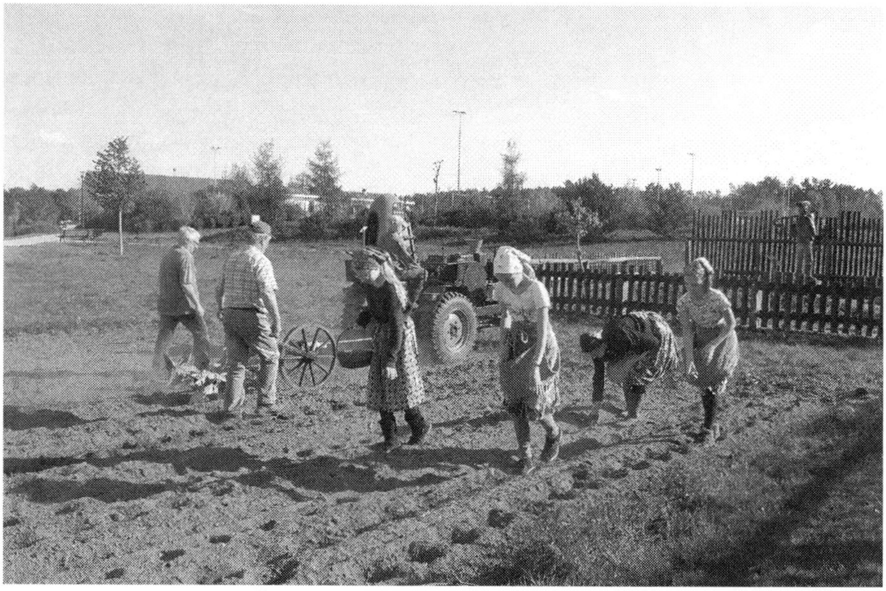

Figure 10.2 Sorbian tradition of picking potatoes at the 'Schusterhof' (shoemaker's yard) in Trebendorf/Trjebin. Photograph by Angelika Balzke.

The rejection of the application in December 2023, however, hints at the limits of reconciling understandings of UNESCO heritage with the ever-changing character of the Lusatian energy landscape. Consequently, the aim of the initiative must be 'to work with change rather than arrest or reserve it,' as observed by DeSilvey and Harrison.[38] This applies to the negotiation of minority rights as well as other aspects, such as re-appropriating the landscape for renewable energy production. Further steps must weigh suitable opportunities for valorisation, which would enable the inclusion of more existing interpretations of the post-mining landscape into the overall narrative, for example, those of miners, resettled people, recultivation practitioners, environmental activists, artists, and others, especially as all of these positions could incorporate Sorbian perspectives and would often overlap with them.

Ultimately, this also means that future uses of the area must be considered when planning its heritagisation. This requires not only balancing economic interests and monument protection but also meaningful interpretation and consideration of Sorbian interests and culture within its legal framework[39] because places like Mühlrose/Miłoraz are located in the official Sorbian settlement area. Here, minority protection can potentially complement monument protection and nature conservation by combining different kinds of protection statuses for a prospective site. It is also apparent that the heritage development of Lusatian post-mining landscapes not only visualises a distant past but also a very tangible present that will have a lasting influence on the future of the local residents.

Participative Heritagisation: Discussion and Outlook

In reflecting on the steps already taken, as elaborated above, we must address the question of whether inclusion of a minority-adjusted perspective in the general narrative amounts to a minority-adjusted heritage initiative. This necessitates reflecting on the initiative through the lens of participation: What does participation mean? Why is achieving it important, and how? Scholars and practitioners alike have been working on the concept of participation for some time now, sharpening its application as a multi-level model through its expansion from obtaining information to self-determination and considering the influence of decision makers as well as the actions of respective stakeholders.[40] This understanding fundamentally informs our work,[41] which is aimed at enriching the local official heritage discourse of the majority population and ultimately diversifying the region's self-understanding to open up more windows of opportunity for the integration of Sorbian interests. To gain access to this majority discourse, between March and October 2021, the SI team was involved in four meetings with district administrators and over 20 meetings with representatives of municipalities, where the aim was to present and discuss the tentative application. During this time, meetings were also held with the two councils for Sorbian affairs in Saxony and Brandenburg, respectively, and with representatives of the DOMOWINA regional associations in Lower Lusatia and Schleife/Slepo.

Clearly, as Di Giovine notes, a World Heritage scheme can only be 'an ethical project, aimed at instilling a sense of collective responsibility towards cultural diversity through governmentality'[42] if all relevant stakeholders are included in the respective negotiation processes. However, as discussed by Fredheim and May more recently, to motivate different stakeholders' participation, official heritage schemes are embedded within a narrative of endangerment, leading to a 'circle of heritage.' According to these authors, for something to be cared for, it has to be valued as endangered. Only if people care for it will they enjoy it, and if they enjoy it, they will want to understand it, and if they understand it, they will want to care for it.[43] This leads directly to the heart of the Lusatian initiative's most influential challenge. During our discussion with Sorbian stakeholders, it became clear that local communities are not concerned about losing visible signs of recultivation techniques. According to May, they do not necessarily need to be concerned, as the sense of ownership of communities participating in heritage schemes stems, first and foremost, from their enjoyment rather than from their awareness of endangerment.[44] However, it is the mining itself, not the post-mining landscape, that is associated with loss, especially for those affected by resettlement and/or those participating in local cultural practices. Members of the Brandenburg Sorbian Council, for example, emphasised the 'great loss of intact village communities combined with the decline of the Lower Sorbian language and culture' as a result of the devastation. Regarding the heritagisation process, they warned against neglecting the 'downfall of the Sorbian people' related to mining. At the same time, they associated the initiative with the hope of 'reaching a broader public and a new understanding among the inhabitants of Lusatia and beyond.'[45]

Adopting the minority perspective may be especially fruitful in freeing heritage from the discourse of endangerment.[46] First of all, Sorbian stakeholders who are consulted have an opportunity to criticise the valorisation of the post-mining landscape in general and, simultaneously, to include minority-adjusted and critical perspectives on the initiative. Nevertheless, as Susemihl concludes, minorities 'must participate in the decision-making process, which involves more than consultation.'[47] In their respective work, both Susemihl and Nicholas have emphasised the importance of basing heritagisation of minoritised or Indigenous culture on participation that leads to what Brown calls 'cultural resilience,' entailing empowerment, capacity building, and community development through rootedness, resourcefulness, and resistance.[48] Specifically, Nicholas proposes six responses to achieve this goal: participation, self-determined decision-making and the flow of benefits, respect for intellectual property, research ethics and developing appropriate new research methods, conjoined policy development and implementation, and corporate responsibility.[49] While the Lusatian initiative does not address all of these recommendations, we elaborated strategies for promoting general participation, dealing with the specifics of autochthony, and marketing/touristic valorisation.[50] These three strategies bring together the legal framework for minority rights, as discussed above, with the updated UNESCO Operational Guidelines, which require the free, prior, and informed consent of local and/or Indigenous groups.[51] Germany also ratified the International Labour Organisation's 1989 Indigenous and Tribal Peoples Convention in 2021 but did not acknowledge the status of Sorbs as Indigenous peoples, as they do not match the ILO's definition of indigeneity, according to the state government.[52]

The ongoing public debate about the advantages and disadvantages of being recognised as Indigenous people calls for a careful adaptation of the concept of indigeneity in the Lusatian context.[53] The implementation of Sorbian interests within the World Heritage initiative is relatively independent of this distinction between 'indigenous' and 'autochthonous.' Both Indigenous people as well as 'rightsholders [...] who have legal or customary rights [and] local communities [...] who possess a direct connection to the heritage place'[54] are defined as important stakeholders within the corresponding UNESCO guidelines. This definition automatically includes Sorbs and other local stakeholders equally.

Therefore, from a legal point of view, the question does not seem to be *whether* Sorbian aspects and interests should be integrated into the process, but rather *how* they should be integrated. Even though the tentative application was rejected at the end of 2022, the initiative remains ongoing, with efforts underway to seek new ways to transfer and increase knowledge about the post-mining landscape as well as new ways to valorise it. This creates opportunities for Sorbian stakeholders to diversify the landscape's narrative and steer the flow of associated benefits back into the community.

However, this effort will only succeed if future steps to implement the initiative succeed in avoiding what can be termed the 'participation paradigm.' Fredheim warns that heritage schemes can reinforce neoliberal forms of governance if they focus solely on *how* to get people involved rather than on *if* people want to get involved, and why.[55] The Lusatian case is also entangled in these logics. Whereas at the end of our discussions in 2021, the two councils for Sorbian affairs in Saxony and Brandenburg gave their free and informed consent to continue the initiative's work, the initiative is still—as we have stressed within our framework—a project of official academic and non-Sorbian political stakeholders. In the case of the Sorbian Institute and the BTU, the project, implemented by academic stakeholders, is aimed at getting local actors to care for the post-mining landscape.

A first step aimed at minimising the negative effects of this top-down dynamic could be to reinforce the transfer of project–intern knowledge. More important, however, would be the provision of resources for an equal exchange with all identified stakeholders as part of the future initiative's design and also to initiate engagement in financial, social, or cultural activities. Concrete steps to include not only a minority-adjusted perspective but also the actual minority, which extends beyond the current exchange with political stakeholders, are yet to be taken. Nevertheless, these steps are crucial to ensure that the initiative becomes a tool for shaping the current regional transformation process in the way that it intends to.

Conclusion

As Harrison argues, 'all heritage needs to be seen as plural, as there are minorities in every society.'[56] Accordingly, integrating the perspectives and interests of minorities into heritagisation processes poses a challenge in many sites, as revealed by recent case studies. In the Lusatian case, this means first overcoming dichotomised understandings of 'Sorbian' and 'industrial culture' within the regional heritage discourse. Those understandings are tightly interwoven with the distinction drawn between tangible and intangible heritage within official heritage schemes. However, a minority-adjusted approach combines both aspects. Through knowledge of landscape-shaping cultural practices and knowledge about their spatial-material basis within the landscape, not only continuities and transformations but also the disappearance of Sorbian life worlds and the entwining of Sorbian and industrial culture can be integrated into the overall narrative and rendered visible within the post-mining landscape.

However, this leads to challenges stemming from the logic of official heritage rather than specifics relating to the minority. World Heritage schemes are always landscape planning processes in which the past is used to negotiate scenarios for the future. Even though the landscape-focused understanding within the UNESCO World Heritage Convention has been criticised by scholars, for example, Brumann et al.,[57] the rejection of the application has foregrounded the difficulties entailed in harmonising the idea of everlasting change—which is part of the OUV, given the nature of the industry that is active here—with understandings of tangibility within UNESCO. In our case, the rejection provides an opportunity to explore further valorisation formats. Could a holistic heritage approach, which works at the different existing levels of regional development, such as tourism, language revitalisation, and municipal politics, be equally useful without the UNESCO label?

In any case, questions relating to participation are at the centre of our future works. Due to their historically evolved minoritised status, autochthonous or Indigenous groups often lack the capacities and competences needed to shape heritage processes in a self-effective way. These include knowledge about planning routines and/or process logics. While capacity building is an important goal emphasised in existing case studies, non-minoritised initiators are just as responsible for creating a minority-adjusted process. Here, elaborating standards, such as the new UNESCO guidelines, and adapting them for local specifics is needed. These should, to return to Nicholas' work, include ethics and research methods for laying out the basis of the scheme as well as developing and implementing effective policies to ensure a flow of benefits towards minoritised stakeholders.[58] The initiative could then be used as an opportunity for the Sorbs themselves to develop collective strategies for the (re)appropriation of the post-mining landscape, not only via consumptive forms of land use, such as tourism and public education, but also through productive forms of land use in the fields of agriculture, forestry, and renewable energies.

Given that renewable energies are increasingly part of a re-appropriation strategy that shapes the post-mining landscape of Lusatia, those forms should be included in its future heritagisation,[59] particularly for raising awareness of associated problems. In the context of the lignite phase-out, the parent company of Lusatia's mining company, Energetický a průmyslový holding a.s. (EPH), is currently being restructured towards the use of renewable energy, raising the general question of how socio-cultural and minority aspects are and should be included in the region's transformation process. Here, the initiative should be seen as part of broader attempts towards minority-adjusted regional planning.

However, strengthening the 'Sorbianness' of the landscape will not undermine other perspectives, such as those of recultivation technicians or mining workers. As we elaborated, Sorbianness is a cross-cutting theme connecting all the disparate aspects of the post-mining landscape in Lusatia. Future concrete valorisation strategies will benefit from a minority-adjusted approach, leading, for example, to more sustainable tourism.[60] In this case, the heritage initiative could serve as a tool for shaping the current transformation process of the region, regardless of its success in the context of UNESCO World Heritage.

Notes

1 Given differences in Sorbian and German names, we will use both names when referring to places within the Sorbian settlement area throughout the text.
2 Hagemann, "Industrie – Kultur – Erbe."
3 Muka, *Statistik der Lausitzer Sorben.*
4 The first part of the project was implemented by three partners: the Institute for Heritage Management (IHM) Cottbus, the Brandenburg University of Technology (BTU), and the Sorbian Institute/Serbski Institut (SI) between July 2020 and December 2021. Additional partners participating in the second part of the project, which will be completed by

February 2025, are the Tourismusverband Lausitzer Seenland (TVLS) and the Institut für Neue Industriekultur (INIK).

5 www.kmk.org/fileadmin/pdf/PresseUndAktuelles/2023/2023_12_04_Abschlussbericht_Fachbeirat_Tentativliste.pdf

6 Susemihl, *Claiming Back Their Heritage*, 331–416.

7 Macdonald, *Memorylands*, 16.

8 Susemihl, *Claiming Back Their Heritage*, 43.

9 Matern et al., "Strategies for Energy Transition."

10 Structural Strengthening Act for Coal Regions ("Strukturstärkungsgesetz"), August 8, 2020. See also: https://medienservice.sachsen.de/medien/news/1048202.

11 Anderson, *Imagined Communities*; Paasi, "The Region, Identity, and Power."

12 Blaschke, *Geschichte Sachsens im Mittelalter*; Schrage, "Entstehung und Entwicklung der Markgrafenschaft Niederlausitz."

13 See: Jatzwauk, *Die Bevölkerungs- und Vermögensverhältnisse der Stadt Bautzen*.

14 Pech, *Ein Staat – eine Sprache?*, 26.

15 Förster, *Um Lausitzer Braunkohle*.

16 For a detailed study on NS-autarky policy and its implications for the lignite sector see: Baumert, *Autarkiepolitik in der Braunkohlenindustrie*.

17 While mining can be viewed as one of the factors influencing the assimilation of the Sorbian languages and self-understandings, further developments during the first half of the twentieth century have to be taken into account. At the broader socio-cultural and political levels, the later National Socialist period, in particular, was marked by increasing discrimination against Sorbian institutions and self-understandings. See: Förster, "Die Nationalsozialistische Wendenzählung von 1939."

18 Elle, *Sprachenpolitik in der Lausitz*; Bednarek et al., "Die Oberlausitz," 231.

19 Wolle, "Energie für die Zukunft."

20 Muka, *Statistik Lausitzer Sorben*, attached map.

21 Steinhuber, "Einhundert Jahre bergbauliche Rekultivierung," 275.

22 Baumert, *Autarkiepolitik in der Braunkohlenindustrie*, 294.

23 Förster, *Verschwundene Dörfer*, 20.

24 Berkner and Gesprächskreis Braunkohlenplanung, *Braunkohlenplanung und Umsiedlungsproblematik*.

25 Elle, *Die Europäische Charta*.

26 Berger et al., *Engpass Arbeit?!*, 13.

27 Nicholas, "Protecting Indigenous Heritage Objects."

28 Faber-Schmidt et al., *Kulturplan Lausitz*; Laschewski et al., *Inwertsetzung des immateriellen Kulturerbes*.

29 Nicholas, "Protecting Indigenous Heritage Objects," 408.

30 Ky Nam Nguyen and Quang Anh Phan, "The Management of Minority Heritage," 746.

31 Ky Nam Nguyen and Quang Anh Phan, "The Management of Minority Heritage," 747.

32 Groschwitz, "Kulturerbe vermitteln."

33 "Tentativantrag 'Lausitzer Tagebaufolgelandschaft,"' Annex 2a.

34 Lähdesmäki, "Politics of Tangibility."

35 Passey and Burns, "The Contested Shift," 1,000.

36 György, "An Attempt."

37 Kittan, "Aus der Dorfgeschichte," 10.

38 DeSilvey and Harrison, "Anticipating Loss," 4.

39 "Sächsisches Sorbengesetz" (Saxonian Sorbs Act) section 3 and "Brandenburgisches Sorben/Wendengesetz" (Brandenburg Sorbs/Wends Act) section 3.

40 For a discussion of the context of Lusatian transformation processes see: Schwan et al., *Kommunale Bürgerbeteiligung*; Lüttringhaus, "Eine Schwalbe."
41 Preliminary thoughts on the adaptation of the concept have been compiled by Tschernokoshewa et al., *Einheit in Verschiedenheit.*
42 Di Giovine, "The Ethics of Participation," 103.
43 Fredheim, "Endangerment-Driven Heritage Volunteering"; May, "Heritage, Endangerment and Participation."
44 May, "Heritage, Endangerment and Participation," 82.
45 Schwelina, *Projekt Lausitzer Tagebaufolgelandschaften.*
46 Fredheim, "Endangerment-Driven Heritage Volunteering," 5.
47 Susemihl, *Claiming Back Their Heritage*, 427.
48 Susemihl, *Claiming Back Their Heritage*, 135; Nicholas, "Protecting Indigenous Heritage Objects," 406–412; Brown, *Resilience, Development and Global Change.*
49 Nicholas, "Protecting Indigenous Heritage Objects," 406.
50 We especially thank our colleague, Hannah Wellpott, the main author of these strategies, for sharing her work with us.
51 United Nations Educational Scientific and Cultural Organisation (UNESCO) and Intergovernmental Committee for the Protection of the World Cultural and Natural Heritage, "Operational Guidelines."
52 Deutscher Bundestag (Wissenschaftliche Dienste), "Mögliche Folgen für Deutschland."
53 Piňosová, "Das Konzept der Indigenität."
54 UNESCO et al., "Guidance and Toolkit," 30.
55 Fredheim, "Endangerment-Driven Heritage Volunteering," 624.
56 Harrison, "Multicultural and Minority Heritage," 197.
57 Brumann and Gfeller. "Cultural Landscapes."
58 Nicholas, "Protecting Indigenous Heritage Objects."
59 For example, there are currently no standards or methods for conserving wind mills. However, the official heritage discourse is slowly shifting towards valorising visible signs of renewable energies as important sites for understanding how humankind is dealing with the challenges of climate change. See: Bockelmann, "Zeit, dass sich was dreht."
60 Global Sustainable Tourism Council, "GSTC Destination Criteria."

Bibliography

Anderson, Benedict. *Imagined Communities: Reflections on the Origin and Spread of Nationalism*. London/New York: Verso, 1991.

Baumert, Martin. *Autarkiepolitik in der Braunkohlenindustrie* [*Autarky Policy in the Lignite Industry*]. Berlin/Boston: De Gruyter, 2022.

Bednarek, Andreas, Jonas Flöter, and Stefan Samerski. "Die Oberlausitz vom Ende des Ersten Weltkrieges bis zur Gegenwart (1918-2000)." ["Upper Lusatia from the End of the First World War to the Present (1918-2000)"]. In *Geschichte der Oberlausitz. Herrschaft, Gesellschaft und Kultur vom Mittelalter bis zum Ende des 20. Jahrhunderts [History of Upper Lusatia. Rule, Society and Culture from the Middle Ages to the End of the 20th Century]*, edited by Joachim Bahlcke, 221–266. Leipzig: Leipziger Universitätsverlag, 2004.

Berger, Stefan. "Vom Nutzen und Nachteil der Nostalgie. Das Kulturerbe der Deindustrialisierung im globalen Vergleich." ["The Benefits and Drawbacks of Nostalgia. A Global Comparison of the Cultural Heritage of Deindustrialisation"]. *Zeithistorische Forschungen/Studies in Contemporary History* 18 (2021): 93–121.

Berger, Wolfram, Gunther Markwardt, Julia Rettig, Jan Schnellenbach, Mirko Titze, and Stefan Zundel. *Engpass Arbeit?! Chance und Risiko für den Strukturwandel in der brandenburgischen Lausitz* [*Bottleneck Labour?! Opportunity and Risk for Structural Change in Lusatia in Brandenburg*]. Study commissioned by the State Chancellery of Brandenburg. Cottbus: Brandenburgische Technische Universität Cottbus-Senftenberg; Leibniz Institut für Wirtschaftsforschung Halle, 2024.

Berkner, Andreas, and Gesprächskreis Braunkohlenplanung, eds. *Braunkohlenplanung und Umsiedlungsproblematik in der Raumordnungsplanung Brandenburgs, Nordrhein-Westfalens, Sachsens und Sachsen-Anhalts: Ergebnisse des Gesprächskreises Braunkohlenplanung [Lignite Planning and Resettlement Problems in the Spatial Development Planning of Brandenburg, North Rhine-Westphalia, Saxony and Saxony-Anhalt: Results of the Lignite Planning Discussion Group]*. Hanover: Verlag der ARL, 2000.

Berkner, Andreas, and Kulturstiftung Hohenmölsen e.V. *Bergbau und Umsiedlungen im Mitteldeutschen Braunkohlenrevier [Mining and Resettlement in the Central German Lignite Mining District]*. Beucha: Sax-Verlag, 2022.

Bockelmann, Leo. "Zeit, dass sich was dreht: Denkmalpflege und Windenergieanlagen" ["Time for Something to Turn: Monument Preservation and Wind Turbines." In *Denkmal_Emotion: Politisierung – Mobilisierung – Bindung [Monument_Emotion: Politicisation – Mobilisation – Commitment]*, edited by Stephanie Herold and Gerhard Vinken, 70–75. Heidelberg: Arthistoricum.net-ART-Books, 2021. https://doi.org/10.11588/arthistoricum.920.c13260.

Bogner, Simone, Gabi Dolff-Bonekämper, and Hans-Rudolf Meier, eds. *Instabile Konstruktionen. Interdisziplinäre Forschungen zu "Identität und Erbe" [Unstable Constructions. Interdisciplinary Research on 'Identity and Heritage']*. Weimar: Bauhaus-Universitätsverlag, 2022.

Brown, Katrina. *Resilience, Development and Global Change*. London: Routledge, 2015.

Brumann, Christoph. "Multilateral Ethnography: Entering the World Heritage Arena." *Max Planck Institute for Social Anthropology Working Papers*, no. 136 (2012): 1–17.

Brumann, Christoph, and Aurélie Élisa Gfeller. "Cultural Landscapes and the UNESCO World Heritage List: Perpetuating European Dominance." *International Journal of Heritage Studies* 28, no. 2 (2022): 147–162. https://doi.org/10.1080/13527258.2021.1941197.

DeSilvey, Caitlin, and Rodney Harrison. "Anticipating Loss: Rethinking Endangerment in Heritage Futures." *International Journal of Heritage Studies* 26, no. 1 (2020): 1–7. https://doi.org/10.1080/13527258.2019.1644530.

Deutscher Bundestag (Wissenschaftliche Dienste). "Mögliche Folgen für Deutschland bei einer Ratifikation der ILO-Konvention 169 zum Schutz der Indigenen Völker" ["Possible Consequences for Germany if it Ratifies ILO Convention 169 on the Protection of Indigenous Peoples"], December 3, 2020. Accessed July 1, 2024. www.bundestag.de/resource/blob/821372/f8d71fa2d34521d943256833de9f6155/WD-2-098-20-pdf-data.pdf.

Di Giovine, Michael A. "The Ethics of Participation, Community Formation and Governmentality in UNESCO's World Heritage Program." In *Between Imagined Communities and Communities of Practice. Participation, Territory and the Making of Heritage*, edited by Nicolas Adell, Regina F. Bendix, Chiara Bortolotto, and Markus Tauschek, 83–108. Göttingen: Universitätsverlag Göttingen, 2015.

Domowina. "Gemeinsame Erklärung der Domowina und Vattenfall – Aus Verantwortung für die sorbische Bevölkerung in den Bergbauregionen der Lausitz" ["Joint Declaration by Domowina and Vattenfall – Out of Responsibility for the Sorbian Population in the

Mining Regions of Lusatia"]. Accessed November 17, 2023, www.domowina.de/forum/gemeinsame-erklaerung-der-domowina-und-vattenfall.

Eiringhaus, Pia. "How Industrial Heritage Became Green. Renaturalisation Narratives in Regional History Culture." In *Boom – Crisis – Heritage. King Coal and the Energy Revolutions after 1945*, edited by Lars Bluma, Michael Farrenkopf, and Torsten Meyer, 257–266. Berlin: De Gruyter, 2022.

Elle, Ludwig. *Die Europäische Charta der Regional- oder Minderheitensprachen und die Sprachenpolitik der Lausitz* [*The European Charter for Regional or Minority Languages and Lusatia's Language Policy*]. Bautzen: Sorbisches Institut, 2004.

Elle, Ludwig. *Sprachenpolitik in der Lausitz. Sprachenpolitik und Sprachenrecht im deutsch-sorbischen Gebiet 1990 bis 2014* [*Language Policy in Lusatia. Language Policy and Language Law in the German-Sorbian Region 1990 to 2014*]. Bautzen: Sorbisches Institut, 2014.

Erler, Annelore, and Dörthe Stein. *Dokumentation bergbaubedingter Umsiedlung. Dokumentacija pśesedlenjow pśez gornistwo* [*Documentation of Mining-Induced Resettlement*]. Forst/Horno: Archiv verschwundener Orte, 2010.

Faber-Schmidt, Brigitte, Franziska Hammer, Frank Kober, et al. *Kulturplan Lausitz – Wir gestalten Kultur! I. Haupttext* [*Culture Plan Lusatia – We Shape Culture! I. Main Text*]. Ministry of Science, Research and Culture of Brandenburg, November 2021. www.lausitz-kultur.eu/images/pdf/01_Kulturplan_Lausitz_Haupttext.pdf.

Förster, Frank. *Um Lausitzer Braunkohle – 1849-1945 [On Lusatian Lignite – 1849-1945]*. Bautzen: Domowina-Verlag, 1990.

Förster, Frank. "Die Nationalsozialistische Wendenzählung von 1939 und der Einfluss ihres Ergebnisses während des Zweiten Weltkriegs" ["The National Socialist Wendish Census of 1939 and the Influence of its Results During the Second World War"]. In *Zwischen Zwang und Beistand. Deutsche Politik gegenüber den Sorben vom Wiener Kongress bis zur Gegenwart [Between Coercion and Assistance. German Policy Towards the Sorbs from the Congress of Vienna to the Present Day]*, edited by Edmund Pech and Dietrich Scholze, 73–101. Dresden: Domowina-Verlag, 2003.

Förster, Frank. *Verschwundene Dörfer im Lausitzer Braunkohlerevier.* [*Vanished Villages in the Lusatian Lignite Mining Area*]. Bautzen: Domowina Verlag, 2014.

Fredheim, Harald. "Endangerment-Driven Heritage Volunteering: Democratisation or 'Changeless Change.'" *International Journal of Heritage Studies* 24, no. 6 (2018): 619–33. https://doi.org/:10.1080/13527258.2017.1399285.

Global Sustainable Tourism Council. "GSTC Destination Criteria. Version 2.0 with Performance Indicators and SDGs," December 6, 2019. Accessed July 6, 2024. www.gstcouncil.org/wp-content/uploads/GSTC-Destination-Criteria-v2.0.pdf.

Groschwitz, Helmut. "Kulturerbe vermitteln. Strategien, Loyalitäten und Ambivalenzen in Beratung und Forschung" ["Communicating Cultural Heritage. Strategies, Loyalties and Ambivalences in Counselling and Research"]. In *Kulturerbe als kulturelle Praxis – Kulturerbe in der Beratungspraxis [Cultural Heritage as Cultural Practice – Cultural Heritage in Counselling Practice]*, edited by Daniel Drascek, Helmut Groschwitz, and Gabriele Wolf, 25–48. München: Bayerische Akademie der Wissenschaften, 2022.

György, Eszter. "An Attempt to Create Minority Heritage: The History of the Rom Som Club (Rom Som Cigányklub) (1972–1980)." *Romani Studies* 29, no. 2 (2019): 205–232. https://doi.org/10.3828/rs.2019.9.

Hagemann, Jenny. "Industrie – Kultur – Erbe. Der Lausitzer Tagebau und seine Folgelandschaft aus minderheitensensibler Perspektive. Essay" ["Industry – Culture – Heritage. Lusatian Open-Cast Mining and Its Post-Mining Landscape from a Minority-Sensitive Perspective.

Essay"]. In *Der Braunkohlenbergbau im 20. und 21. Jahrhundert. Geschichte – Kultur – Erinnerung [Lignite Mining in the 20th and 21st Century. History – Culture – Memory]*, edited by Jan Kellershohn, 261–281. Halle (Saale): Landesamt für Denkmalpflege und Archäologie Sachsen-Anhalt – Landesmuseum für Vorgeschichte, 2023.

Harrison, Rodney. "Multicultural and Minority Heritage." In *Understanding Heritage and Memory*, edited by Tim Benton, 164–201. Manchester: Manchester University Press, 2010.

Jaramillo, George S., and Juliane Tomann. "Introduction." In *Transcending the Nostalgic. Landscapes of Postindustrial Europe beyond Representation*, edited by George S. Jaramillo and Juliane Tomann, 1–17. New York/Oxford: Berghahn, 2022.

Kittan, Udo. *Aus der Dorfgeschichte von Petershain [On the History of the Village Petershain]*. Neu-Petershain: Eigenverlag, 2001.

Lähdesmäki, Tuuli. "Politics of Tangibility, Intangability, and Place in the Making of a European Cultural Heritage in EU Heritage Policy." *International Journal of Heritage Studies* 22, no. 10 (2016): 766–780.

Laschewski, Lutz, Gregor Schneider, Daniel Häfner, and Fabian Jacobs. *Inwertsetzung des immateriellen Kulturerbes im deutsch-slawischen Kontext – Teilprojekt AP3 "Entwicklung von Inwertsetzungsstrategien" Aktionsplan 2021 / Plan póstupowanja 2021* [*Valorisation of the Intangible Cultural Heritage in the German-Slavic Context – Subproject AP3 "Development of Valorisation Strategies" Plan of Action 2021 / Plan póstupowanja 2021*]. Cottbus: Sorbisches Institut, 2021.

Lausitzer Braunkohle AG, ed. *Ortschronik Wolkenberg [Chronicle of Wolkenberg]*. Niederlausitzer Heidemuseum, 1993.

Lüttringhaus, Maria. "Eine Schwalbe macht noch keinen Sommer – Grundvoraussetzungen für Aktivierung und Partizipation" ["One Swallow Does Not Make a Summer – Basic Requirements for Activation and Participation"]. *Newsletter Wegweiser Bürgergesellschaft* 20 (2009): 1–8.

Macdonald, Sharon. *Memorylands. Heritage and Identity in Europe Today*. New York: Routledge, 2013.

Matern, Antje, Martin Špaček, Jessica Theuner, Robert Knippschild, and Julius Janáček. "Strategies for Energy Transition and Regional Development in European Post-Coal Mining Regions: Ústí Region, Czechia, and Lusatia, Germany." *Territory, Politics, Governance* (2023): 1–22. https://doi.org/10.1080/21622671.2023.2231972.

May, Sarah. "Heritage, Endangerment and Participation: Alternative Futures in the Lake District." *International Journal of Heritage Studies* 26, no. 1 (2020): 71–86. https://doi.org/10.1080/13527258.2019.1620827.

Muka, Arnošt. *Statistik der Lausitzer Sorben. Aus dem Obersorbischen übersetzt und herausgegeben von Robert Lorenz* [*Statistics on the Lusatian Sorbs. Translated from Upper Sorbian and edited by Robert Lorenz*]. Bautzen: Domowina-Verlag, 2019.

Nguyen, Ky Nam, and Quang Anh Phan. "The Management of Minority Heritage: Critical Challenges to Vietnamese Catholic Heritage Seen from the Case Study of Bui Chu Cathedral." *International Journal of Heritage Studies* 27, no. 7 (2021): 734–751. https://doi.org/10.1080/13527258.2020.1858139.

Nicholas, George. "Protecting Indigenous Heritage Objects, Places, and Values: Challenges, Responses, and Responsibilities." *International Journal of Heritage Studies* 28, no. 3 (2022): 400–422. https://doi.org/10.1080/13527258.2021.2009539.

Paasi, Anssi. "The Region, Identity, and Power." *Procedia in Social and Behavioral Sciences* 14 (2011): 9–16.

Passey, Emma, and Edgar A. Burns. "The Contested Shift to a Bicultural Understanding of Place Heritage in Aotearoa New Zealand." *International Journal of Heritage Studies* 29, no. 9 (2023): 988–1003. https://doi.org/10.1080/13527258.2023.2231911.

Pech, Edmund. *Ein Staat – eine Sprache? Deutsche Bildungspolitik und autochthone Minderheiten im 20. Jahrhundert. Die Sorben im Vergleich mit Polen, Dänen und Nordfriesen* [*One State – One Language? German Education Policy and Autochthonous Minorities in the 20th Century. The Sorbs in Comparison with Poles, Danes and Northern Frisians*]. Bautzen: Domowina-Verlag, 2012.

Piňosová, Jana. "Das Konzept der Indigenität als Scharnier zwischen Natur- und Minderheitenschutz." In *Minderheit – Macht – Natur. Verhandlungen im Zeitalter des Nationalstaats*, edited by Jana Piňosová, Susanne Hose, and Marcel Langer, 83–105. Bautzen: Domowina-Verlag, 2022.

Schwan, Gesine, Katja Treichel, and Julian Kilchling. *Kommunale Bürgerbeteiligung in der Lausitz. Geeignete Formate zur nachhaltigen Unterstützung des Transformationsprozesses* [*Municipal Citizen Participation in Lusatia. Suitable Formats for Sustainable Support of the Transformation Process*]. Berlin: Humboldt-Viadrina Governance Platform GmbH, 2022.

Schwelina, Kathrin. *Projekt Lausitzer Tagebaufolgelandschaften "Land-Innovation-Lausitz" (LiL). Letter of the Council for Sorbian Affairs at the Landtag Brandenburg to the Institute for Heritage Management*. September 22, 2021.

Smith, Laurajane. *Uses of Heritage*. New York: Routledge, 2006.

Steinhuber, Uwe. "Einhundert Jahre bergbauliche Rekultivierung in der Lausitz. Ein historischer Abriss der Rekultivierung, Wiederurbarmachung und Sanierung im Lausitzer Braunkohlenrevier" ["One Hundred Years of Mining Recultivation in Lusatia. A Historical Outline of Recultivation, Reclamation and Rehabilitation in the Lusatian Lignite Mining Area"]. PhD diss., Palacky University Olomouc, 2005.

Susemihl, Geneviève. *Claiming Back Their Heritage: Indigenous Empowerment and Community Development through World Heritage*. Cham: Springer Nature, 2023.

"Tentativantrag 'Lausitzer Tagebaufolgelandschaft'" ["Tentative Application 'Lusatian Post-Mining Landscape'"]. Unpublished manuscript, October 28, 2021.

Tschernokoshewa, Elka, Ines Keller, and Fabian Jacobs, eds. *Einheit in Verschiedenheit. Kulturelle Diversität und gesellschaftliche Teilhabe von Minderheiten auf dem Prüfstand* [*United in Diversity. Cultural Diversity and Social Participation of Minorities Put to Test*]. Münster/New York: Waxmann, 2015.

UNESCO, ICCROM, ICOMOS, and IUCN, eds. "Guidance and Toolkit for Impact Assessments in a World Heritage Context," 2022. Accessed September 19, 2024, https://whc.unesco.org/en/guidance-toolkit-impact-assessments/.

United Nations Educational Scientific and Cultural Organisation (UNESCO). "The 2005 Convention on the Protection and Promotion of the Diversity of Cultural Expressions," October 20, 2005. Accessed September 19, 2024, www.unesco.org/creativity/en/2005-convention.

UNESCO and the Intergovernmental Committee for the Protection of the World Cultural and Natural Heritage. "Operational Guidelines for the Implementation of the World Heritage Convention," July 31, 2021. Accessed September 19, 2024, https://whc.unesco.org/en/guidelines/.

Wolle, Stefan. "Energie für die Zukunft. Das Kombinat 'Schwarze Pumpe' und die sozialistische Wohnstadt Hoyerswerda" ["Energy for the Future. The 'Schwarze Pumpe' Combine and the Socialist Town of Hoyerswerda"]. *Aus Politik und Zeitgeschichte* 70, no. 6–7 (2020): 8–14.

Concluding Remarks

Landscapes and Energyscapes

Petra Dolata

This book is about landscapes that are changing and how they are remembered. Individual chapters trace industrialising and deindustrialising landscapes while addressing the politics surrounding the construction, closure, and reclamation of energy sites. They contextualise these energy sites as integral parts of industrial as well as post-industrial landscapes, thereby addressing the spatiality and temporality of energy systems. In highlighting what the lived experiences of people in transitioning energy systems look like, they help us to understand what we may be up against as we embark on a path to decarbonise our energy systems. The contributions in this volume describe how change impacts communities in these energy-producing regions, including their sense of place, feelings of belonging, and their identities, reminding us that a landscape is not only something to be observed, administered, or controlled from a distance; it is also a space inhabited by communities and energy workers who give it meaning. It provides what Andrew Demshuk, in his contribution, refers to as 'anchors poignant for personal memory [and] communal identity.' As these energy sites are decommissioned and no longer provide energy for the nation and beyond, and as jobs for local and transient workers disappear, their industrial heritage becomes contested, particularly at a time when hydrocarbon energy systems symbolise a dirty past, which is responsible for climate change, and societies are planning for an energy transition towards greener energy systems in the Anthropocene. When landscapes change due to industrialisation, deindustrialisation, or energy transitions, existing meanings are challenged and renegotiated or even erased, as illustrated by the villages in East Germany in Andrew Demshuk's chapter that were destroyed and relocated to make room for lignite mining. While grappling with what these changes entail, the authors in this volume write about landscapes that have been shaped by the economic desire to produce energy. They explore the ways in which work and habitation created specific cultural landscapes in these industrial regions, which are now undergoing significant transitions, necessitating a reconsideration of their industrial heritage. This reconsideration also entails political discussions around national identities, state survival, and citizen participation.

This book brings together two strands of landscape research. Firstly, it puts emphasis on the geographies and spatial cultures of energy-producing sites.

DOI: 10.4324/9781003473534-15

Secondly, because cultural landscape forms a central aspect of heritage studies, it focuses on how to remember these industrial sites of energy production. Because the focus of this book is on cultural landscapes of *energy*, discussions of energy systems are also addressed, albeit implicitly. These become particularly pertinent when authors refer to the detrimental environmental impacts and human costs of energy production in what are often considered pristine or peripheral regions. What all of the contributions share is an interest in human agency in these places and, more specifically, in those who inhabit or work in energy landscapes. Notwithstanding their examination of the lived experiences of *energy* landscapes, these analyses transcend the local scale, as energy systems intertwine global, national, regional, and local sites of energy production, consumption, and infrastructure. Energy not only connects these various scales but also their many logics. These include national political goals of energy security, autarky or independence, the global capitalist strategies of multinational companies and international energy markets, the transnational mobility of labour, and the movement of money as workers send remittances to their families back home.

Whose Landscape? People and Place

Landscape is a concept that includes both the material and ideational aspects of relating to the environment. Various conceptualisations of landscape that are based on different philosophical traditions have emerged, especially within cultural geography but also within anthropology, history, and the creative genres. As a landscape can be 'the world we are living *in*, or a scene we are looking *at*, from afar,'[1] it covers diverse human experiences from the perspectives of both observers and those inhabiting a landscape, thus combining different scales and bringing together distance and proximity, embeddedness and aloofness, detachment and connection. While artists, researchers, or institutions may look upon landscapes from afar, others live in, experience, and shape these landscapes. To the former, a landscape may only become apparent from a distance, with smaller details merged into one coherent, homogenising image, viewed as a visually distinguishable entity with clear boundaries. This unbroken, all-encompassing, and continuous view of a scene from an elevated vantage point is also reflected in the artistic concept of the panorama, which Sarem Sunderland mentions when discussing the aesthetics of Swiss alpine landscapes. Seeing a landscape—as opposed to inhabiting it—distances the observer from place, confirming the argument made by Raymond Williams that 'a working country is hardly ever a landscape. The very idea of landscape implies separation and observation.'[2] For many artists, landscapes evoke ideas of beauty nourishing 'aesthetic discourses such as the pastoral, picturesque and sublime.'[3] Landscapes are to be painted, photographed, or otherwise creatively represented. During the period of industrialisation, these discourses included notions of the industrial sublime, which turned industrial architecture into 'cathedrals of work.'[4] More recently, industrial ruination has equally turned post-industrial landscapes into places of aesthetic enjoyment, a shift that has been criticised by deindustrialisation scholars as 'ruin porn.'[5] In her chapter on the post-industrial landscape of oil shale extraction in north-eastern Estonia, Saara Mildeberg quotes a filmmaker who turned the local observation of the ugliness of the abandoned site into an opportunity, exclaiming, 'Thank God everything hasn't been fixed up here. Thank God you have paint chipping on the walls.' Equally, as Sarem Sunderland argues, empty lakebeds and unsightly reservoir shorelines in the Swiss Alps have become aestheticised in recent years, turning wastelands into 'objects of fascination.' And while this fascination may lead to local reappropriation of industrial sites, it is also clear that these redefinitions of post-industrial landscapes are often framed by those situated outside them, whether they are tourists, visitors, corporate actors, national institutions of governance, journalists, or politicians. These examples also underline the challenges of reanimating industrial landscapes either through gentrification or heritage tourism, which, through its distanced and top-down approach, disregards local needs and voices.

The experiences of those who inhabit a landscape differ, and their involvement with its topographies and terrains is much more immediate. Embedded in the landscape, they live in, work on, and travel through its terrain, not always observing it as a whole. They actively construct the landscape through their practices and values, turning it into a cultural landscape full of social markers and cultural meanings. Through the lived experience of those who inhabit energy sites, cultural landscapes emerge. It is the interaction between human activity and nature that creates these distinctive landscapes. As a consequence, cultural landscapes have a history, or as Rune Frandsen insists, 'a right to history.' They can change over time due to human interaction and social, cultural, political, or economic processes. And change over time or 'landscape as process' constitutes the very meaning of cultural landscape along with place and practice.[6] Not surprisingly then, in a volume that focuses on 'cultural landscapes of energy,' humans play a central role and so does change over time.

As Rune Frandsen reminds us, humans need not even be residing in the landscape for long to contribute to its cultural significance. Transient workers equally leave their marks and give meaning to the landscape. Through their movement between energy sites, which they construct, they connect various locations and, by doing so, transcend the boundedness of cultural landscapes. These transient workers are not only involved in the construction of energy infrastructure but also in actual energy production. Using more recent examples in Northern Scotland, Ewan Gibbs reminds us that within the oil industry, workers and managers are often flown in and out of drilling sites all over the world, thus connecting the local and global scales of energy systems and creating transnational energy landscapes. It is this complex and at times tense relationship between various spatial and temporal levels that most of the authors highlight in their analyses. They are acutely aware of the difference and connection between seeing landscape from a distance and inhabiting it. The story of a cultural landscape of energy cannot be told unless the narrator looks at both vantage points and links their different ways of seeing landscape. Equally, these energy landscapes cannot be understood without tracing their history and recent changes or without addressing the multiple ways energy landscapes are narrated, represented, and remembered. Most of the chapters in this book combine the two ways of seeing to highlight the tension between those who work and live in the energy landscape and those who view it from a distance, often bestowing it with national political or economic significance or criticising its detrimental environmental effects. This complexity is needed to evaluate energy heritagisation efforts. In their respective chapters, Rune Frandsen and Sarem Sunderland caution that attempts to commemorate dams as significant energy landscapes leave out other dimensions of such landscapes, namely reservoirs, which, while engineered for the purpose of hydropower, are often misrecognised as being part of the natural environment or the cultural landscapes of those who created these material artefacts.

Industrial landscapes, such as energy landscapes, are made up of a built environment, which supports the energy industry; the natural environment upon or into which these were built, for extracting its subterranean riches; and its cultural landscape. Yet, despite this diversity, energy landscapes are often exclusively remembered or heritagised through their material remnants, such as dams, factory buildings, pitheads, derricks, or pumpjacks. And often, these do not even include the physical homes of workers. This is not to say that examples to the contrary do not exist. In her chapter, Saara Mildeberg describes how former miners in Estonia work as guides in a mining museum, leading tourists through underground tunnels, thus retaining their agency as energy workers in the transformed energy landscape. They also contribute to the oral histories of the site's mining past through clubs. However, as Jenny Hagemann, Fabian Jacobs, and Lutz Laschewski caution, not everyone is equally involved in creating this energy heritage, and some may not even want to get involved. If landscape does indeed have a 'right to history,' we still need to ask *whose* history. Rune Frandsen and Ewan Gibbs entreat us to look for the stories of workers, reminding us that it is through stories and histories that natural landscapes become cultural and energy landscapes. This may also necessitate connecting narratives across borders. Furthermore, it may entail uncovering the emotive, sensory, and bodily experiences of landscapes.

The Many Scapes

Using a landscape approach to understand changes in energy systems enables the authors in this book to draw attention to specific scapes. A scape denotes a geographical 'area that has a particular character because of the type of things that can be seen, heard, smelled, or experienced in it.'[7] In his contribution on transient energy workers, Rune Frandsen borrows Tim Ingold's concept of a 'taskscape' to combine a local and a transnational logic of dam construction in the Swiss Alps. By combining different geographical scales and introducing different temporal perspectives, he shows how the specific interactions of transient energy construction workers with nature in the Swiss Alps are informed by an 'industrial prehistory' of migrating Italian workers. Ingold uses his concept to discuss landscape as a 'process of becoming.'[8] In turn, Rune Frandsen employs it to plead for an industrial heritage of dam building 'as part of a continually evolving organism,' which cannot be tied to one particular location or time period.

In his chapter, Andrew Demshuk describes the East German lignite landscape as a moonscape. This description evokes an image of a sacrifice zone, a concept that is discussed further below and one that underlies artistic representations of surface mining, whether lignite, shale oil, or oil sands mining. He also explains that this moonscape has now been partly reclaimed as a system of lakes and forests and thus turned into a recreation area, leaving no physical traces which could provide an anchor for remembering the communities that were erased to make room for lignite mining. In addition, the flooding of the former mining landscape leaves no traces for energy workers to remember their place and role in the lignite industry. Instead, new meanings are inserted into the landscape. Jenny Hagemann, Fabian Jacobs, and Lutz Laschewski begin their chapter by taking the reader along vineyards which now adorn the former lignite mining region of Lusatia.

Moving beyond these normative scapes, Merve Neziroğlu, Hanna Lempinen, and Leyla Sayfutdinova employ politically connoted scapes in their respective chapters. Focusing on memories of violence, Merve Neziroğlu describes the Romanian Iron Gates region, which was devastated and flooded to make room for hydropower. She views it as a traumascape, which is remembered as a landscape in which traumatic events took place. Memories of loss prevail, and in the context of post-socialist states, they are often strongly associated with a negative memory of life under a dictatorship. Depending on the political trajectory after the end of socialism, they also remain hotly contested. The chapter by Hanna Lempinen demonstrates powerfully how forced energy transitions can lead to 'cultural trauma' and how changing energy landscapes may be interpreted very differently when viewed from a distance as opposed to inhabiting the landscape, which is producing an energy source no longer considered desirable. After peat was established as a source of climate change and the Finnish government made the subsequent decision to phase out its production, once-respected peat workers became 'outcasts' and 'environmental criminals.' Competing visions of the future of energy thus inform the ways in which energy pasts are remembered. Leyla Sayfutdinova's description of a 'geo-architectural assemblage' in her chapter on Baku resembles Carola Hein's concept of a petroleumscape,[9] connecting all physical infrastructure and the built environment supporting the financial flows of the industry and hence capitalism. She uses the ideas of financial and energetic flows, money, and power to expose the political power underlying the very specific energy landscape in Baku.

Interestingly, apart from Rune Frandsen, who writes about 'landscapes of labour,' and Ewan Gibbs' chapter on oil-related industries in Scotland, there is little discussion of labourscapes, focusing on energy landscapes as sites of work and labour relations. Such discussions bring out competing visions and conflict while highlighting the significance of these landscapes through their provision of economic livelihoods and homes, both material and ideational. Marin Kuijt and Gertjan Plets imply such a scape in their reference to an energy landscape that 'reshaped labour relations,' alluding to Rhenish capitalism and the corporatist response of labour, industry, and government in the Ruhr District in the second half of the twentieth century. A labourscape highlights what Ewan Gibbs calls 'worker mobility and labour agency' and includes built environments dedicated to energy workers, some of which were temporary. From a heritage perspective, this redirects our attention away from artefacts of energy production to include material remnants of what sustained those who produced energy, their ways of life, and cultural markers. In addition, it sheds light on exploitation and equity issues. Ewan Gibbs' discussion of women in the industry alerts us to remember that energy labourscapes are often extremely gendered and may become sites of sexual violence that expose 'toxic masculinity.' In addition, from an energy history perspective, a labourscape underscores the fact that despite the mechanisation and industrialisation of energy production, most of these energy landscapes still need a considerable amount of human labour and manual work, an often-forgotten fact that Ewan Gibbs calls attention to and which labour and energy historians have begun to remedy in recent years. It takes energy to construct and run energy production sites. Energy labourscapes also bring new values, norms, and practices to existing landscapes that have been the location of very different and often more traditional lifestyles, for example, in the alpine valleys of Switzerland. They may also allow for more nuance in describing these different ways of life. Not all workers in energy landscapes are the same. Jenny Hagemann, Fabian Jacobs, and Lutz Laschewski point out how some of the lignite miners in Lusatia 'maintained their rural lifestyle.'

Another scape is the energyscape that closely relates to the idea that energy systems are not just economic and technological arrangements for producing energy; they are also always social and cultural structures, and, as a consequence, their landscapes contain material energy technologies and infrastructures as well as social and cultural meanings, norms, and values.[10] Energyscapes have a history and are constituted through human practices. This is important for our understanding of the social dimension of energy systems and has some significant ramifications. First of all, this means that everything that supports a specific energy industry at a particular site should be considered as constitutive of that energyscape. Thus, related industries such as the ones that Ewan Gibbs discusses in his chapter are an integral aspect of the energy system and are not just associated with it, as are the transient workers in Rune Frandsen's chapter. Beyond this, even people who may not be employed by the energy industry but who live in these landscapes are contributing to the energyscape. They are contributing, whether by supporting those who are energy workers, such as housewives who provide reproductive labour, or through identification with the industry, as we have seen in the examples of the Ruhr District and Baku. An energyscape is where the influx of workers into remote landscapes creates what Ewan Gibbs calls 'fluid social relations.' Beyond the scalar and spatial nuances of energyscapes, the historical emergence of industrial landscapes in the wake of energy production adds an important temporal dimension, while highlighting economic and political rationales and practices. Energyscapes are closely linked to modernity, economic development, and the history of industrialisation. Industrialisation, deindustrialisation, and sometimes reindustrialisation are different stages through which a landscape may pass, providing it with a multitude of possible pasts competing for remembrance.

In a more geological and geographic sense, energyscapes are located both above and below ground; they are terrestrial and aquatic, as the examples of rivers and lakes for hydropower and the offshore production sites alluded to in Ewan Gibbs' chapter show. Baku's energyscape in Leyla Sayfutdinova's analysis is mainly reflected in its related architecture, both above and below the ground. Of course, the subterranean landscapes are much less visible and seeable from a distance. Her approach redirects the gaze to connect two verticalities, below and above ground, through the metaphor of fire. Not only did fires in the landscape indicate the existence of subterranean fields of oil, which needed to be drilled up, but they also stood for wealth and national pride. In Leyla Sayfutdinova's analysis, Baku is also unique as an energy space, as it hosts oil extraction and connected industries, combining 'direct and indirect energy landscapes,' which in other jurisdictions are often spatially separated. In addition, all of its urban infrastructure serves the overarching logic of oil production.

However, the most compelling insight emerging from Leyla Sayfutdinova's chapter relates to what we mean by 'seeing' landscape. In the Western world, those who inhabit cultural landscapes have been trained to see landscape in a particular way. Western visual culture has taught them to see the world as a 'scenery to be visually observed.'[11] Landscape then becomes a way of seeing. Yet, many energyscapes include locations that are below the surface and thus cannot be seen; they are invisible or unobservable. The same would apply to the hydropower reservoirs that Sarem Sunderland discusses in his chapter and which also become invisible as energy infrastructure, especially when aesthetically embedded in the existing landscape. Leyla Sayfutdinova argues that these extractive energy landscapes 'are co-produced by invisible matter, which is often hidden underground.' She encourages us to replace the horizontal line of seeing a panoramic landscape with a vertical plane and suggests that a focus on fire seems promising, as this 'volumetric approach' allows for the connecting of the subterranean with the surface landscapes of energy. The birth of the modern petroleum industry is visually connected to a landscape of derricks (technology) as well as uncontrolled gushers (nature) when oil was struck. Both of these symbols interrupt landscapes vertically. Leyla Sayfutdinova's historical perspective reveals how these gushers, like fire, came to symbolise wealth and industry. As will be discussed below, more recently gushers have been reinterpreted to symbolise the environmental impact and ecological disasters, thus contributing to the concept of an energy sacrifice zone. Whether visible or invisible, energyscapes can be seen differently, either because their economic and environmental impacts are connected in the present or because they are reinscribed with new meaning over time, sometimes because our ways of seeing landscape change.

Energy Systems and Energy Transition

The above discussion of energyscapes also prompts us to ask to what extent energy systems are delineated through energy landscapes. Energy systems are not only anchored in energy landscapes, but they also transcend these landscapes by connecting at larger scales, whether through markets, transport infrastructure, or their (geo)political significance. Their character and scenery are derived from the actual energy resource that they are primarily producing. For example, coal miners or peat diggers define themselves through their work and hence also through that particular energy resource, which has a unique materiality. This does not mean, however, that energy systems are exclusively based on one particular energy source; this is, after all, an energyscape and not a coalscape, shalescape, or oilscape. Often different energy resources coexist within a single energy system. As mentioned above, even the most mechanised energy productions still require a fair amount of manual labour. Conceptualising cultural energy landscapes as energy systems also allows for the integration of the entire suite of practices connected with energy, from producing to refining, transporting, and ultimately using energy. Separation of the sites of production and consumption has long been a problem in energy studies. This conception loses sight of the fact that communities are always involved in both sites. Any human being inhabiting an energy landscape is part of both processes. From a distance, onlookers will often only ascribe the landscape its energy meaning by singling out its function as a location for energy production. Yet, everyone who inhabits that landscape also consumes energy. This does not mean that those who inhabit energy landscapes always consume the very energy that they produce. Hanna Lempinen notes that peat extracted in Finland is not destined for residential use. Peat's local significance is not based on its energetic value; rather, it becomes a symbol of the survival of rural regions and ways of life and of energy dependency.

Hanna Lempinen also reminds us that what counts as energy resources are human constructions. Organic and inorganic matter becomes a 'resource' through the values that humans ascribe to this matter. In the same way that landscape entails a cultural approach of seeing landscape, coal, peat, and oil need to be framed as valuable resources to move to the centre of an energy landscape. Thus, energy sites have to be imagined. Sarem Sunderland describes in detail how planning documents and proposals for a hydropower site in the late nineteenth century created the very idea of an energy landscape by highlighting the economic rationalisation and utilisation of nature, in this case, comprising a valley that was prone to flooding. As plans for the reservoir shaped up, romanticised ideas of alpine lakes were employed to prove how aesthetically pleasing this human-engineered lake would be, thus beautifying the existing natural landscape.

Energy systems are also socio-economic systems. Energy industries provide employment and income. Because energy resources cannot be found everywhere and do not always coincide with population centres, given that transport routes and energy infrastructure can easily bring them to markets, they often provide the only source of livelihood for communities. And while energy transitions may turn existing energy landscapes into less environmentally damaged sites, the concomitant social changes due to phase-out have detrimental effects on the cultural landscape. Hanna Lempinen describes how peat workers not only lost their jobs but also lost their sense of place, identity, and community belonging. Understanding how these changes impact the landscape and the people who inhabit it helps to provide a nuanced view on the lived experience of energy transitions, highlighting the complexities of inhabiting an energy landscape, especially one that is changing. While energy workers identify with those energy sources which they produce, they are much more ambivalent when it comes to the impact of these sources on their health and their environment.

Despite the coexistence of various energy sources within one energy system, there are normative differences between them and, by extension, the ways in which energy landscapes may be viewed from afar and also remembered after their disappearance. Because the burning of fossil fuels, such as oil, gas, and coal but also peat, contributes to carbon dioxide emissions and thus to climate change, the landscapes that produce these energy resources are viewed in a much more negative light than hydropower, which generates clean electricity with much lower emissions. This is not to say that the generation of hydroelectricity does not create a CO_2 footprint, nor does it imply that hydropower is an entirely environmentally friendly energy source. The chapters on hydropower in this volume provide some inkling of how the construction of hydro infrastructure leaves considerable environmental destruction in its wake. Equally, a lifecycle assessment of hydroelectricity generation from its inception to its endpoint would show that it also entails greenhouse gas emissions. Still, normative discussions on energy landscapes differentiate between fossil fuels and renewables and pose a challenge to the heritagisation of such contentious energy sites.

It is also not the first time that energy systems are transitioning. History is full of examples of one dominant system of energy use replacing another, and this volume references a number of these transitions. As Sunderland shows, in the case of Lake Sihl, hydropower replaced peat. This example also shows that energy transitions in the industrial age do not always follow the dominant historical narrative of societies moving from an organic to a mineral energy system in which coal, and later oil, replace wood and other organic energy forms.[12] In Sunderland's case study presented in this volume, water, a renewable energy form, replaced peat, a hydrocarbon. Saara Mildeberg and Ewan Gibbs describe how, more recently, former energy landscapes of oil are (potentially) being replaced by ones that feature wind turbines. Jenny Hagemann, Fabian Jacobs, and Lutz Laschewski even go so far as to maintain that renewable energies have become part of a 'reappropriation strategy.' At the same time, Hanna Lempinen cautions that what she observed in her study of a national peat phase-out in Finland does not leave much room for agency as it has little to do with calls for a just energy transition. Both Hanna Lempinen and Andrew Demshuk explain how, through the discourse on energy transition, the national significance of the same energy source has changed over time. First pursued by the respective states to ensure their survival, the same states have now decided that peat and lignite need to be phased out. If these fossil fuels previously provided solutions to address national concerns about energy supply and security, now these very same sources find themselves at the heart of what constitutes a problem—that of climate change.

Energy Landscapes as Sacrifice Zones

The normative discussion of energy landscapes rests on their detrimental local impacts as well as their contribution to global greenhouse gas emissions. Interpreting energy landscapes as sacrifice zones, a term which almost none of the contributions in this volume use but somehow imply, relies on a particular way of seeing that landscape. Landscapes are culturally constructed; the same energy landscape can simultaneously be a symbol of wealth and jobs as well as one of environmental disaster. It can be celebrated nationally and loathed locally. It may initially be seen as having a mainly positive impact, but with the passage of time, it takes on many more negative connotations, changing into something very different. This change in meaning is especially pertinent as energy transitions and the move toward decarbonised economies are on national and international agendas. Sites of energy production often bring destruction in their wake, as their construction, for example, the dam featuring in Merve Neziroğlu's chapter, constitutes 'drastic intervention into the landscape,' which can lead to the loss of artefacts and the displacement of people. Because lignite mining, like oilsands mining, presents an immense and visible scarring of the earth's surface, images of lignite mines have come to represent the concept of the Anthropocene. The multidisciplinary Anthropocene project developed by artists Edward Burtynsky, Jennifer Baichwal, and Nicholas de Pencier is one of many artistic representations of open-cast mining.[13] Burtynsky, a photographer and environmental activist, has become famous for his large-scale images of industrial landscapes, which visualise the large-scale effects that human activities have on the Earth while creating an almost sublime aesthetic for energy sacrifice zones. Saara Mildeberg's description of the history of the oil shale industrial site in north-eastern Estonia equally details the environmental destruction in this energy landscape. Waste and pollution impact the local community, and the extraction of oil shale below and above ground physically damages and scars the landscape. She also distinguishes between these local detrimental impacts and the national political significance of oil deliveries to the rest of the Soviet Union. Here is a peripheral region with little political clout, which is exploited for the benefit of a national whole and a central government. This energy landscape became 'a sacrifice zone for maintaining energy independence.'

In their respective chapters, Merve Neziroğlu, Andrew Demshuk, Sarem Sunderland, and Jenny Hagemann with Fabian Jacobs and Lutz Laschewski lament the loss of home as human beings and their cultural artefacts have to make room for industrial energy infrastructure, such as hydropower dams or lignite mines. This interaction between human activity and nature appears to be more of a sacrifice in some energy landscapes than in others. Constructing dams in what are considered pristine landscapes, such as the Alps or picturesque villages around Leipzig, seems to convey a much more critical perspective on energy production than those energy sites that are located in peripheral regions with few aestheticised views. The same would potentially apply to picturesque coastal communities, which have become the sites of either tidal energy projects or onshore and offshore wind parks. Sites of waterpower energy open up another interesting aspect of energy landscapes as sacrifice zones. It is not only the relocations and flooding of cultural landscapes that have influenced the ways in which these landscapes are considered sacrifice zones but also competing uses of water. Because rivers, such as the Danube, serve as important transport routes that are less destructive to these villages than hydropower, the economic rationale of using water to generate energy could at least be countered by pointing to already existing economic uses of the water, which did not incur the destruction of entire villages and the flooding of islands.

Are energy sites always destructive, turning landscapes into sacrifice zones? For Andrew Demshuk they clearly are, as they erase villages to make room for lignite mining for economic purposes and to ensure state survival. If an energy industry erases a village entirely, its significance will vanish after the single generation that experienced that loss. The main reason for this obscurity is that no one inhabits that landscape anymore. It is only through the presence of a local community that any memory, no matter how much it has been challenged by energy production, will survive into the next generation. For cultural landscapes to change and adapt, they need to be inhabited. However, what is remembered may also be precisely the fact that it was a sacrifice zone, which in itself becomes a challenge if heritage is about envisioning futures. Jenny Hagemann, Fabian Jacobs, and Lutz Laschewski equally grapple with the question of how big a role the environmental destruction should play in heritage. Mölbis, once 'the filthiest village in Europe,' has been restored as a hopeful example of a post-industrial future. The challenge here is to make sure that everyone is heard, including the workers, who may have a much more ambivalent position on the industrial past, which takes account of its environmental imprint on the landscape. In general, the response of community members is much more complex and nuanced. As Andrew Demshuk explains, some of the younger villagers welcomed the modern amenities that they had access to in their new homes after their forced resettlement. This also raises the question of which of the many pasts—feudal, democratic or communist, pre-industrial, industrial, or post-industrial—may be more deserving of remembering and the most politically convenient for imagining a future.

Not all the contributors to this volume consider energy landscapes as sacrifice zones. The example of Baku, which Leyla Sayfutdinova examines in her chapter, shows how energy production gave political power to the capital of Azerbaijan. Ewan Gibbs also notes how Sullom Voe changed Shetland and brought 'positive outcomes,' including 'long-term increases in the population and employment' and the 'benefits that oil-related activity brought to municipal budgets.' Jobs and economic development thus become the measures of success. However, some of the people interviewed by Gibbs also recounted how rapid industrialisation led to sacrifices relating to social cohesion, community welfare, and the natural environment. In another unexpected twist, sacrificing settlements to energy production allowed archaeologists to dig up other much older pasts, as was the case in the East German lignite region examined by Andrew Demshuk and the Iron Gate hydropower plant, which Merve Neziroğlu describes.

Thus, there is no simple equation that associates industrial energy sites with environmental sacrifice zones. An energy landscape is not just about loss and environmental degradation; it is also about a workplace and way of life. Post-mining landscapes may have less pollution after closure, as the example of Möbius shows, but as post-industrial landscapes, they also signify the loss of a particular way of life, which drew its meaning from energy production. During Soviet times, the industrial towns which had been constructed for the purpose of producing energy drew pride from their energetic significance. At the same time, the industrial nature of these towns marginalised them culturally, as they stood in direct opposition to prevalent rural values and the prioritisation of nature. When remembering this landscape in post-socialist times, it is not the controversial Soviet-era industrial heritage that is prioritised; rather, it is memories of better and greener times before the Soviet Union. Unlike former open-pit mines in lignite-producing Germany, these oil shale mines were not flooded but were instead reforested. Adding temporality to the concept of energy sacrifice zones, Saara Mildeberg describes how post-industrial sites also openly carry 'sociocultural scars.'

The Politics of Energy Landscapes

Politics in and of energy landscapes is probably the most important common thread that runs through all the chapters in this volume, and it plays out in various ways. In industrial energy landscapes, it helps to explain the construction and siting of energy production while underlining the power differentials between geographical centres and peripheries and informing the violence and destruction connected with energy production. In post-industrial landscapes, political considerations often frame what will be remembered and heritagised. The coming energy transition is also highly dependent on ambitious political decisions, as the case of the German *Energiewende* (energy transition) shows. Marin Kuijt and Gertjan Plets argue that there is a 'political economy in which heritage sites are embedded,' and this is particularly complicated in the case of energy because former sites of fossil fuel production are not only remembered but their heritagisation also invites comments on the future of energy systems. They maintain that politics have led to separate 'civic and corporate uses of industrial heritage in the Anthropocene.' Examining the case of the German energy company, RAG, they encourage us to follow the money, not in the way that Carola Hein's petroleumscapes connect sites of production with financial flows, but more directly to establish who pays for heritage sites and what kind of an agenda is being pursued. Their critical analysis reveals that the exemplary heritage sites in the Ruhr District, foremost amongst them being Zeche Zollverein, have a much darker role to play in current debates in the Anthropocene as they have become sites of 'art-washing' and the rewriting of history. Through heritagisation, companies may not only try to gloss over their historical role in the industrial landscape but they may also actively use the post-industrial site to promote new fossil imaginaries and thus support a fossil economy. This is all the more surprising given that the German government has pursued its *Energiewende* quite adamantly, and the turn to renewables was also featured in the highly acclaimed Internationale Bauaustellung Emscherpark (IBA) exhibition, which has been considered crucial for the successful heritagisation of coal mining in the region.

Replicating the all-too-familiar corporatist approach, which in the past celebrated governments, industry, and labour working together in the Ruhr District to weather the storms of declining coal demand, RAG has a considerable financial stake in the regional heritage foundation. Marin Kuijt and Gertjan Plets insist that this allowed the company to control the foundation's activities. A more sympathetic view would be that this was part of their legal obligations to deal with all follow-up costs from their industrial pursuits. However, in direct negotiations over the content of the exhibition, the corporate actors did demonstrate their steadfast belief in the future of a fossil economy despite the exhibition's motto which was 'coal goes, and the sun comes.' Coal miners themselves supported this position, showcasing their attachment to the energy source, which gave them jobs and a place-based identity.

Other authors in this volume highlight the national significance of energy landscapes and the meaning that is transferred on these landscapes from a distance. National significance may be derived from a government's wish to establish energy independence or autonomy and often serves the purpose of achieving energy autarky or energy security. Energy independence helped to boost the Etzelwerk hydropower project discussed by Sarem Sunderland in his contribution. He explains how after World War I, Switzerland, which was looking for ways to become less dependent on coal imports, found a domestic energy alternative in its many rivers and lakes. Where a strong national narrative about the significance of energy resources for ensuring the independence, security, and survival of the state exists, workers not only derive livelihoods and local identities from energy production, but they also feel pride and patriotism, thus connecting the local and national scales.

Industrial energy sites in socialist countries also took on ideological meaning and were instrumental in showcasing national and engineering grandeur, proving to the world that these regimes were at the forefront of industrial production. In the case of Baku, the centrality of its energy landscape was reflected in the many state symbols it provided for the Soviet Union but also for post-socialist Azerbaijan, especially through the construction of the Flame Towers. Whether in socialist or capitalist states or historical empires, governments shared similar state rationales for creating energy sites. Framed as ensuring state survival while launching regions into economic modernity, these energy landscapes were seen as contributing to both national security and local development. Outside of Europe, hydropower facilities were often constructed on a very large scale and pursued as high modernist projects, linking energy and development strategies primarily to address the economic challenges of remote regions but also for nation-building purposes. This kind of industrial modernity is also reflected in the case of Baku, Azerbaijan, which is examined in the chapter by Leyla Sayfutdinova. Here, fire becomes a potent metaphor for petroleum and belonging. Unlike most of the other case studies, in Baku, the national significance of energy coincides geographically with a national centre of power. Baku is not only the site of hydrocarbon extraction but also the state capital, even though it was located at the periphery of the Russian empire when oil production began in the nineteenth century. According to Leyla Sayfutdinova, its long existence as a centre of energy production has normalised or even naturalised the link between resource and nation, proving that the temporal dimensions should not be neglected when attempting to understand the cultural landscapes of energy. Nor should the kind of energy resource that defines an energy landscape be discounted. As Timothy Mitchell has argued in his book *Carbon Democracy*, there are different carbon pasts. Even though both petroleum and coal are subterranean energy resources, their differing materiality have created very different political economies and cultural landscapes.[14]

The national significance of energy landscapes raises questions about political participation, especially that of marginalised groups and those who are not citizens. Landscapes of energy are sites of extraction and industrial production, often located at spatial peripheries, such as the Alps, or inhabited by more powerless sections of the population, such as the small villages in Communist East Germany or north-eastern Estonia. They attract workers who migrate to these energy sites, and while some do so only temporarily, others, like in the Scottish examples described by Ewan Gibbs, settle permanently in these regions. At the same time, highly mobile, foreign, middle-management professionals sojourn in energy sites, representing international companies and also bringing new practices to the sites, as the example of the introduction of Sunday work in the Scottish Highlands illustrates. Through seasonal or long-term migration, energy landscapes connect very different energy sites, transcending geographical boundedness and national boundaries. Because these workers were not citizens of the landscapes they worked in, they did not have the right to reside in the energy landscapes indefinitely. The Italian construction workers in Switzerland whom Rune Frandsen writes about needed annual permits, and their bodies were regulated through medical requirements. As seasonal workers, they did not have the same access to social and labour benefits. The cases of East German lignite mining and the Romanian Iron Gates region raise another challenge relating to participation. If some of these energy sites were national projects, what would happen if these landscapes changed jurisdiction? What happens when former imperial or socialist sites become part of a different state? Who remembers these former industrial sites, what is remembered, and in what language? Is it enough for Iron Gates museums to simply provide stories and histories in Romanian and English? Should heritage sites in Lusatia provide information in Sorbian, German, and English?

Political power is also reflected in the locations that are chosen for energy development projects. Andrew Demshuk reminds us that the sheer existence of energy resources does not automatically lead to the construction of energy landscapes and the erasure or resettlement of communities. There is always an element of power involved. It is often the powerless and the peripheries that will become industrial sacrifice zones. Thus, the city of Leipzig, which was at the centre of a cultural landscape that included significant buildings and identity-building monuments, was not turned into a coal-mining area. Andrew Demshuk's concept of 'architecture beyond ideology' implies that only those spaces that did not hold a particular political meaning for the nation and the state could be transformed into new energy landscapes and ultimately into energy sacrifice zones. His concept pits the different functions of local landscapes against each other: economic versus cultural survival of the Communist state. Merve Neziroğlu provides a different political perspective in her examination of the transnational dimension of the Iron Gate power plant at the border between Romania and Serbia. She describes the displacement of villages and the disappearance of the island of Ada Kaleh, formerly inhabited by Muslim communities, who were offered the option of emigration and resettlement in Turkey, an offer that several dozen families decided to accept.

As spelled out in the above section on energy sites as sacrifice zones, violence and exploitation are an integral part of 'cultural landscapes of energy.' Power in its double meaning is also central to some of the energy projects that we encounter in this book. Merve Neziroğlu writes of the political power that prestigious water energy projects brought to socialist leaders such as Tito and Ceaușescu. In particular, hydropower mega projects became symbols of national grandeur and technological superiority at a time when the Eastern and Western blocs were competing over ideologies. The struggle over who decides what happens to former energy sacrifice zones is especially complex in post-socialist landscapes, as the chapter by Saara Mildeberg shows. In her view, industrial landscapes are 'heritage sites in economic transition from mining to tourism.' Transition and transformation take on a whole new meaning in post-socialist energy landscapes, whether in Romania, East Germany, or Estonia. Post-industrial landscapes are highly politicised. Jenny Hagemann, Fabian Jacobs, and Lutz Laschewski explain how the 'heritagisation of landscapes' is a 'field for negotiating local and often regional identities.' Economic restructuring is a highly political and often contested aspect of post-industrial landscapes. It raises the question of agency and power differentials. Who gets to sit at the table and who decides? And what about minorities such as the Sorbs in the case of the post-industrial lignite landscape of Lusatia? There is also an activist role that we can play as researchers, especially in support of minoritised heritage. However, Jenny Hagemann, Fabian Jacobs, and Lutz Laschewski caution that while capacity building is crucial to ensure the inclusion of minority stakeholders in heritagisation, we also need to allow locals to make the decision of not getting involved. Not everyone may feel the need to be part of the heritagisation initiative.

However, these power differentials did not always automatically translate into political apathy. The contributions in this volume are full of examples of political agency, whether through worker mobility, environmental protests, or political autonomy. An example is the case of Shetland, described by Ewan Gibbs, which, despite its peripheral geographical position at the northernmost tip of the UK, had been at the centre of an extensive economic maritime network. A 'cultural landscape of energy' is not just any kind of landscape. These landscapes carry political significance as societies plan for an energy transition. Thus, any heritagisation is also about making a political statement about our energy futures. Because the stakes are so high, everyone needs to be involved in deciding how best to remember the industrial past of energy sites. As researchers and practitioners, we need to find ways to give voice to those not represented at the table, while humbly accepting that some choose not to be involved in these conversations. In the process, we may learn to accept the ambivalence and inconsistencies. We may embrace Saara Mildeberg's 'scavenger ethnography' as we 'chew [...] on someone's leftovers, trying to make sense of lost places like post-industrial complexes or industrial wastelands, while striving for a more practical, future-oriented way of engaging with them, which [is] not based on nostalgia.'

Notes

1 Wylie, *Landscape*, 1.
2 Williams, *The Country and the City*, 120.
3 Wylie, *Landscape*, 8.
4 Ebert and Bednorz, *Kathedralen der Arbeit.*
5 Strangleman, "Smokestack Nostalgia."
6 Brown and Goetcheus, *Routledge Handbook*, 2.
7 Cambridge Dictionary online, "Scapes."
8 Ingold, "The Temporality of the Landscape," 164.
9 Hein, "Oil Spaces."
10 Howard et al., "Energyscapes."
11 Wylie, *Landscape*, 3.
12 Wrigley, *Energy and the English Industrial Revolution.*
13 "The Anthropocene Project."
14 Mitchell, *Carbon Democracy.*

Bibliography

"The Anthropocene Project." Accessed 24 September 2024. www.edwardburtynsky.com/projects/the-anthropocene-project.

Brown, Steve, and Cari Goetcheus. *Routledge Handbook of Cultural Landscape Practice.* Oxford: Routledge, 2023.

Cambridge Dictionary online, "Scapes." https://dictionary.cambridge.org/dictionary/english/scape?q=scapes.

Ebert, Wolfang, and Achim Bednorz. *Kathedralen der Arbeit. Historische Industriearchitektur in Deutschland.* [*Cathedrals of Work: Historical Industrial Architecture in Germany*]. Berlin: Wasmuth Verlag, 1996.

Hein, Carola. "Oil Spaces: The Global Petroleumscape in the Rotterdam/The Hague Area." *Journal of Urban History* 44, no. 5 (2018): 887–929.

Howard, D. C., J. Burgess, S. J. Butler, S. J. Carver, T. Cockerill, A. M. Coleby, G. Gan, C. J. Goodier, D. Van der Horst, K. Hubacek, R. Lord, A. Mead, M. Rivas-Casado, R.A. Wadsworth, and P. Scholefield. "Energyscapes: Linking the Energy System and Ecosystem Services in Real Landscapes." *Biomass Bioenergy* 55 (2013): 17–26.

Ingold, Tim. "The Temporality of the Landscape." *World Archaeology* 25, no. 2 (1993): 152–174.

Mitchell, Timothy. *Carbon Democracy: Political Power in the Age of Oil.* London: Verso, 2013.

Strangleman, Tim. "'Smokestack Nostalgia,' 'Ruin Porn' or Working-Class Obituary: The Role and Meaning of Deindustrial Representation." *International Labor and Working-Class History* 84 (2013): 23–37.

UNESCO. *The Operational Guidelines for the Implementation of the World Heritage Convention*. Paris: UNESCO, 2021.
Williams, Raymond. *The Country and the City*. London: Chatto & Windus, 1973.
Wrigley, E. A. *Energy and the English Industrial Revolution*. Cambridge: Cambridge University Press, 2010.
Wylie, John. *Landscape*. New York: Routledge, 2007.

Index

Note: Page numbers in *italics* indicate figures and in **bold** indicate tables on the corresponding pages.

Made in the USA
Monee, IL
12 November 2026

25d5434a-c431-4c42-a85c-109485c24934R01